Geometrical Objects

Archimedes

NEW STUDIES IN THE HISTORY AND PHILOSOPHY OF SCIENCE AND TECHNOLOGY

VOLUME 38

Archimedes has three fundamental goals; to further the integration of the histories of science and technology with one another: to investigate the technical, social and practical histories of specific developments in science and technology; and finally, where possible and desirable, to bring the histories of science and technology into closer contact with the philosophy of science. To these ends, each volume will have its own theme and title and will be planned by one or more members of the Advisory Board in consultation with the editor. Although the volumes have specific themes, the series itself will not be limited to one or even to a few particular areas. Its subjects include any of the sciences, ranging from biology through physics, all aspects of technology, broadly construed, as well as historically-engaged philosophy of science or technology. Taken as a whole, *Archimedes* will be of interest to historians, philosophers, and scientists, as well as to those in business and industry who seek to understand how science and industry have come to be so strongly linked.

For further volumes:
http://www.springer.com/series/5644

Anthony Gerbino
Editor

Geometrical Objects

Architecture and the Mathematical Sciences 1400-1800

Springer

Editor
Anthony Gerbino
Art History and Visual Studies
University of Manchester
Manchester, UK

ISSN 1385-0180 ISSN 2215-0064 (electronic)
ISBN 978-3-319-35923-6 ISBN 978-3-319-05998-3 (eBook)
DOI 10.1007/978-3-319-05998-3
Springer Cham Heidelberg New York Dordrecht London

Printed on acid-free paper

Springer is part of Springer Science+Business Media (www.springer.com)

Acknowledgments

This volume originated as a conference held at the Museum of the History of Science and at Worcester College, Oxford, on 19–20 March 2007. It has been a collaborative venture from the start, with a number of participants. I would like to thank the contributors, each of whom has produced a fascinating account, as well as Mario Carpo and Marco Panza, who helped organize the conference on which this edited collection is based. Jim Bennett, former Director of the Museum of the History of Science, and Stephen Johnston, Assistant Keeper, invited us to hold our proceedings in the august space of the museum and helped generously in the preparations. I am also grateful to Richard Smethurst, Provost of Worcester College, for his support and to Martin Kemp and Henk Bos for their participation. Jed Buchwald has guided the book through the review process, patiently waiting for its completion. We are proud to have it included in such an eminent series. I have benefited greatly from conversations with Matthew Cohen and Robert Bork, who generously read drafts of the Introduction, while Darin Stine provided much-needed editorial assistance at the finish line. Finally, I would like to thank our funding bodies, each of whom contributed to various aspects of this project. The conference was generously supported by the Paul Mellon Centre for Studies in British Art, the Samuel H. Kress Foundation, and the John Fell Fund of the University of Oxford. The Graham Foundation for Advanced Studies in the Fine Arts provided a generous publication subsidy. I am also grateful to the Leverhulme Trust, whose grant award in 2013–2014 gave me the time away from my teaching responsibilities at the University of Manchester to finalize the manuscript. I am, as always, enormously indebted to my wife, Mary Bosworth, for her constant kindness and encouragement.

Contents

Editor and Contributors

Editor

Anthony Gerbino Art History and Visual Studies, University of Manchester, Manchester, UK

Contributors

Kirsti Andersen Emerita, Centre for Science Studies, Aarhus University, Aarhus, Denmark

Francesco Benelli Department of Art History and Archaeology, Columbia University, New York, NY, USA

Bernard Cache School of Architecture, Civil, and Environmental Engineering (ENAC), École Polytechnique Fédérale de Lausanne, Lausanne, Switzerland

Filippo Camerota Museo Galileo, Florence, Italy

Pascal Dubourg Glatigny Centre Alexandre Koyré, Paris, France

David Friedman Emeritus, Department of Architecture, School of Architecture, Massachusetts Institute of Technology, Cambridge, MA, USA

Jacques Heyman Emeritus, Department of Engineering, University of Cambridge, Cambridge, UK

Ann C. Huppert Department of Architecture, University of Washington, Seattle, WA, USA

Jason M. Kelly Department of History, Indiana University, Carmel, IN, USA

Jeanne Kisacky Independent Scholar, Ithaca, NY, USA

Susan Klaiber Independent Scholar, Winterthur, Switzerland

Frank Salmon Department of History of Art, University of Cambridge, Cambridge, UK

David Yeomans Independent Scholar, Banbury, UK

About the Authors

Kirsti Andersen is Associate Professor (Emeritus) at the Centre for Science Studies, Aarhus University, Denmark, where she taught the history of mathematics since 1967. She has been a member of the board of editors for *Centaurus* since December 1983 and is a member of the Académie Internationale d'Histoire des Sciences. With Henk J.M. Bos, she coauthored the chapter "Pure Mathematics" in Vol. 3 of *The Cambridge History of Science* (Cambridge University Press, 2006). She has also published *The Geometry of an Art. The History of the Mathematical Theory of Perspective from Alberti to Monge* (Springer, 2007) and entries on Guidobaldo del Monte and Piero della Francesca for the *New Dictionary of Scientific Biography* (Scribner's, 2008).

Francesco Benelli is an Associate Professor in the Department of Art History and Archaeology at Columbia University. His interests and publications focus on issues of design, style, building materials, structures, the interpretation and acquisition of the Antiquity during the Renaissance, as well as the relation between structure and decoration and the theoretical discourse from Leon Battista Alberti to Andrea Palladio. His book *The Architecture in Giotto's Painting* (Cambridge University Press, 2011) explains how Giotto created a new way of conceiving pictorial space and of representing architecture through an unexpected knowledge of classical sources. His current project is a book on the education of the Renaissance architect through the case study of Antonio da Sangallo the Younger and his reading of Vitruvius.

Bernard Cache developed the concept of non-standard architecture in his book *Earth Moves* (MIT Press, 1995). This concept was given the name "*objectile*" by Gilles Deleuze in *The Fold: Leibniz and the Baroque* (University of Minnesota Press, 1993). Bernard Cache founded the company Objectile together with his partner Patrick Beaucé in order to conceive and manufacture non-standard architecture components. He is currently dedicated to the reading of classical texts (such as Vitruvius's *De Architectura* and Dürer's *Underweysung der Messung*) with the help of CAD/CAM software. After teaching nomadically he is now Professor at the École Polytechnique Fédérale de Lausanne

Filippo Camerota is Deputy Vice Director and Collection Manager of the Museo Galileo in Florence and a research associate at the Istituto Universitario di Architettura di Venezia (Department of Architectural History). His research has focused on the history of scientific representation, with a particular interest in the Renaissance perspective and measuring instruments. His principal published works include critical editions of Giorgio Vasari the Younger's unpublished treatise on instruments (1997), Fabrizio Mordente's treatise on the proportional compass (2000), Ludovico Cigoli's *Prospettiva pratica* (2010), and a full account of the art and science of the Renaissance perspective (2006).

Pascal Dubourg Glatigny is a researcher in art history at the CNRS. He has published many works on practical perspective in the Renaissance, most recently *Il disegno naturale del mondo. Saggio sulla biografia di Egnatio Danti con l'edizione del Carteggio* (Aguaplano Biblioteca, 2011). With Matthias Bleyl he has coedited the volume *Quadratura: Geschichte, Theorien, Techniken* (Deutscher Kunstverlag, 2011). He is currently studying the role of technique in architecture in the early eighteenth century.

David Friedman is Professor Emeritus in the History, Theory, and Criticism Section of the Department of Architecture at MIT, where he has taught since 1978. He is best known for his work on medieval Italian urbanism, in particular for his landmark book of 1989, *Florentine New Towns.* His most recent study along these lines, "The Mercanzia and its Palace" published with Antonella Astorri in *I Tatti Studies* of 2006, is an institutional and architectural history of the Florentine merchant court. His most recent work is concerned with the history of urban cartography. He has published on the fifteenth-century view of Florence, the so-called View with a Chain, in the *Zeitschrift fur Kunstgeschichte* of 2001 and with Hilary Ballon, a survey on Renaissance city views for the third volume of David Woodward's *History of Cartography* (University of Chicago Press, 2007). An article discussing how the absence of maps affected urban planning in the late Middle Ages has appeared in the acts of the I Tatti conference "Arnolfo's Moment" of 2005 (Olschki, 2009).

Anthony Gerbino is a historian of early modern architecture in France and England. His research focuses on the role of architecture in seventeenth-century scientific and academic circles and on the technical and mathematical background of early modern architects, engineers, and gardeners. His more general interests lie in the interaction of art, science, and technology; the professional and intellectual world of early modern artisans, architectural treatises and the culture of the printed book, cartography and its relation to landscape, and the urban history of Paris. He has published *François Blondel: Architecture, Erudition, and the Scientific Revolution* (Routledge, 2010) and is coauthor of *Compass and Rule: Architecture as Mathematical Practice in England 1500–1750* (Yale University Press, 2009). He is a Lecturer in History of Art at the University of Manchester.

Jacques Heyman is a Fellow of the Royal Academy of Engineering, of the Institution of Civil Engineers and, of the Society of Antiquaries and is an Honorary

Fellow of the RIBA. He was Professor of Engineering at Cambridge University and Head of Department from 1983 until his retirement in 1992. He is the author of 13 books and coauthor of two others and has written extensively on the plastic behavior of steel structures, on masonry structures, on the history of structural engineering, and on general structural theory. He has acted as consulting engineer for work to several cathedrals and churches and served on the Cathedrals Fabric Commission for England.

Ann C. Huppert is an Assistant Professor in the Department of Architecture at the University of Washington. Her recent publications have explored the position of perspective drawings in the building workshop and designs for the new St. Peter's Basilica at the Vatican (*Journal of Architectural Historians*, 2009) and Renaissance practices of mapping ancient Rome (*Memoirs of the American Academy in Rome*, 2008). Her forthcoming book, *Becoming an Architect in Renaissance Italy: Art, Science, and the Career of Baldassarre Peruzzi* (Yale University Press, 2015), examines the role that drawing played in the sixteenth-century practitioner's transition from figural artist to architect within the broader context of architecture's emergence as a distinct profession.

Jason M. Kelly is an Associate Professor in the Department of History, Indiana University. He is a historian of eighteenth-century Britain, specializing in antiquarianism, literary networks, and political culture. Recent articles have appeared in *The British Art Journal* and the *Journal of British Studies*. His monograph *The Society of Dilettanti: Archaeology and Identity in the British Enlightenment* was published by the Paul Mellon Centre and Yale University Press (2010).

Jeanne Kisacky trained as an architect at Princeton and as a historian at Cornell. She has taught at Syracuse, Cornell, and Binghamton universities. Her research focuses on how ideas of health interact with environmental design, particularly on the influence of medical and scientific discoveries on architectural design theory and practice. She was the managing editor of *Isis* and has published articles in the *Bulletin of the History of Medicine* and the *Journal of the Society of Architectural Historians*.

Susan Klaiber is an architectural historian based in Winterthur, Switzerland. She trained at Columbia University and at the universities of Bonn and Münster. She has held fellowships from the American Academy in Rome and the Mellon, Kress, Whiting, and Getty foundations. Her publications include the book *Guarino Guarini* (Allemandi, 2006), coedited with G. Dardanello and H.A. Millon. She serves on the governing committee of the European Architectural History Network and was founding editor of the EAHN newsletter (2007–2010). She is currently coauthoring the corpus of Guarini's drawings with Giuseppe Dardanello.

Frank Salmon is Senior Lecturer in History of Art at the University of Cambridge and a Fellow of St. John's College, Cambridge. He was Head of the Department of History of Art from 2009 to 2012. His prize-winning book *Building on Ruins: The Rediscovery of Rome and English Architecture* was published in 2000. He has subsequently worked on Greek archaeology and antiquarianism, making a major contribution to *James 'Athenian' Stuart, 1713–1788: The Rediscovery of Antiquity* (Yale University Press, 2006). In 2008, he published "C.R. Cockerell and the Discovery of Entasis in the Columns of the Parthenon" in *The Persistence of the Classical*, a book he also edited. More recently, he has been working on William Kent in connection with the monographic book and exhibition being held at the Bard Center, New York and the Victoria and Albert Museum, London in 2013–2014. Frank Salmon is a Fellow of the Society of Antiquaries.

David Yeomans trained as an engineer and a historian. He taught architecture at the universities of Liverpool and Manchester and, as a Visiting Professor, at the universities of Washington and Oregon. He practices as an engineer, specializing in the design and restoration of timber structures. His research interests are in the development of building technology and its effect on architecture. He has written extensively on the development of roof carpentry in the seventeenth and eighteenth centuries and on twentieth-century reinforced concrete architecture. He is Vice Chairman of the Construction History Society.

Introduction

Anthony Gerbino

This volume explores the mathematical character of architectural practice in diverse pre- and early modern contexts. It draws together two nominally distinct disciplines; the history of architecture is here seen through the prism of the history of science, and one subfield of that discipline in particular. Our theme concerns the role of practice in the scientific revolution. This subject—sometimes expressed in more anachronistic terms as the relationship between science and technology—has burgeoned in recent years, and our contributions here are premised on the results of important recent work in this area.[1] In contrast to the oppositional and hierarchical categories that used to mark the historiography of this subject, scholars now tend to emphasize the jumble of intellectual, scientific, and technical factors associated with various forms of practice and, conversely, how practical and material factors were implicated in the process of actually doing science.

One of the most fruitful innovations of this approach is that it levels the artificial disparity between the mental and the manual, knowledge and know-how, theory and "application". Even where such categories are evident, our challenge is to show how they depend on and reinforce each other, not in a process of top-down "vulgarization" but rather in something like a reciprocal cycle or feedback loop. We emphasize, likewise, a micro-historical focus. In architectural as in scientific practice, various forms of knowledge—whether explicit and codified as in "science" or implicit and tacit as in "craft"—meet, interact, and augment each other in local, embodied ways. Such a focus is perhaps not unfamiliar to architectural historians, who are used to working at a fine-grained level of the individual project. To the extent that our approach entails a change in perspective, it is one that sees the designer's studio, the stone-yard, the drawing floor, and construction site not merely

A. Gerbino (✉)
Art History and Visual Studies, University of Manchester,
Oxford Road, MP13 9PL Manchester, UK
e-mail: anthony.gerbino@manchester.ac.uk

A. Gerbino (ed.), *Geometrical Objects: Architecture and the Mathematical Sciences 1400-1800*, Archimedes 38, DOI 10.1007/978-3-319-05998-3_1,

as places where the architectural object takes shape, but where knowledge itself is deployed, exchanged, and amplified among various participants in the building process.

Mathematics provides an obvious disciplinary framework for this kind of investigation. In contrast to its academic counterpart, natural philosophy, early modern mathematics was partly defined by its orientation toward practice. In the Aristotelian tradition, this connection was most clearly marked in the "subalternate", intermediary, mixed, or composite sciences, as they were variously known. Those fields—which included optics, astronomy, harmonics, and mechanics—all treated geometrical quantity abstracted from sensible matter. That is, they treated the properties of the physical world with a level of demonstrative rigor comparable to that of "pure" geometry.[2] This classification of mathematical disciplines echoed that of the traditional *quadrivium*, but by the mid-seventeenth century, the mixed sciences had grown immensely in importance. Not only did they provide a strictly mathematical rationale for the investigation of nature, they also served as an umbrella category for newly resurgent technical arts of virtually any geometrical character. Early modern surveyors, cartographers, engineers, instrument-makers, gunners, navigators, and even painters routinely identified their activities with the various "mathematical sciences" and themselves as part of an all-embracing culture of both pure and practical mathematics.[3] Moreover, these practitioners created an important intellectual and technological context for the work of scholars and mathematicians—the kinds of figures whose names and discoveries feature more prominently in the history of early modern science. Galileo stands out here, but many others would also fit the bill. Historians have been increasingly attentive to the way practical and theoretical concerns were imbricated in their work as well.[4]

This volume proceeds from a conviction that architectural history, too, can benefit from an approach that contextualizes design and construction in terms of contemporary mathematical knowledge, attendant forms of mathematical practice, and relevant social distinctions between the mathematical professions. This perspective is intended to respond, in the first place, to the actual character of the art; geometrical and arithmetical operations of some form or another lay at the heart of early modern architectural practice. Indeed, the process of design was largely defined by the application of proportional or compass-based rules. These protocols were more-or-less pervasive, potentially controlling the design in both plan and elevation, from the concept to details. Mathematics was indispensable in other ways. Measurement and scale conversion—particularly important when fitting the proportions of a design to on-site dimensions—surveying, cost estimates, bookkeeping, and even the use of routine graphic techniques all presupposed a certain amount of mathematical training.

Architecture was also connected to learned or theoretical traditions of mathematics, those associated not with the workshop or building site, but rather with the university and the humanist's library. The profession was after all largely shaped by scholar-practitioners working in or alongside a tradition of classical commentary. Leon Battista Alberti is the obvious touchstone here, but it is important to point out how broad and long-lasting this tradition was. Alongside its purely architectural content, Vitruvius's *De Architectura* contains a wealth of information on ancient theories of proportion, an anecdotal knowledge of Greek mathematics, and several chapters on technical and engineering-related subjects.[5] From the earliest printed editions, Renaissance commentators found the text to be a compelling stimulus, both as a source for ancient science and in many areas of recent research. This interest is perhaps epitomized by Daniele Barbaro, whose translation and commentary, appearing in two editions of 1556 and 1567, set new standards for the interpretation of the book's technical and mathematical contents, particularly those parts dealing with astronomy and sun-dialing (Book 9). Barbaro's famous description of the Vitruvian analemma—reconstructed with the help of Federico Commandino's then-recent edition of Ptolemy's *De Analemma* (1562)—was quickly recognized as a milestone both in the understanding of ancient astronomy and in the study of gnomonics.[6] In this respect, Barbaro was an exemplary figure, but not an uncommon one. A similar conjunction of mathematics, technology, and architecture is evident in the work of Bernardino Baldi, Guillaume Philandrier, François Blondel, Teofilo Gallaccini, Nicolaus Goldmann, and Giovanni Poleni.[7]

There was clearly a manifold bond between mathematical and architectural practice, yet historians have only partially explored this relationship. The most relevant research focuses on the design methods of medieval architects and on the proportional and geometrical layout of medieval buildings, an important vein of scholarship that I will have occasion to review below. More recently, there have been a growing number of specialist subfields—in the history of structural mechanics or building technology, for example—that sometimes overlap with architectural history.[8] Our volume builds on these approaches, but it is worth pointing out that they have remained largely peripheral to the discipline. The relative lack of scholarly engagement is not fully explained by the technical nature of the subject. Research into the mathematical basis of architectural design has a long pedigree, even if it has recently fallen off. Nor have other disciplines run afoul of the same hurdles. The historiography of early modern mechanics—to take one example relevant to architecture—has shown how technical content can be fruitfully combined with broader hermeneutic and historical concerns.[9]

The obstacles to further study in this area are several, but among the most challenging—and paradoxical—may be those posed by the physical reality of the building itself. Although mathematical practice was integral to the making of architecture, it is often subsumed and concealed by the finished object. In many cases, the designer's intentions have to be reconstructed, and—in the absence of original drawings or written records—with often partial or even unreliable results. The most valuable scholarship in this area hews closely to the measurements of the built work, to known drawing practices, and to rare written sources about the design process. Even with these controls, however, conclusions can remain speculative, and, too often, scholars have done without such checks. Indeed, older scholarship on design methods—until well past mid-twentieth century—is frequently characterized by fanciful, complex diagrams overlaid onto building plans. Recent research in this field is finally beginning to overcome the stigma of these earlier attempts.[10] Aside from being difficult to recuperate, the designer's point-of-view is only one of the perspectives that informs the history of architecture. Scholars are equally and justifiably attentive to the desires, resources, and input of both patrons and users of a building. Any reconstruction of a project's genesis will naturally involve a host of issues, from planning and style to patronage, iconography, and reception. Mathematics, in other words, is merely one part of a complex and multifaceted process.

The current state of scholarship reflects a normative conception of the design process that sees it solely as the preliminary to the building, but it is worth noting that this view involves some unintended drawbacks. To the extent that the built work remains the privileged object of study, it hinders a fuller understanding of the figure of the architect, particularly those aspects of his intellectual culture and expertise that were non-architectural. Indeed, this approach unintentionally restricts the wide range of mathematical practices that went hand-in-hand with building. Early modern architects built fortifications, drew maps, used instruments, and designed machines, but that is not what they are remembered for. To be sure, practitioners themselves did much to establish architecture as an autonomous discipline. The touchstones provided by Vitruvius and by the built remains of the ancient world allowed early modern architects and theorists to separate themselves very clearly from the wider strata of mathematical practitioners, even when their practical activities were almost indistinguishable. Yet, from a modern disciplinary perspective, the often-uncritical adoption of such a focus is nonetheless distorting, for it relies implicitly on a relatively narrow, twentieth-century definition of what the architect is, one that defines the profession retrospectively in terms of distinct socio-professional boundaries, where none existed at the time.[11]

The alternative is certainly not to ignore the built work, but rather to shift our view slightly. Instead of a restricted focus on buildings as the outcome of the design process, we might rather see design itself as part of a constellation of related activities that were no less central to the architectural culture of the period. We might try to contextualize, in other words, not only the design of specific projects, but rather the process of design itself. That process was and remains, of course, historically contingent. It is subject both to technological constraints and the level of knowledge available at a given time. Such an approach would consider not merely whether a given reconstruction fits the measured dimensions, but also how projects are conceptualized and executed within a horizon of existing practices, abilities, tools, and techniques.

Crucial to any attempt of this sort is to recognize the distance between educated elites and the traditional craft culture of the building trades. The mathematical notation, number systems, and methods of calculation that we take for granted cannot often be assumed for early modern practitioners until well into the period. Building craftsmen appear to have begun receiving education in the *abaco*—commercial arithmetic and simple geometry—only from the Trecento, and such schools were not widespread outside of central Italy. Nor did formal mathematics of this sort always take precedence over the practical training of the traditional apprenticeship, which tended to provide graphical and rule-of-thumb solutions to most problems.[12] In any given case, the most likely design methods were those that produced simple arithmetical and geometrical relationships, generated by physical manipulations of the compass and ruler or, alternatively, of the yardstick or stakes and string. It is not simply that such methods kept calculations as straightforward and unobtrusive as possible. They also reflected the manual and instrumental character of the setting-out process, in which elements of the project are treated not in terms of abstract number or dimension but as linked properties of the full-scale design.[13]

The relationship, in other words, between drawing and building has hardly been constant or universal. Indeed, the recent advent of computer-generated design and the almost complete disappearance of pen-and-paper drawing from architectural offices and schools bring home the mutability of this relationship like nothing else. Architectural practice in the early modern period was no less dependent on changes in contemporary mathematical culture and available technology, and it is this theme that ties together the essays presented in this volume. In the first place, our contributors explore the various *uses* of mathematics by early modern architects. The emphasis here is on practice, on activities as basic to architecture as drafting, calculating, measuring, surveying, composition, and design. In this sense, our papers will present a picture of architects as "consumers" of mathematics, dependent both on the level of mathematical knowledge available at the time and the degree to which they were able to understand and employ it in their own work. We also explore the opposite side of this issue, that is, the extent to which architects were themselves "producers" of mathematical knowledge or the degree to which they collaborated with mathematicians and natural philosophers in the production of such knowledge. These groups had long been associated and we know that in many cases they overlapped, particularly in the seventeenth century. Figures such as Christopher Wren, Robert Hooke, François Blondel, Girard Desargues, and Guarino Guarini spring easily to mind.

As these names suggest, the role of the sciences already colors our understanding of early modern architecture—at least for the Baroque. In fact, the entire period bounded by this volume was one of increasing mathematical and scientific expertise among architects, and this transformation was largely characterized by the reciprocal relationship between the two phenomena described above. On the one hand, mathematical and technological advance in architecture was often frustrated by the limited educational background and conservative practical outlook of the average builder. This "advance" is not a whiggish story of unimpeded progress. Yet, on the other hand, we also see continual instances in which architectural practice was both deepened and enriched by coeval advances in mathematics from both practical and learned spheres. This kind of influence was not unusual and kept the two fields closely connected throughout our period. It accounts for the live curiosity among architects in what mathematicians and mathematical scholars were actually doing and, contrariwise, for the fact that so many scholars found in the art a natural outlet for their own interests.

Scholars and Practitioners at Milan Cathedral

As an introduction to the papers that follow, this chapter presents three examples of the interaction of "high" and "low" mathematics, to illustrate its transformative effects over the period covered in this volume. I mention these instances not to suggest that they are canonical nor to imply any causal relationship between them, but to outline the range of issues inherent in our subject. Indeed, their geographical and chronological separation serves to highlight the evolution of some common themes across several centuries. These include: the interdependence of geometrical design and arithmetic calculation; the embeddedness of architecture in other mathematical disciplines; and the link between drawing and instruments. My examples also serve to illustrate a methodological point: the mathematical content of architectural practice has often been relegated as either peripheral or merely preparatory to the form-generating process of design. Making that content visible requires a change in perspective that places it in its own historical schema.

The late fourteenth-century conferences held by the cathedral workshop of Milan, made famous in a classic article by James Ackerman, still offer a useful baseline from which to compare similar events.[14] The preserved records of these meetings comprise one of our most extensive sources for medieval architectural theory in relation to an actual project. More importantly for our purposes, they also offer telling evidence for the relative level of mathematical knowledge among at least one group of medieval masons. As Ackerman recounts, the Lombard masters had fixed the ground plan and even began construction on the foundations before deciding on what the upper profile of the building was going to look like. They had envisaged an elaborate Gothic decoration for the cathedral—which still sets it apart from other comparably sized Italian churches—and began the work on a much more ambitious scale than they were normally used to working with. As a result, they were forced to call in a succession of outside experts to advise them both about decorative matters as well as the optimal height and form to give to the cross-section of the nave and aisles.

The workshop's initial intentions for the nave section are preserved in a drawing by a visiting Bolognese architect, Antonio di Vincenzo (Fig. 1). Typically dated to early 1390, some four years after work began, the sketch combines the measured plan of the cathedral with a section of the nave as it was then projected. The plan is presumably based either on a model or on a survey of the cathedral's rising walls, but the design of the upper parts was still very much in flux. The vertical elements are not drawn to scale, but Di Vincenzo's annotated dimensions suggest that the section was based on a simple modular schema, in which a basic unit of 10 Milanese *braccia* (about 5.95 m) served to establish the height of the various vertical elements (Fig. 2). In this early scheme, the springing of the outer aisle vaults were to be 30 *braccia* high, that of the inner aisles 50 *braccia* high (including tall capitals of 10 *braccia*), and the springing of the nave vault 60 *braccia* high. The ground plan, in contrast, had been laid out according to a different module. Corresponding to the aisle bays, it formed a square 16 *braccia* to the side. With the nave two modules wide, the total width of the cathedral section was 96 *braccia*.[15]

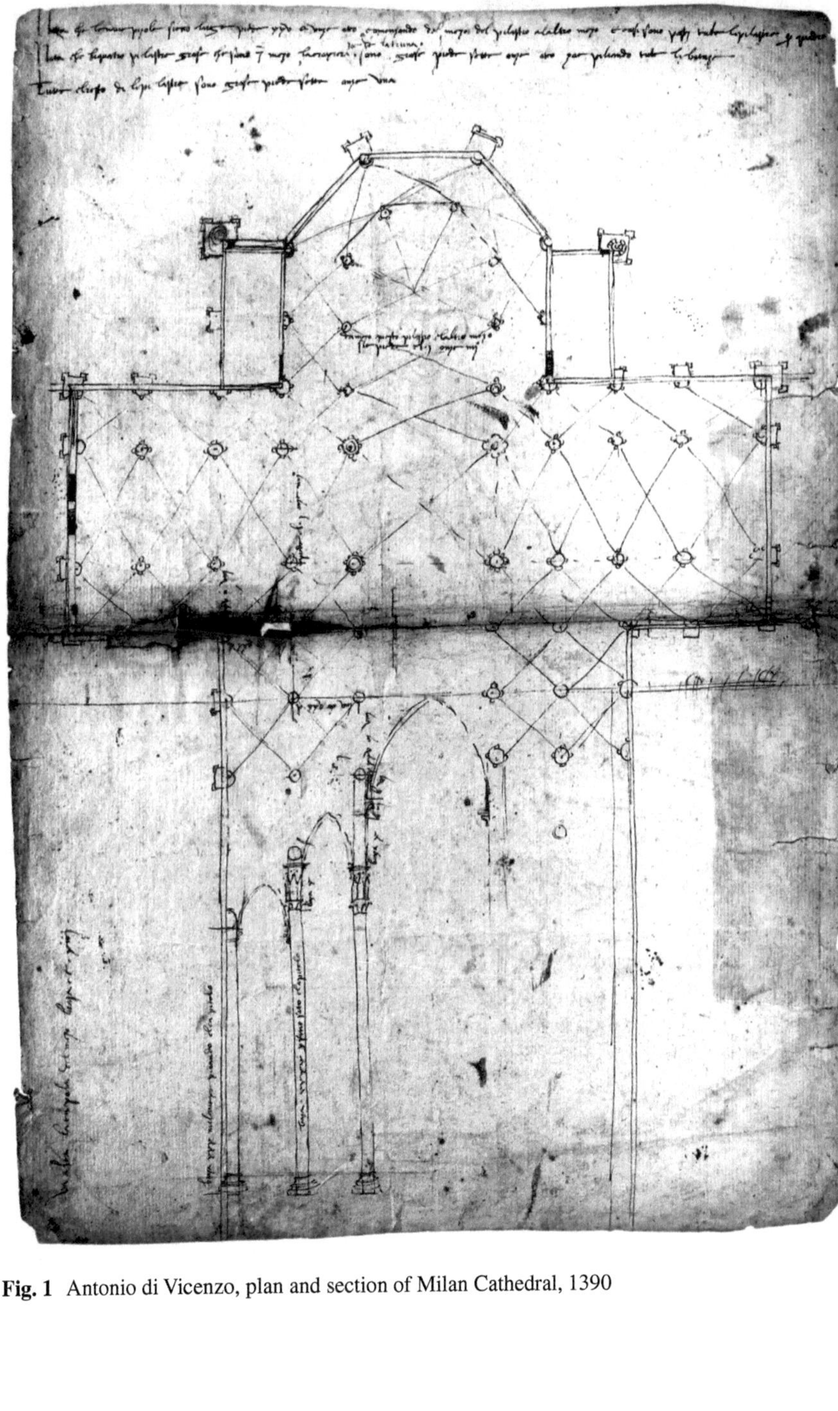

Fig. 1 Antonio di Vicenzo, plan and section of Milan Cathedral, 1390

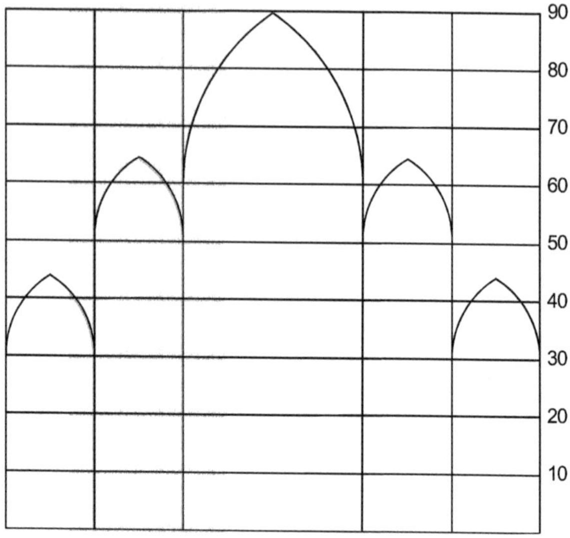

Fig. 2 Modular schema for the first design of Milan Cathedral

At an early point in the story, sometime after March 1391, the Milanese masters decided to adopt a revised design for the cross-section, based on an equilateral triangle. This idea, put forward by the new consultant engineer, Annas de Firimburg, was presumably intended to lower the profile of the building and to inscribe it in a regular geometric shape, avoiding the somewhat arbitrary relationships between the aisles and the nave that had governed the earlier design. In accepting the new scheme, however, the building council encountered a difficult problem, for they needed somehow to calculate the projected height of the new structure, now incommensurable with its width. The workshop appealed, not to another consulting engineer, but to a mathematician from Piacenza, one Gabriele Stornaloco. Described as an "expert[us] in arte geometriae," he was asked to "discuss the questions concerning the height and about other things with the engineers of the said *fabbrica*," that is, to determine the height and reconcile it with a modular system based on the Milanese *braccio* for the rest of the cathedral.[16]

Stornaloco's solution to the problem is known from the reply that he sent to the committee, which was decoded by Paul Frankl and Erwin Panofsky in another famous article from the 1940s.[17] Panofsky showed that Stornaloco solved the problem by translating it into a four-step algorism—that is, a calculation using Hindu-Arabic numerals—involving the multiplication and division of sums to three and four places. The formula served to approximate the irrational term that was central to the solution of the problem, namely the square root of three.[18] Panofsky surmised that Stornaloco had employed an inherited formula, invented by Leonardo Fibonacci, but in wide use by the late fourteenth century. Adjusting the algorism to suit the circumstances of this particular problem (to express the value in terms of a whole number divisible by 8, that is, half that the module length), Stornaloco was able to determine

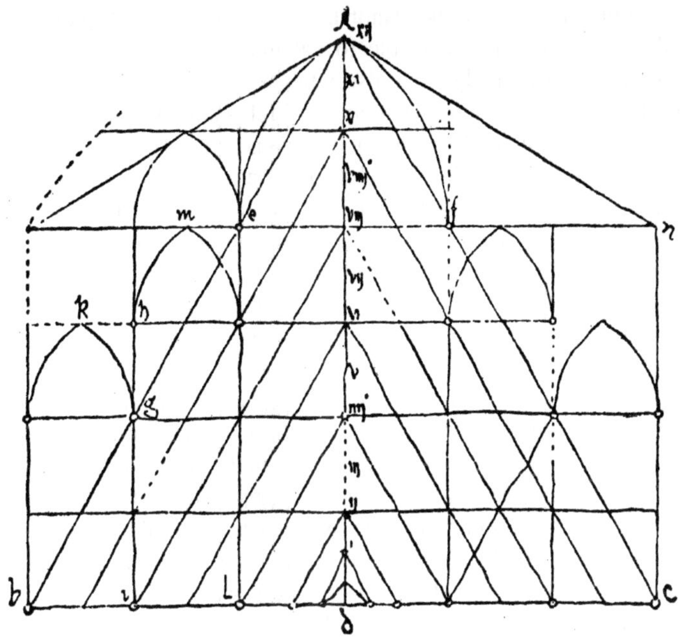

Fig. 3 Stornaloco's design for Milan Cathedral (From Beltrami 1887)

the height of the nave within .04 % of the true value (83.168 *braccia*). Indeed, he went farther. Using 84 *braccia* as a convenient approximation (giving him a module height of 14 *braccia*, as opposed to 16 in width), he was also able to correlate the nave height with the springing of the outer aisle and nave vaults. As Stornaloco pointed out in his letter, the scheme implies a series of concentric, similar triangles that link the module widths at the base of the section with the module heights along the centerline (Fig. 3).[19] The two largest of these inscribed triangles establish the heights of the principal vertical elements, namely the nave and aisle piers.

Stornaloco's solution is characteristic of an academic mathematician, not a practicing builder. Indeed, the strict formal regularity and thorough internal consistency of his design are unparalleled in Gothic architecture. These qualities may partially explain the project's immediate impact. The workshop adopted his proposal not only for the nave height, but for the entire cross-section, using it to direct construction for the next several months. This was enough to dictate the height of the existing outer aisles and to provide the basis for a revised, compromise design for the rest of the work in May 1392.[20]

What does this episode tell us about the mathematical abilities of medieval masons? The answer is not straightforward. The incident is typically seen in terms of the masons' limitations, that is, as an example of the kind of mathematical problem that medieval architects could not solve. While that reading is true in broad terms, the details need some unpacking in light of what we have learned about medieval design since Frankl and Panofsky's day. The relevant questions are not only whether the workshop was able to calculate the height of the nave *à la* Stornaloco—which is unlikely—but also whether the masons were able to determine that dimension using their own methods. In this latter respect, the fact that the workshop felt compelled to call on a mathematician is indeed surprising. Geometrical design was a mainstay of medieval architectural practice, as were the "irrational" relationships it produced. The masons' techniques were largely instrumental, graphic, and empirical—not demonstrative. Yet, combined with on-site testing and verification, they were more than enough to achieve very high levels of building precision. This was true not only for equilateral triangles. Gothic builders appear to have been fully capable of incorporating even more complicated shapes into their buildings: octagons and pentagons were not unusual.[21]

Looking at the problem from a different angle—arithmetically—we might again ask whether the workshop strictly needed an outsider to determine the nave's height. Assuming the cathedral masons could not calculate the solution in the same manner as Stornaloco, is it possible that they could have approximated it with an arithmetic rule-of-thumb, one that could relate the base of the cathedral to its height in terms of a commensurable ratio? Procedures of this kind are believed to have been used widely. In the first place, geometrical relationships were not always possible to set out physically. Existing buildings or other obstructions might easily block the sweep of a long diagonal, and such operations were particularly difficult to perform in elevation, as Frankl himself noted with respect to the problem at Milan. Practical difficulties alone would suggest the occasional need for rational approximations of geometrically derived dimensions, and scholars have, in fact, found some evidence for the use of such ratios by medieval builders. Indeed, recent discoveries by Matthew Cohen have thrown this practice into sharp relief. At Santa Maria del Fiore in Florence, the first bay of the nave arcade, designed by Francesco Talenti around 1357, incorporates a ratio of 29:41 *braccia*. These dimensions approximate the relationship between the side and the diagonal of a square with an error of only .03 %. The same ratio appears in an even more sophisticated form at San Lorenzo, built by Brunelleschi from 1421, where the width of the bays and height of the six westernmost columns measure 9⅔ and 13⅔ *braccia* respectively.[22] This approximation and others were known in antiquity. Theon of Smyrna reports an arithmetic method for generating increasingly accurate whole-number approximations for the side-to-diameter ratio of a square, that is, a series that progressively converges toward the square root of two. Although a route of transmission remains elusive, it is possible that this method was known to ancient builders and handed down through the Middle Ages. Indeed, some scholars have gone farther, arguing that medieval builders would have had similar formulas for calculating ratios that progressively converge toward any desired square root, thereby approximating proportions inherent in the equilateral triangle, the root-5 rectangle, and even the "golden section".[23]

These broader claims have not been verified, but in the case of Milan cathedral they are probably not relevant. A simple ratio like 8:7—the same used by Stornaloco in his own design (96 × 84 *braccia* for the width of the cathedral to its height)—could have been handed down as part of the masons' oral tradition. The workshop would, in any case, have been fully capable of working out an adequate ratio for the cross-section of the church using scaled drawings or cords set out at full scale. What the masons could not have known is the degree of divergence between their physical approximation and the closest possible numerical value, but this level of accuracy was for all practical purposes unnecessary, as Stornaloco himself seems to have recognized. In this light, the decision to call on the mathematician appears less as evidence of an intellectual failing on the part of the masons than as an artifact of the very unusual circumstances at Milan. As historians have long recognized, the Lombard masters were unused to the Gothic design system of their northern consultants. This clash of sensibilities—which would only intensify with Heinrich Parler's arrival in late 1391 and Jean Mignot's in 1399—may have led the two parties to see Stornaloco as a useful go-between.[24]

Even this charitable interpretation, however, should not obscure the essential fact about this incident. As "unnecessary" as Stornoloco may have been, his solution to the problem was nonetheless diametrically opposed to the methods then available to architects and masons. Instead of working out a sequence of increasing number pairs for the sides and height of the triangle (assuming the masons had such a technique), Stornaloco recognized the root that lay at the heart of the problem and expressed it, moreover, in terms suitable to the particular circumstances he found at Milan cathedral. If Panofsky is right—his is still the most convincing explanation—Stornaloco used a formula that expressed the side of the triangle in units divisible by eight (half the module width) and multiplied all the terms by ten to avoid fractional remainders. He then adjusted his approximation for $\sqrt{3}$ to be both easier to manipulate and more accurate (175/101 instead of 173/100).[25] In comparison to Theon's method, which calls for nothing more than basic addition, Stornaloco's is a true algorism that requires multiplication and division of terms to three and four places.

Stornaloco's approach to the problem, in other words, was thoroughly arithmetized, and it is this quality that sets it apart from the masons' techniques. Indeed, his explanatory letter suggests that it may not have been simply the translation of the problem into a formula that lay beyond the capabilities of the builders, but also the manipulation of the numbers themselves. Why else would he have expressed the coefficients used in the solution with Roman numerals, rather than the Hindu-Arabic ones necessary to calculate it? It is also worth noting that Roman numerals predominate in the conference minutes. Calculation using Hindu-Arabic numerals was still not universal in 1391, and Stornaloco may have had some reason to assume that the masons were not familiar or comfortable with them.[26]

These conferences are very well-trodden ground, but the great advantage of this material is that it offers a concrete historical link between the mathematics of the worksite and that of the classroom. Indeed, these conferences remain the only source we have for the direct interaction between medieval mason-practitioners and a university- or *abaco* trained mathematical scholar. At the same time, the Milan affair also makes clear the distance between these two worlds, and in this respect, reinforces the work of other historians in this area. As Lon Shelby has shown, the "geometry" of the medieval mason appears distant from most practical geometrical texts of the time. The few and scattered sources we have of the mason's art—texts by Villard de Honnecourt, Matthias Roriczer, and Hanns Schmuttermayer—reveal that it consisted largely of rule-of-thumb procedures involving the construction and manipulation of simple geometrical forms. Unlike the methods taught in the schoolman's *Practica geometriae*, the mason's techniques were often imprecise and approximate. They utilized no arithmetic calculations of the kind known at the time, nor did they reveal any understanding of the Euclidian theorems and proofs that would justify the operations involved. Although medieval masons certainly possessed a sophisticated intuitive grasp of spatiality and of spatial forms, their design and construction methods appear to have been essentially pragmatic and empirical, not mathematical or analytic.[27] What is immediately striking about the Milan story, particularly about Stornaloco's involvement, is that the two parties involved correspond, almost perfectly, to the two modes of geometrical thought that Shelby describes.

"High" and "Low" Mathematics: Antonio da Sangallo the Younger

Did medieval builders ever benefit from contact with "higher" mathematics, that of the university classroom or the humanist's library? The record is indeed scarce, but historians have unearthed some isolated examples that point to increasing interaction between the two domains. The Florentine new towns of the early fourteenth century, brought to light by David Friedman, provide unique evidence of an entirely innovative and sophisticated application of geometry to the problems of urban design.[28] The street plans of San Giovanni and Terranuova, founded in 1299 and 1337 respectively, are laid out in such a way that the widths of the residential blocks decrease in proportion to the chords of a circle advancing at set intervals. As Friedman points out, these designs presuppose a knowledge of trigonometry that could only have originated in a textbook tradition far removed from the working knowledge of most mason-builders. Matthew Cohen's work on San Lorenzo in Florence has brought to light an analogous example. The nave arcade appears to embody a complex set of interdependent proportional relationships that may incorporate a Boethian number system.[29] Without these examples, we might be entirely justified in believing that the level of mathematical knowledge at Milan cathedral was representative of medieval masons in general. In reality, there may have been particular workshops, cities, or regions, where graphical, numerical, and technical ability were more advanced and where individual craftsmen were more receptive to influences from parallel or analogous fields. Given the prevalence of the *abaco* curriculum in central Italy, the influence of a highly educated merchant culture, and the sophistication of local surveyors, it is perhaps not surprising that innovations of this kind first appear there.

The Stornaloco incident is worth scrutinizing not to belittle medieval masons, but to underline one of the central premises of this volume, namely that Renaissance architectural practice was characterized by a new orientation toward both the speculative tradition of ancient geometry and to the advances in practical mathematics that accompanied its revival. Indeed, Renaissance architects often distinguished themselves from their predecessors on these very grounds. In the opening pages of the *Primo Libro* (1545), for example, Sebastiano Serlio complained of those "who today bear the title, 'architect' but who do not know how to give a definition of a point, a line, a plane or body, or say what correspondence and harmony are."[30] He explicitly identified his approach with the theoretical principles derived from Euclid's *Elements*. This transformation has typically been laid to the influence of Neoplatonism, but it was in fact impelled by a number of factors, including the rediscovery of linear perspective and the general resurgence of the practical, mathematical sciences in the fifteenth and early sixteenth century. This influence is evident in many areas. As several recent studies have affirmed, Renaissance architects show a newfound awareness of the dynamic properties of structure, a greater familiarity with techniques of arithmetic calculation, and a growing interest in new mathematical sciences, such as trigonometry.[31] This change is also reflected in the art's renewed status as an intellectual discipline. Sixteenth-century divisions of knowledge—ramified disciplinary "trees"—often list architecture under the general heading of mathematics, usually alongside mechanics.[32]

It is against the background of this transformation that I want to set a second example, a series of drawings of the 1520s and 1530s from the recently published corpus of Antonio da Sangallo the Younger.[33] Ann Huppert's chapter below discusses these documents at length and in a more specific architectural context. They serve here simply as a contrast with the picture suggested by the Milan cathedral workshop and to illustrate the enormous sea change that was entailed in the transition to the humanist architectural culture of the Renaissance. In that light, the drawings are remarkable, because they show the architect engaged in a purely personal study of just the kinds of problems that appear to have stumped the Milanese masters and that Shelby describes as being outside the normal working methods of medieval masons in general. Indeed, the great fascination of these notes and sketches is that we have very few earlier examples for this kind of interest or ability among architects. It is important to note here that we are not dealing here with an Alberti or a Barbaro, but with a building practitioner trained in a traditional—if not to say medieval—apprenticeship system.[34]

In the first place, the drawings evince an ease and facility with arithmetic calculations, examples of which cover large portions of the sheets concerned. These are all the more striking in light of the Milan episode, since the multiplied sums often include regular fractions in an attempt to find approximate values for square- and cube roots (Fig. 4). In a general sense, the figures show the importance of the *abaco* in the period. Indeed, the editors of the volume make a point of noting Sangallo's mastery of technique.[35] Not only was he apparently taught by such a master, he was also able to think in terms of algorism, adapting it to new problems thrown up by craft practice and his theoretical interests. In this respect, Sangallo was far from alone among contemporary architects. As Ann Huppert remarks in her essay below, his mathematical abilities were matched, if not exceeded, by those of Baldassarre

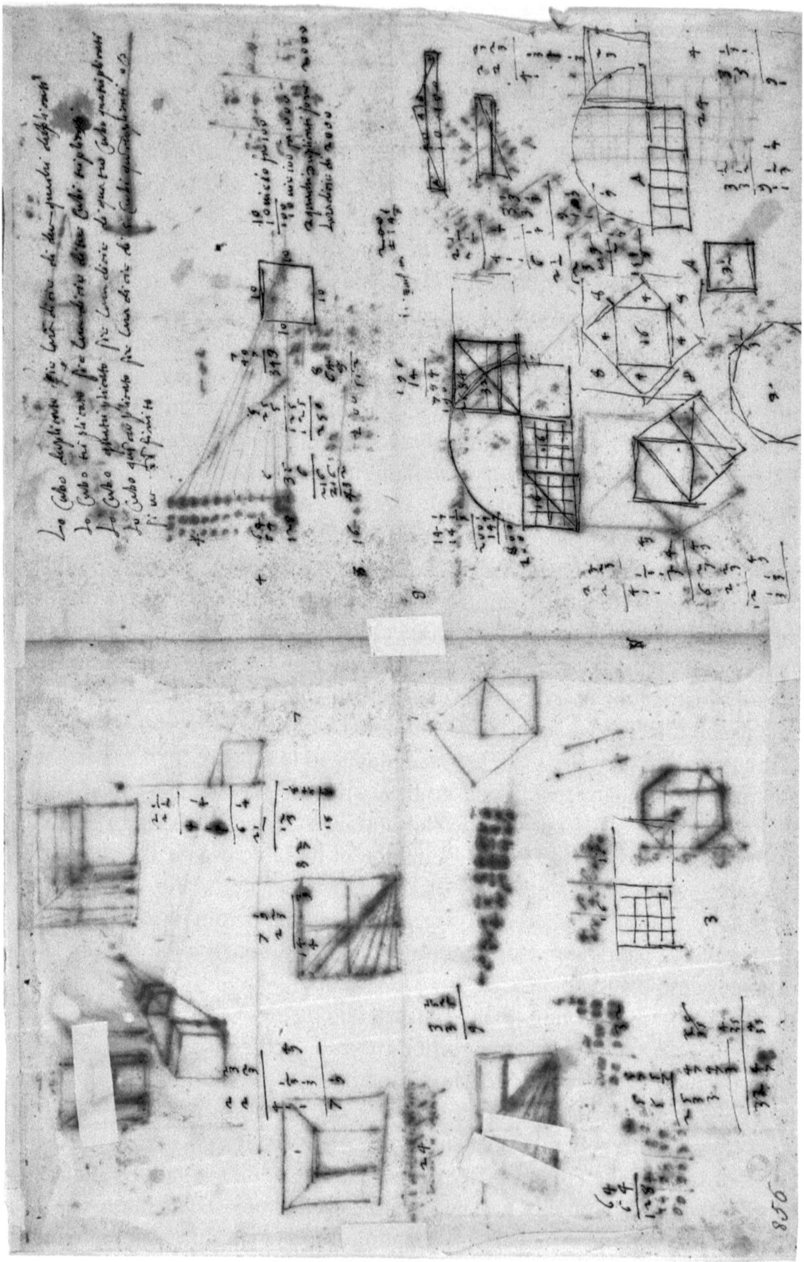

Fig. 4 Antonio da Sangallo the Younger, sheet of calculations, UA 856v

Peruzzi, whose own use of algorism was even more closely intertwined with his architectural practice.

In many respects, these drawings are characteristic of what Carlo Maccagni has described as the intermediary or "vulgar" science of Renaissance artisans, technical experts, and practitioners. Works of this kind typically take the form of *taccuini* and *zibaldone*, written in the vernacular with a mixture of notes and diagrams. Like Leonardo's notebooks—the most emblematic of the genre—they represent an *ad hoc* process of learning adapted from the experience of the workshop, where activity tends to follow not a formal curriculum, but rather the meandering course of an apprenticeship or the unpredictable demands of a clientele. Such works investigate individual problems as they are encountered and worked through on a case-by-case basis. In this respect at least, they tend to mirror the format of contemporary abacist manuals and practical geometries.[36]

Sangallo's drawings illustrate the way in which his *abaco* education provided routes into his theoretical and quasi-scholarly interests. In the case noted above, the architect's calculations seem to have arisen from an exploration of geometric constructions relating to the doubling of the square and the cube, as mentioned in Vitruvius (Book 9, Introduction). Other drawings also seem related to those parts of the *abaco* curriculum concerned with solid geometry. One sheet, for example, reflects attempts to find the volume of pyramids of different form (Fig. 5). This kind of problem was famously and rather more expertly explored by Piero della Francesca and Luca Pacioli in their own *abaco* instruction manuals, but here Sangallo seems to have again been inspired by Vitruvius. As the editors point out, the interspersed diagrams of stepped pyramids may reflect an attempt to reconstruct the Mausoleum of Halicarnassus as described by the Roman author (Book 2, Chapter 8.10–11).[37] The figure in the upper right portion of the sheet, showing the doubling of the square, is likely stimulated by the same source.

The range and variety of Sangallo's graphic *oeuvre* are remarkable. Like the drawings of his older contemporary, Francesco di Giorgio Martini, Sangallo's encompass an array of technical and engineering-related activities that far exceeds modern notions of the architect's role. Drawings of artillery, instruments, and machines of all kinds are perhaps the most unexpected, precisely for their lack of any specific connection with building. Yet, it is also clear that these sketches formed part of a common disciplinary constellation. The same range of interests—especially in the fields of astronomy and cartography—would characterize the principal concerns of mathematical practitioners throughout the sixteenth and seventeenth centuries.

Among the most striking examples of Sangallo's curiosity is a proposal—"my opinion" he calls it—for a system of curved panels or gores for the construction of a globe (Fig. 6). This technique of cartographic projection was relatively new, having been published for the first time by Martin Waldseemüller in 1507. Sangallo might have come across the idea in an intermediary text—several other examples had appeared over the intervening twenty-odd years—but his awareness of an innovation so far outside his own training is nevertheless surprising. The projection is composed of twelve gores, dividing the globe by 30° intervals of longitude. The figure

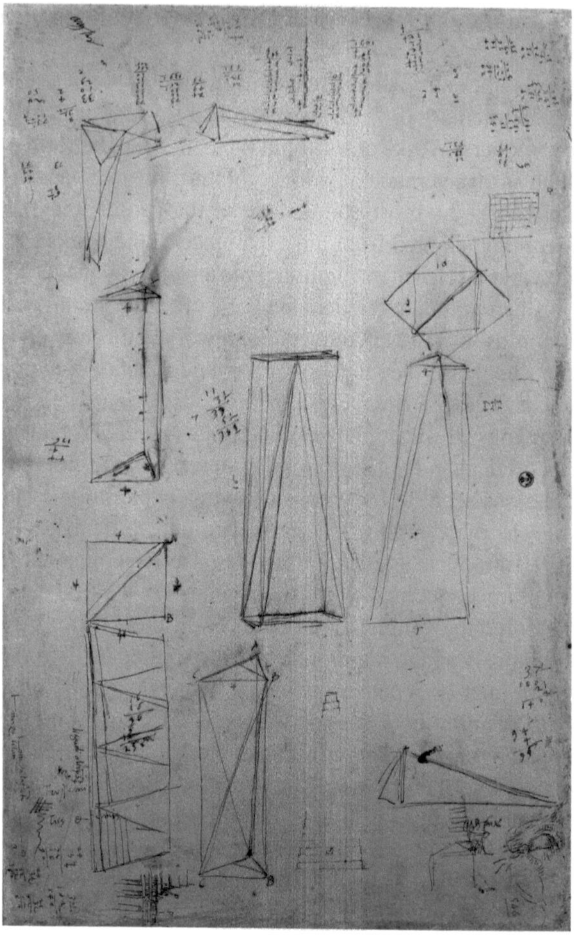

Fig. 5 Antonio da Sangallo the Younger, calculations and geometrical diagrams, UA 857r

1865 is inscribed between each gore, calculated (incorrectly) as one twelfth of the total circumference of the Earth, which Sangallo gives at the top of the sheet as 22,500 miles. As the editors point out, no other world maps are found among the architect's drawings, but that is not to say that the work is entirely isolated. Antonio executed several topographical surveys—typically involving fortifications—and still other drawings related more broadly to cosmography. A geometrical study of the constellations in the northern hemisphere, for example, is conceptually not far removed from Sangallo's globe gores.[38]

Several drawings of mathematical instruments reveal another dimension of this concern for terrestrial and celestial measurement. The most extraordinary of this group—indeed, of the corpus as a whole—is an astonishingly faithful replica of an Arabic astrolabe, complete with its Kufic inscriptions (Fig. 7). Sangallo surely could not read them, but this did not diminish his fascination with the object, as is

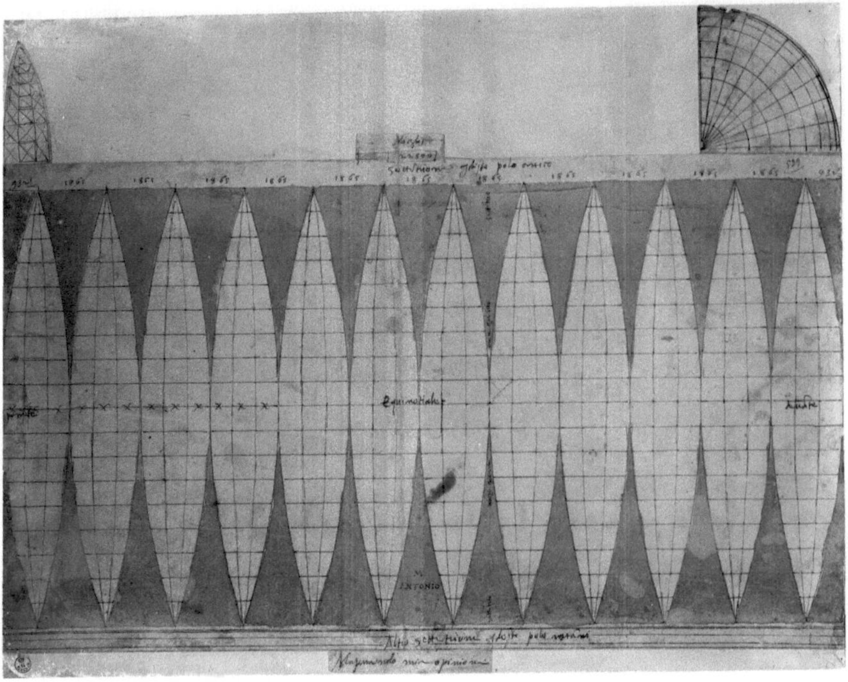

Fig. 6 Antonio da Sangallo the Younger, globe gores, UA 850r

evidenced by his painstaking method of description. The drawing stunned the editors of the volume. "We marvel," writes Gustina Scaglia, "at his determination to make a facsimile of all the letters." Indeed, the transcription is legible enough to determine the maker's signature, as well as the place and rough date of origin: Morocco, probably—on the basis of its similarity to other known instruments—in the tenth or eleventh century.[39] Not only was Sangallo's study precise, it is also systematic. The astrolabe's front and rear faces are juxtaposed, and its component parts—alidade, rete, even a replacement screw and nut—presented separately and distinctly. As Scaglia points out, Sangallo evidently knew something about such instruments. The inscribed notes suggest that he was familiar with the names, arrangement, and function of the parts, including the use of the sight holes in the alidade. He evidently knew how to engrave the scales and geometric constructions on the two faces. Astrolabes featured commonly in medieval and quattrocento practical geometries, where they are typically shown as helpmates for rudimentary surveying problems. Sangallo's knowledge of such instruments seems to have gone much deeper.

One last image from the group reveals Sangallo's more-than-passing interest in instruments. In terms of contemporary advances in mathematical practice, it is one of the most noteworthy of the entire corpus: the earliest known representation of a flat-sided, proportional compass—or sector—with scales incised on the face of each leg (Fig. 8). It is a hinged rule used with dividers to transfer dimensions to and from

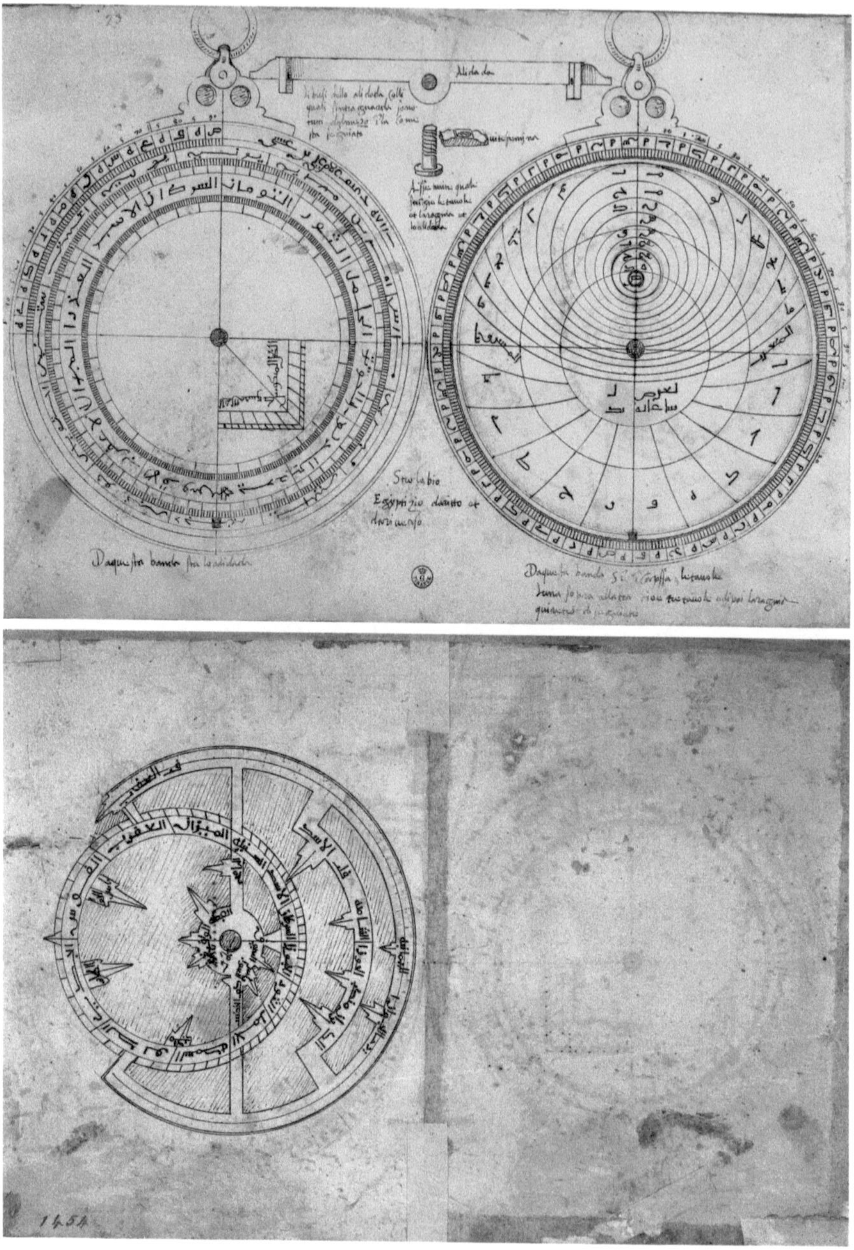

Fig. 7 (**a, b**) Antonio da Sangallo the Younger, front and rear faces of an Arabic astrolabe, with rete below, UA 1454r-v

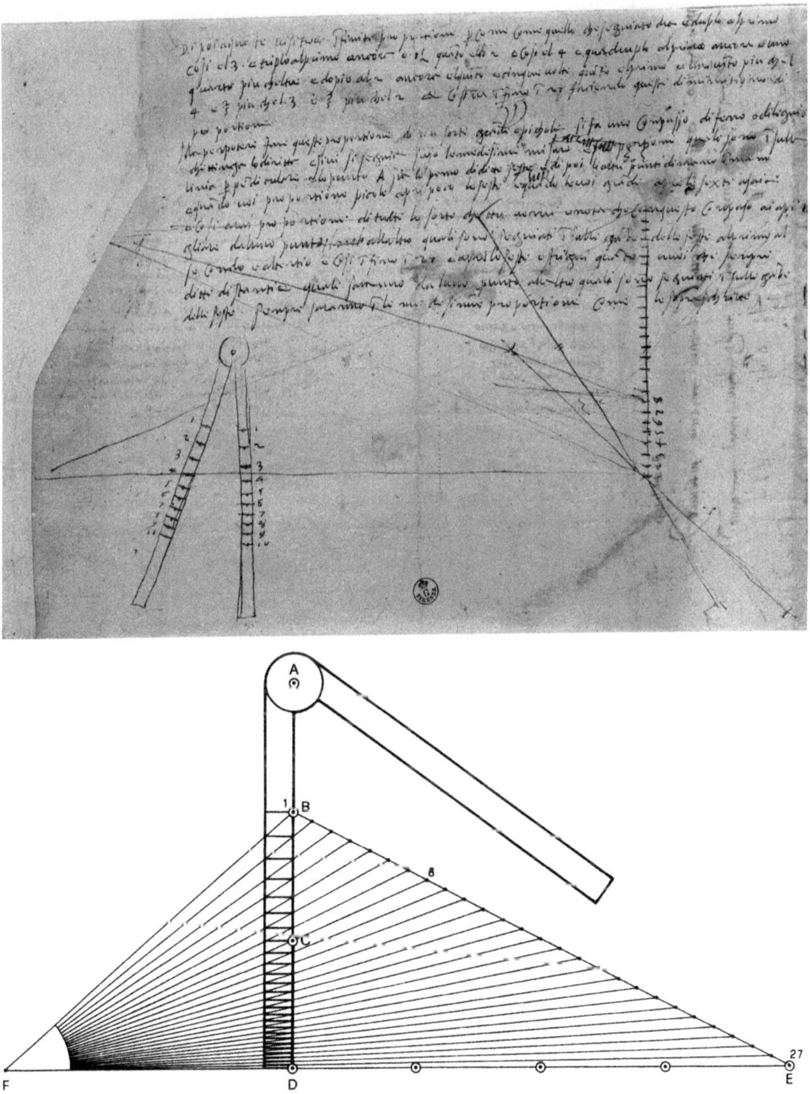

Fig. 8 (**a, b**) Antonio da Sangallo the Younger, sheet of notes for a proportional compass or sector (UA 1491r), with modern reconstruction (from Camerota 2000)

the pairs of identical scales on its two legs. Opened to a given length, the matching scale divisions provide additional dimensions continuously proportioned to that initial length. Sangallo designed the instrument as a rough graphic method of approximating cubes and cube roots. As Filippo Camerota has shown, theorists of perspective—Alberti, Piero, and Dürer, in particular—had earlier demonstrated the basic principle, but it was Sangallo who gave this instrumental form to it. The idea may have been inspired by his study of the problems in Book 9 of Vitruvius; the drawing appears to have been preceded by a number of preparatory studies exploring analogous methods of approximating square and cube roots. In practical terms, the instrument could be used to easily rescale architectural elements, but the important innovation was conceptual: the sector provided a mechanical and geometrical approach to problems that resisted a quick arithmetic solution. The same principle governed the many more sophisticated forms of the instrument that appeared throughout the century, culminating in Galileo's own version, invented in 1597 and published in 1606. That the device had originated in an architectural context had been largely forgotten, but it makes sense in light of later attempts to adapt the sector to the rules of classical design, first by Ottavio Revesi Bruti in the early seventeenth century and later in eighteenth-century England.[40]

Sangallo's mathematical drawings make a blunt, but telling, contrast to the Milan cathedral debates—the Stornaloco episode in particular. In light of the latter, the former reveals the profound transformation that had taken place in the mathematical culture of elite building practitioners during the preceding century. Sangallo's training was traditional, but it was bolstered by a sound mathematical education and by a rich humanistic culture that linked his art with the latest advances in the practical mathematical sciences. Sangallo's inspirations were not only textual, but also personal and professional. The drawings as a group reflect two important influences. The first was the Della Volpaia family of clock- and instrument makers, with whom the Sangallos had been linked since the late fifteenth century. Antonio appears to have been close to Benvenuto della Volpaia, in particular. The two men are believed to have shared drawings of instruments and mechanisms; several duplicates are found among their respective manuscripts. The second—and more important—influence was likely the learned architect, engineer, and antiquarian Fra Giovanni Giocondo of Verona. Among contemporary humanists, Giocondo had a virtually unique ability to marry erudition and practice, an admixture evident both in his groundbreaking 1511 edition of Vitruvius—the first with illustrations—and in his work as a manuscript hunter. Giocondo was an assiduous collector of medieval practical geometries, including several rare French versions, acquired presumably during his time in Paris as the royal engineer to Charles VIII. Sangallo's connection with Giocondo probably dates to sometime after June 1514, when both men were employed at the workshop of St. Peter's. In terms of the direction of influence, we are here on more solid ground. A note on one of the geometrical studies in Antonio's collection attributes it explicitly to the Franciscan friar.[41]

From the Mathematical to the Physical Sciences: Pierre Bullet

It would be difficult to link Sangallo's mathematical notes and drawings directly to the design of any one of his buildings. As Ann Huppert demonstrates, below, they bear rather on the history of design itself. Architects of Sangallo's generation were among the first to incorporate Hindu-Arabic algorithms into the design process,

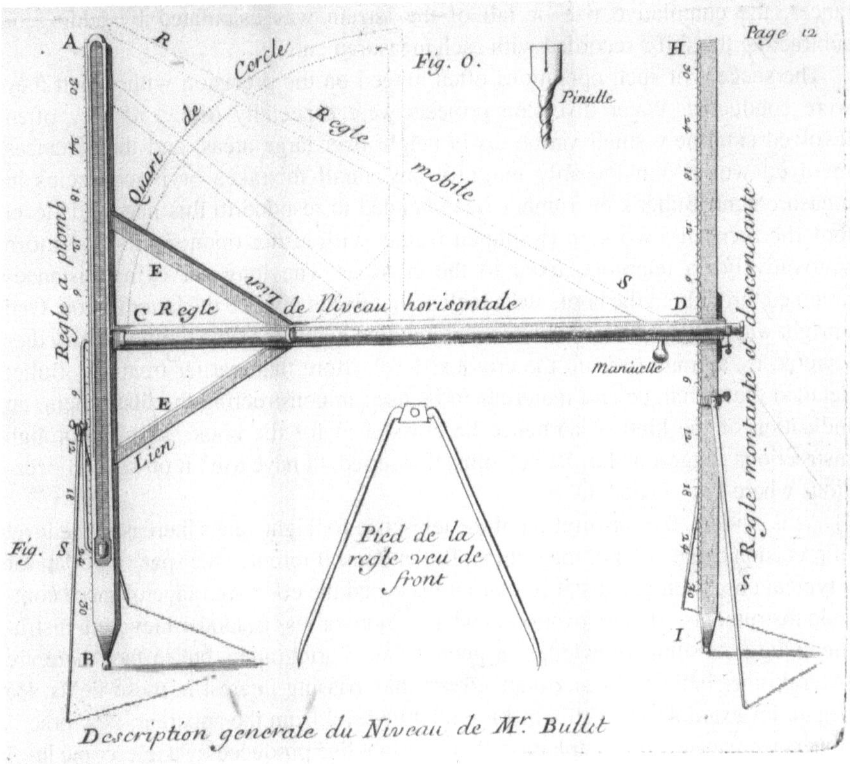

Fig. 9 Long-distance level (from Bullet 1688)

with all the advantages that entailed for cost estimates and bookkeeping, not to mention site surveying and the ability to build from accurate drawings.[42] Sangallo's notes also say something about his broader intellectual culture, particularly his receptivity to forms of mathematics that lay, strictly speaking, outside his craft. In this respect, the influence of figures like Giocondo must have been decisive. This kind of encounter is also significant in a longer historical perspective, for it was typical of the kind of intellectual relationships that architects would continue to exploit throughout the early modern period.

It is with this idea in mind that I want to adduce a third case-study. This one comes from a later moment, in 1688, toward the end of our period. In that year, the architect Pierre Bullet published a treatise on the long-distance level (Fig. 9).[43] This was an instrument used to determine precise gradients over large tracts of land, to cut slopes and terraces, for example, or to align canals or divert waterways. The principle was simple enough. One team would stabilize the device horizontally and aim it into the distance. Communicating with visual signals, a second team at the other end of the area to be surveyed would raise or lower a cardboard marker until it met the sightline. Changes in elevation were determined simply by subtracting the height of the eyepiece from the height of the marker or vice versa. Over longer distances, the cumulative rise or fall of the terrain was calculated by adding or subtracting the shifts recorded with each measurement.

The success of such operations often hinged on the precision with which they were conducted. Water diversion projects were especially tricky, as they often involved extremely small variations in height over large areas, and the distances involved would considerably magnify any small mistakes or inaccuracies in measurement. Bullet's instrument was intended to respond to this need. His level took the form of a wooden H-shaped frame, with either open sights or—more innovatively—a telescope fixed to the crossbar. The long surveying distances assumed throughout the book make it clear that Bullet favored the latter option. One upright was hollowed out to contain a plumb line, while the other could be raised or lowered by a crank to keep the crossbar level. More than earlier treatises, Bullet detailed the technique and materials to be used in constructing the instrument, an indication of the kind of audience he envisioned for the book. As his thorough instructions suggest and as Bullet himself asserted, "I have used it on several occasions where I've needed it."[44]

As is evident, the substitution of the telescope for sight vanes increased the level of precision by an order of magnitude. Bullet himself refers to the span of 400 m for a typical observation, a distance that far exceeded the effective range of more common instruments. This innovation in what is more or less a standard leveling instrument suggests some knowledge of geodesy and cartography, but in fact there are several other indications of Bullet's more-than-passing interest in these fields. He began, for example, by distinguishing the true level from the apparent. Few practitioners were aware, he complained, that the sight line produced by a telescopic level does not follow the curved surface of the earth but actually forms a tangent to that surface (Fig. 10). The longer the distance surveyed, the more the visual ray actually rises away from the true level, which follows a circular arc with the radius toward the center of the earth. He then provided a rule-of-thumb to compensate for the deviation. In distances over 200 m (100 *toises*) the apparent level has to be lowered by about 21 mm (1/12 *pouce*) to remain accurate. After 600 m, it has to be lowered by 190 mm, and so on. More accurate values for the deviation, Bullet noted, could be obtained by using sine- or log tables. In contrast, his rule was "more mechanical but also quicker and easier to understand." [45]

That a guild-trained architect would be familiar with the basic trigonometry of the problem is surprising enough. More impressive is the fact that the effective value that he employed for the radius of the Earth was taken from the most up-to-date work of "les astronomes modernes." That was not all. Bullet also warned of the refraction of the visual ray caused by water vapor in the air. "This," he explained to his readers, is "what mathematicians call parallax." In describing the phenomenon, Bullet articulated a fully-fledged mechanical theory of evaporation, noting that the rays of the sun on the surface of the earth "excite and cause large particles of water to rise with the more subtle matter."[46] His method of correcting such errors called for two leveling teams, each compensating for the discrepancy in the other's measurement. The technique was borrowed directly from astronomical practice for verifying the built-in deviation caused by non-centered lenses.

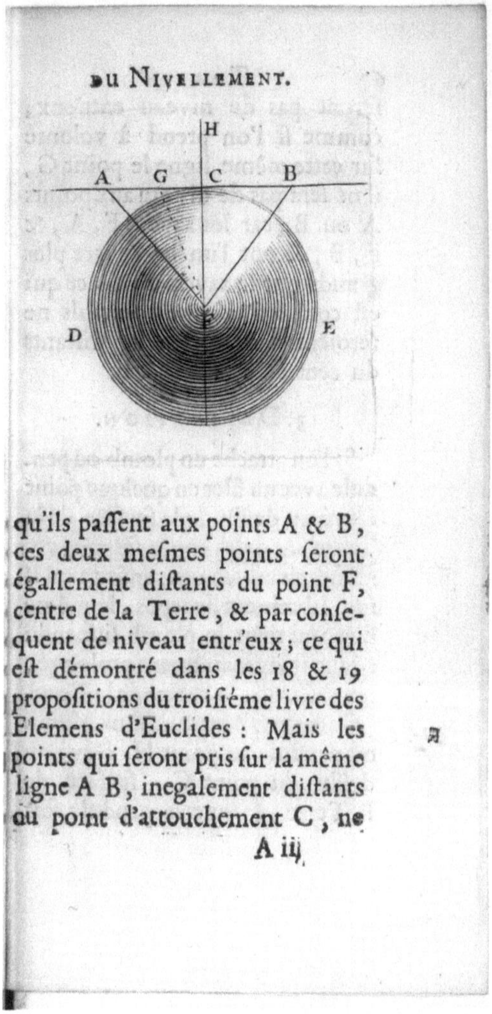

DU NIVELLEMENT.

qu'ils paſſent aux points A & B, ces deux meſmes points ſeront égallement diſtants du point F, centre de la Terre, & par conſequent de niveau entr'eux ; ce qui eſt démontré dans les 18 & 19 propoſitions du troiſiéme livre des Elemens d'Euclides : Mais les points qui ſeront pris ſur la même ligne A B, inegalement diſtants du point d'attouchement C, ne

A iij

Fig. 10 The apparent level (from Bullet 1688)

Bullet's source for this information is not difficult to find. His interest in the subject of large-scale leveling followed a number of recent advances in surveying by the Académie royale des sciences, in particular by the astronomer Jean Picard, as part of his research in geodesy and large-scale cartography. As reported in his official account, Picard's procedure involved establishing an arc of meridian between two distant localities by connecting between them a series of triangles formed by prominent landmarks—typically hilltops and church steeples. Triangulation had been used for this purpose since the early sixteenth century, but Picard revolutionized the technique. By incorporating telescopes into his surveying device—an astronomical quadrant adapted to take horizontal measurements—he greatly expanded the distance between stations (Figs. 11 and 12). The resulting meridian ran through

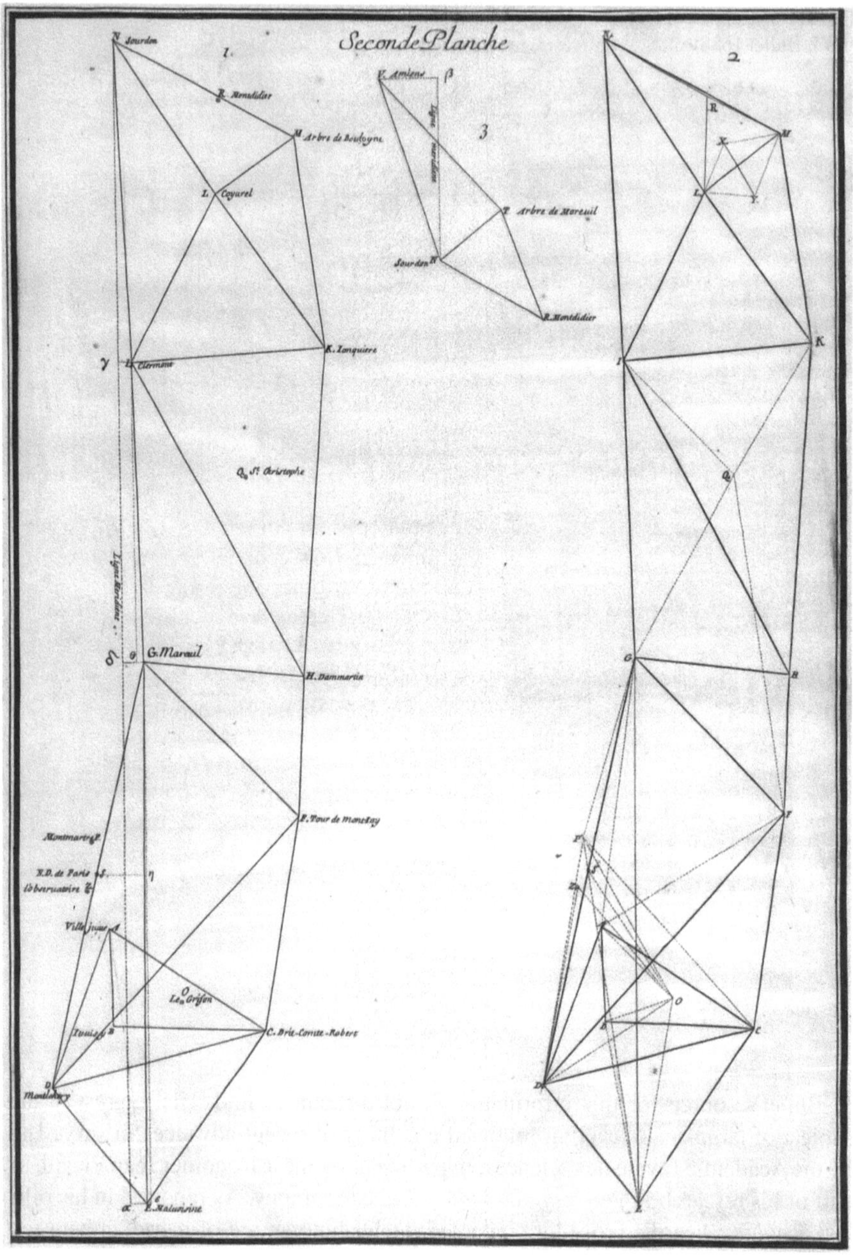

Fig. 11 Triangulation of the meridian between Malvoisine and Amiens (from Picard 1671)

Fig. 12 Triangulation of the Paris meridian, showing a night-time observation (from Picard 1671)

Paris between Amiens and Malvoisine, a distance of about 154 km, calculated to the nearest third of a meter. By comparing the latitude at both ends of the meridian, Picard used this measure to determine the length of a terrestrial degree. His results for the Earth's dimensions were more accurate than any previously attained and were cited well into the eighteenth century. In fact, they have been shown to be within 0.15 % of the latest values for this latitude (111.38 km).[47]

As a by-product of his meridian project Picard developed a second instrument: a telescopic level—the progenitor of Bullet's—for measuring changes in elevation over long distances (Fig. 13). Unlike the telescopic quadrant, Picard's level played no role in the construction of the meridian, which was established using only angular, horizontal measurements.[48] The level was, therefore, purely a practical tool, brought about as an incidental outcome of the meridian project. Yet it was precisely this connection that legitimized it, for the results of the operation made it possible to utilize the device with previously unimaginable precision. As Picard pointed out, the newly derived value for the radius of the Earth now made it possible to calculate the rise of the apparent level over the true level, that is, the rate at which the visual ray of the telescope diverged from the curved surface of the globe. Anticipating Bullet, he even provided a table of measurements to compensate for that deviation in distances beyond 100 m. The telescope also prompted another important "philosophical" consideration: the problem of water in the atmosphere, which subjected the visual ray to downward refraction as it passed from thinner to denser air, thus raising the apparent level. In this case, rather than tinkering with readings Picard recommended the use of middle stations to avoid the problem altogether.[49] What Bullet took from Picard's work was not merely a newfangled instrument, but an optical-physical theory—partly adopted from astronomical practice—to explain and justify its use.

Picard's *Mesure de la terre*—prestigious as it was—would hardly have been enough to give the telescopic level a broad appeal. People did notice, however, what

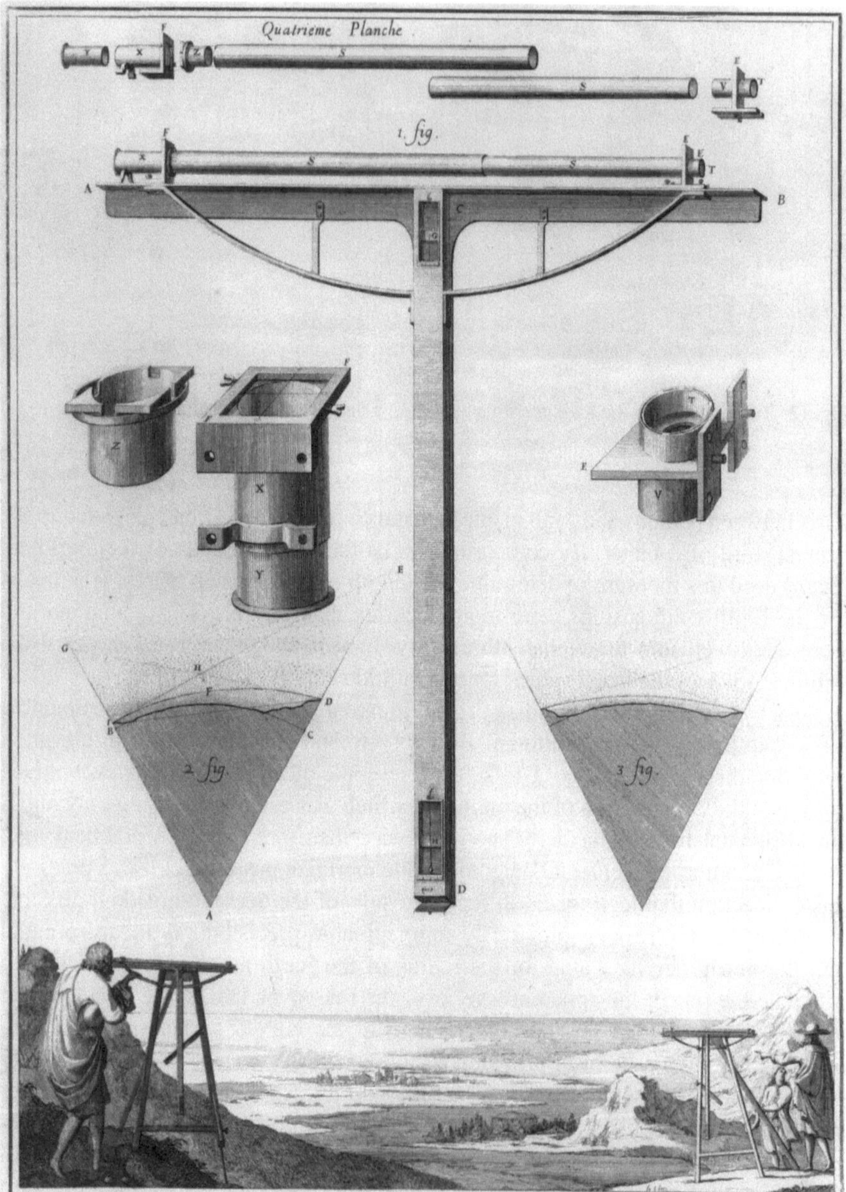

Fig. 13 Telescopic level (from Picard 1671)

the academicians did with it. In spring 1668, the academy began a long involvement with the gardens of Versailles, when they used the new instrument to establish the level of the Grand Canal. In the early 1670s, Picard and his colleagues systematically surveyed the terrain around the chateau and, from October 1677, developed the first branches of an extensive rainwater collection system to feed the garden's many fountains. Picard himself set out the channels and determined their rate of descent. Six years later, the marquis de Louvois commissioned Philippe de La Hire to extend Picard's system to the west, by damming several additional plains at consecutively higher elevations and linking them to the already existing conduits. In 1684–1685, the monumental—if ultimately aborted—project for the canal de l'Eure offered the Académie even greater scope for their abilities.[50] By the time Bullet published his own treatise, the telescopic level had been used to reshape the area around Versailles for 20 years.

The academicians knew that their work in this area had a potentially large and untapped audience. In 1684, La Hire edited and published Picard's manuscript treatise, *Traité du nivellement*, and in 1689, he wrote his own work dealing substantially with the subject, *L'Ecole des arpenteurs*. Unlike Picard's sumptuous folio volume, *Mesure de la Terre*, these were inexpensive, pocket-sized books oriented explicitly toward practitioners. The Académie's efforts to publicize their work paid off. Indeed, the added combination of efficacy and glamour associated with such devices could hardly fail to have had a wider effect. References to them began to appear in the *Journal des savants* from 1677 and in popular books on practical geometry from 1685.[51]

Bullet may have learned about the instrument from one of these texts, but a direct route is more likely; as a royal architect, he travelled in circles close to the Académie. The architect-academician François Blondel, with whom Bullet had collaborated since the late 1660s, is one possible source. From 1672, the two men worked together on a new urban plan of Paris, the first since 1652 to coordinate the streets and landmarks in a rigorous geometrical survey. For an architect interested in the latest advances in surveying and cartography, the Académie's work in this area would have been a natural focus of curiosity. More direct knowledge probably dates to 1685, during the planning of the canal de l'Eure. In spring of that year, Louvois summoned the Académie royale d'architecture to submit proposals for the aqueduct at Maintenon. A long elevation for the project is held among Bullet's papers in the Stockholm Nationalmuseum.[52] Whatever the ultimate source of Bullet's knowledge, it is the result of this interaction that stands out. His treatise represents a serious effort to respond to the Académie's discoveries in terms of everyday use. He was the first practitioner to digest these developments systematically and re-present them in a manner suitable for professional gardeners and architects.

A comparison between Bullet and Sangallo—artificial though it may be—serves to highlight some salient themes. Both men subscribed to a disciplinary identity that transcended their background in the building trades and that linked architecture with analogous—but quite distinct—forms of mathematical practice. Indeed, one of the striking similarities of this comparison is how prominently geography and astronomy figure in the work of each. Mathematics provided not only the basis of the craft, but also the means by which the architect might exceed the limitations of the builder's traditional background. New instruments were a particular focus of interest, both for the routes that they opened in to more prestigious, "theoretical" fields and for the greater control and effectiveness that they promised in everyday practice. The ambitions and motivations of both men, finally, were stimulated by personal and intellectual relationships, formed in active communities of both scholars and "enlightened" practitioners.

The comparison suggests a broad continuity of mathematical expertise and interests between elite building practitioners in the early sixteenth and late seventeenth century, but some significant differences are also worth noting. The first is the role of institutional authority in fostering a culture of both "high" and "low" mathematics. Not only did the crown sponsor Picard's high-status work in geodesy, the commissions for Versailles also spurred the Académie to develop the practical benefits of the research. The court provided, moreover, a pervasive patronage context that served to drive interest in the telescopic level. The instrument was effective and it had prestigious scientific origins, but it also contributed dramatically to the self-presentation of the king in an area that was both deeply important to him personally and central to contemporary notions of absolutist monarchy. The gardens at Versailles represented, above all, the control of nature as an expression of royal power. This patronage context—fully attuned to the material and political value of "utility"—contrasts markedly with that of Sangallo, who was able to pursue a broader interest in mathematical practice only in the shadow of his "official" architectural work.

A second difference involves the role of print. Whereas Sangallo's mathematical interests were sustained by a humanistic and erudite print culture—witness his continued engagement with Vitruvius—manuscripts were evidently still an important source of information. The extent to which he traded, copied, and collected drawings underlines both the personal nature of his collaborations and their origins in the workshop. Bullet, in contrast, had recourse to a stream of inexpensive publications, directed towards like-minded practitioners. Indeed, Bullet himself participated in this extended community, publicizing the new surveying techniques to a wider audience of architects, builders, and gardeners.

The third and most important difference has to do with the broadening scope of mathematical practice in the late seventeenth century, transformed not only by new inventions and techniques, but by novel physical claims about the natural world. These arguments—like Picard's about the accurate size of the Earth—were both derived from instrumental practice and consequently fed back again into it. For Bullet, the telescope made the "theory" of surveying—that is, its optical and physical content—integral to its use. As a result, his directions for using the instrument were replete with "scientific" content wholly novel for a surveyor's manual. What is perhaps most unexpected is the role that architects played in this cycle. Although mathematical instruments do not feature in the scholarship on early modern architecture, they were nevertheless central to the profession. They also show how the art was connected to the broader intellectual currents of the early modern period, indeed, to the scientific revolution itself.

Comparing Bullet's example to that of Stornaloco turns us back 180 degrees. The later episode—in which theory and practice directly informed each other—is the inverse of the earlier. By the late seventeenth century, Shelby's distinction between the schoolman's textual understanding of practical geometry and the mason's empirical, "constructive" approach no longer held. Indeed, they two traditions of mathematics had folded into one another.

Aims and Scope of the Volume

In addition to this long narrative trajectory, the three examples adumbrated above also suggest something of the diversity inherent in early modern architectural and mathematical practice. In the Milan cathedral episode, we see how mathematical knowledge—unequally distributed across disparate mathematical communities—informed the design of an actual project. In Sangallo's notes and drawings, we see mathematics shaping the wide-ranging interests of an individual architect, while Bullet's publication on the long-distance level reveals how innovations in mathematical and instrumental technology drove architectural practice. All three cases involve the interaction—sometimes incidental, sometimes concerted—of scholars and practitioners engaging in, for want of a less value-laden distinction, "high" and "low" forms of mathematical practice. Juxtaposing these case studies also serves a valuable end, for it evokes themes that broadly represent the historical evolution of this relationship. These transitions, though blunt, do highlight a number of broad realignments in the period covered by this volume. These include a marked increase in mathematical competence among practitioners, the shift from a culture of manuscript to one of print, and the transformation in institutional and patronage structures.

The essays in this volume may also be considered as case studies of a kind. As discrete examples of the use of mathematics in architectural practice and discourse, they reveal the diverse forms this relationship took, while also greatly expanding on the issues presented above. Setting these historical episodes against each other is likewise intended to generate questions about development and change. Our contributions are arranged chronologically and thematically into sections that follow a familiar subdivision into four broad historical "moments": Antiquity, Renaissance, Baroque, and Enlightenment. We use these terms not to reify them, but merely as convenient shorthand, to group papers into coherent sections, while also marking out some basic historical shifts. Although the contributions cover a wide time span, they are linked by a basic premise: the use of mathematics was a defining feature of early modern architectural practice, one that both characterizes the period as a whole and helps to explain developments within it.

Notes

1. See, for example, Long (2001, 2011); Ash (2004); Smith (2004); Roberts et al. (2007); Smith and Schmidt (2007); Ash (2010) and Smith et al. (2014). My thanks to Suzy Butters for alerting me to this last volume.
2. This classification originated in Aristotle's discussion of demonstrative science in the *Posterior Analytics*. See Apostle (1952, 131–136) and McKirahan (1978). On the medieval evolution of the mixed sciences, see Weisheipl (1965); Gagné (1969); and Laird (1983). For the seventeenth century, see Lennox (1986); Dear (1995, 34–40); and Remmert (2009). For the eighteenth century, see Brown (1991).
3. For an overview, see Bennett (1986, 2006). For a now-classic study of mathematical professions in Italy, see Biagioli (1989). For a recent monograph on an exemplary mathematician-practitioner, see Marr (2011). On classical geometry in practical situations, see Camerota (2006) and Brioist (2009).
4. On the practical background to Galileo's mathematics, see Lefèvre (2000); Renn et al. (2000); Renn and Valleriani (2001); and now Valleriani (2010). Fokko Jan Dijksterhuis teases out the mix of "manipulation and contemplation" in the seventeenth-century study of dioptrics. See Dijksterhuis (2007).
5. On ancient technical culture as reflected in Vitruvius, see Romano (1987, 195–219) and Rowland (2002). On his Renaissance "rediscovery", see Pagliara (1986). Also see Bernard Cache's contribution in this volume (Proportion and Continuous Variation in Vitruvius's *De Architectura*) and the Introduction to Part I (Part I: Foundations).
6. The literature on Barbaro has grown considerably in recent years. For his translation of and commentary on Vitruvius, see Barbaro (1997). On Barbaro's intellectual context, see Tafuri (1987, 1989, 114–38). On Barbaro's reconstruction of the Vitruvian analemma, see Losito (1989a, b). More recently, see Mitrovic (2004); Cellauro (2004); and Howard (2011).
7. On Alberti, see Wassell et al. (2010); Carpo and Furlan (2007); and March (1998). On Baldi, see Serrai (2002); Becchi (2004); and Nenci (2005). On Philandrier, see Lemerle (2000, 2011). On Gallaccini, see Payne (2012). On Goldmann, see Goudeau (2005, 2006–07). On Blondel, see Gerbino (2010). On Poleni, see Lenci (1975) and Soppelsa (1988).
8. See, for example, the work of the *Avista Forum*, the *Nexus Network Journal*, the *Associazione Eduoardo Benvenuto*, and the *Construction History Society*. Each of these bodies promotes research that falls partially under the rubric adopted here.
9. See, for example, Keller (1976); Gabbey (1993); and Meli (2006). For examples of the impact of mechanics in architecture, see Schlimme (2006) and Gargiani (2008).
10. On the abuse of proportions in the nineteenth and early twentieth century, see Wittkower (1960). Also see the prefatory remarks in Wilson Jones (2000, 1–6) and Bork (2011, 11–20). For some methodological guidelines, see Fernie (1990, 2002). For an exemplary recent study, which promises to set new methodological standards for the subject, see Cohen (2008), now expanded in Cohen (2013).

11. Early modern architects also worked within other broad disciplinary matrices. Engineering, for example, involved a similar combination of drawing, mathematics, print culture, antiquarian study, and large-scale construction. For different approaches to this tradition, see Vérin (1993); Conforti (2002); Long (2008); and Maffioli (2010). As an actors' category, the concept of *disegno* is also relevant here, particularly for the way it united different art forms, including architecture, with the practice of drawing and mathematics. See Barzman (2000, 143–80) and Marr (2011, 167–76).
12. On the *scuole dell'abaco*, see Goldthwaite (1972). On the implications of the new system, see Swetz (2002). On the mathematical education of craftsmen, see Zervas (1975) and Shelby (1970, 1972, 1977).
13. On classical and medieval design methods respectively, see Jones (2000) and Bork (2011). For recent scholarship on drawing conventions and setting-out, see Toker (1985); Ousterhout (1999, 58–85); Davis and Neagley (2000); Wu (2002); Carpo (2003); Rossi (2004); Gerbino and Johnston (2009, 17–44); Hadjitryphonos (2009); and, most recently, Yeomans (2011). On the relationship between architectural practitioners and scholars in an Islamic context, see Özdural (1995).
14. See Ackerman (1949), reprinted in Ackerman (1991) and Romanini (1964, vol. 1, 351–415). There is still some disagreement about whether the Lombards' discomfort with the Gothic design system was characteristic of Italian architects in general. For two recent interpretations, see Ascani (1997, 115–27) and Bork (2011, 411–20).
15. Older literature concentrates on the metrical content of the two superimposed drawings and their role in the project's history. See Ghisalberti (1994), with a round-up of earlier studies, and Ascani (1991). More recently, scholars have become attuned to the drawings' unusual visual conventions. See Sakarovitch (1998, 45); Ackerman (2002, 45–46); and Bork (2011, 413–14).
16. "Deliberaverunt, quod discreto viro Gabrieli Stornalocho de Placentia experto in arte geometriae, pro quo missum fuit parte deputatorum dictae fabricae juxta deliberationem in consilio dictae fabricae factam die 24 septembris p.p. et Mediolanum venit… causa discutendi cum inzingneriis dictae fabricae de dubbis altitudinis et aliorum de quibus dubium erat inter dictos inzignerios… dentur… dono pro recognitione et recumpensatione expensarum per eu factarum veniendo…" Cantù (1877–85, vol. 1, 55) (13 October 1391), quoted in Ackerman (1949, 90).
17. Frankl and Panofsky (1945).
18. Stornaloco's cryptic, but crucial, sentence reads "Erit […] altitudo summitatis eclexie radix dc dcc mxx XXVII/[sesquialtera] quia tregesime, quod est aliquid minus de LXXXIIII." Including emendations and in modern notation, Panofsky's paraphrase would read, "the maximum height of the church will be [a] root [approximated by] 600×700÷1010." This gives 415.842, which is 10 times the height divided by two, or 83.168. The actual height of an equilateral triangle with a side of 96 is 83.138, which diverges from Stornaloco's result, as Panofsky pointed out, by less than .04 %. The second part of the sentence can be paraphrased "[the height is also approximated by] 10×(27×1½), because

[27] is one thirtieth [of ten times the height], which is somewhat less than 84...." The result, 405, is only roughly approximate to 415.841, but if 28 represents the upper whole-integer limit for "one thirtieth" of 10 times the height, 27 is the lower limit for that value. See Frankl and Panofsky (1945, 61–64).

19. The drawing accompanying the letter no longer survives, but it is known from later copies, including a version made by Cesare Cesariano, which he published in his 1521 edition of Vitruvius. A simplified, nineteenth-century copy by Luca Beltrami, shown here, is generally accepted as an adequate substitute. See Frankl and Panofsky (1945). Beltrami's diagram was originally published in Beltrami (1887), reprinted in Ramelli (1964, 37–91, esp. 73).
20. Ackerman (1949, 93–95).
21. Although the methods used to set out such shapes at full scale remain unknown, surviving architectural drawings often show a very close match to the built work. See Bork (2011) for several examples, including the equilateral triangles in the nave and façade of Strasbourg cathedral (64–76) and the octagons in the section of Prague cathedral (205–14). For the use of pentagonal geometries, see his remarks on 32–33.
22. See Cohen (2010) and, now, Cohen (2013). For an exploration of similar issues in a northern context, see Nussbaum (2011).
23. Peter Kidson, in particular, has long argued for the advanced numeracy of medieval masons on just these grounds. He has pointed to metrical evidence, for example, to suggest that various local measures were related to Roman ones as more-or-less accurate approximations of $1:\sqrt{2}$ and of the "golden section" (in geometrical terms, the ratio of the length of a 2:1 rectangle to the sum of its diagonal and its width, or $1: (1+\sqrt{5})/2$). See Kidson (1990). He has also argued for similar relationships in medieval buildings. His analysis of Salisbury Cathedral, based on a measured survey, emphasizes dimensions that appear to be related in ratios of 4:9 and 5:11 (both for $1:\sqrt{5}$), as well as 5:8 (for the golden section). See Cocke and Kidson (1993, 62–82). More recently, see Kidson (2008).
24. Peter Kidson has offered an analogous explanation: that Stornaloco was called not to provide the cathedral workshop with a solution to the problem, but merely as a security measure. In this interpretation, someone in authority felt it necessary to call on an "expert" to confirm a decision that had already been made. The same author also proposes an alternative interpretation of Stornaloco's letter (see below, note 25): Kidson (1999). For similar skepticism about Stornaloco's role, see Beaujouan (1975).
25. See Frankl and Panofsky (1945). Peter Kidson's analysis of Stornaloco's letter—which differs considerably from Panofsky's—proposes that the mathematician's formula was based implicitly on the ratio 256/153, as an approximation for $\sqrt{3}$. However, this interpretation, while ingenious, is not as convincing as the one it is intended to replace. Panofsky's has several advantages: it accounts for the lack of decimal notation in the period; it is more clearly related to the wording in Stornaloco's description; it provides several parallels in contemporary mathematics; and it is generalizable for any given input (as Panofsky was

careful to point out, the algorism needed to be formulated in terms of a any given module). See Kidson (1999).

26. For the conference minutes for this period, see the transcription in Ackerman (1949, 108–11). Documents of the *Opera* of Santa Maria del Fiore in Florence show that Hindu-Arabic numerals begin to overtake Roman ones from 1411. See Cohen (2010, 16, 2013, 244). On Renaissance architects' knowledge of arithmetic, see Carpo (2003) and Ann Huppert's contribution in this volume (Practical Mathematics in the Drawings of Baldassarre Peruzzi and Antonio da Sangallo the Younger).
27. See Shelby (1972). On the medieval *Practica geometriae*, see Saint-Victor (1991) and Victor (1979). For two recent attempts to challenge Shelby's rather strict distinction between mathematical and masonic practices, see Zenner (2002) and Liefferinge (2010).
28. Friedman (1988).
29. Cohen (2008).
30. Serlio (1996–2001, vol. 1, 5). Although purely practical in aim, the *Primo Libro* is nevertheless organized according to a deductive, "Euclidean" structure, in which preceding constructions furnish the concepts necessary to complete subsequent ones. Alberti begins *De pictura* with a similar formulation. This method contrasts with Albrecht Dürer's *Underweysung der Messung* (Nuremberg, 1525), which is otherwise taken as Serlio's model. See Lorber (1989).
31. On Renaissance innovations in the theory and practice of structural design, see Sanabria (1982, 1989); and Betts (1993). On the rise of numeracy and arithmetic calculation, see Carpo (2003). For examples of the application of trigonometry to town planning, see Friedman (1988, 117–48) and Jäger (2004).
32. There were ancient and medieval precedents for such classifications. For one Hellenistic classification system, see Downey (1948). For medieval schemes and early Renaissance schemes involving fortification, see Wilkinson (1988). Alina Payne mentions several sixteenth-century examples in Payne (1999). For the English context, see Bennett (1993, 23–30).
33. The drawings are held in the Gabinetto dei Disegni e Stampe degli Uffizi, Florence, Architettura (hereafter UA). See Frommel and Adams (1994–2000).
34. See the still-useful overview by Ackerman (1954, 3–11). For a basic biography, see Bruschi (1983). On the role of drawings in Antonio's career, see Frommel and Adams (1994–2000, vol. 1, 1–60).
35. Pier Nicola Pagliara and Gian Luca Veronese, "U 856A *recto*" and "U 856A *verso*," in Frommel and Adams (1994–2000, vol. 1, 156–58). William E. Wallace, however, notes that there is also a degree of literalism in Sangallo's arithmetic, which appears rather alien to modern modes of calculation. See Wallace (1995).
36. See Maccagni (1993, 1996). An earlier example of such a manuscript from the hand of an architect is Francesco di Giorgio Martini's practical geometry, focusing largely on surveying problems. See Martini (1970).
37. Pier Nicola Pagliara and Gian Luca Veronese, "U 857A *recto*," in Frommel and Adams (1994–2000, vol. 1, 158). Also see Veronese, "U 851A *verso*," and "U

1478 *recto*," in Frommel and Adams (1994–2000, vol. 1, 154, 240–41). On Piero's mathematics, specifically his *Trattato d'abaco*, see Davis (1977) and Field (2005, 6–32, 119–28). On Pacioli, See Jayawardene (2008) and Baldasso (2010). On Sangallo's study of Vitruvius, see Pagliara (1986, 46–55, 1988).

38. Nicholas Adams, "U 850A recto" in Frommel and Adams (1994–2000, vol. 1, 153–54). On the star chart, see Maria Losito, "U 1459A *recto* and *verso*," in Frommel and Adams (1994–2000, vol. 1, 231–32). Also see Snyder (1993, 40–43) and Shirley (2001).
39. In addition to recording instruments himself—see his similarly detailed study of a trecento quadrant (U 1455A recto)—Antonio also collected such drawings. Scaglia cites anonymous reproductions of two other Moorish astrolabes among the architect's former papers. See Gustina Scaglia, "U 1454A *recto* and *verso*," in Frommel and Adams (1994–2000, vol. 1, 227–29). Sangallo's transcription skills appear to go well beyond those of his contemporaries. On Italian artists' fascination with Kufic script, see Mack (2002).
40. See Nicholas Adams, Pier Nicola Pagliara, and Gian Luca Veronese, "U 1491A *recto and verso*," in Frommel and Adams (1994–2000, vol. 1, 246–47). For Sangallo's preparatory studies, see 856A*r*, 1456A*r*, 1457A*r*, 1466A*r*, 1499A*r*, 1500A*r*, 3949A*r*. On the history of the proportional compass, see Camerota (2000, 5–19). On Galileo's sector, see Drake (1978). On the later adaption of the sector to architecture in the seventeenth and eighteenth century, see Bruti (1627) and Gerbino and Johnston (2009, 111–51).
41. See Gustina Scaglia, "1463A recto and verso," in Frommel and Adams (1994–2000, vol. 1, 233–34). For the most recent biography, see Pagliara (2001). On Giocondo's Vitruvius, see Juřen (1974), Ciapponi (1984), and Rowland (2011). On his mathematical manuscripts, see Tura (2008).
42. On this subject, also see Thoenes (1990) and Carpo (2003, 463, 468–9). On Sangallo's drawing conventions, see Lotz (1977) and Lefèvre (2004).
43. Bullet (1688). More broadly on this subject, see Gerbino (2008).
44. Bullet (1688, Unpaginated Preface).
45. Bullet (1688, 29–35).
46. Bullet (1688, 60–65); Gerbino (2008, 89).
47. Picard (1671). See also Taton (1987). For the comparison of modern results with Picard's, see Levallois (1987). For a longer perspective, see Gallois (1909). More recently, see Pelletier (2002).
48. Changes in elevation between stations were disregarded, as these were considered to be minimal in relation to the circumference of the Earth. Picard (1671, 16).
49. Picard ignored refraction while establishing the meridian, as this phenomenon affected the direction of the visual ray only on the vertical plane, leaving the angular measurements between stations unaltered. Picard (1671, 27–28).
50. For Picard's work on the reservoir system, see Loriferne (1987). For a still-useful overview of the whole water management system, see Barbet (1907). For the Académie's later involvement, see Gerbino (2008).

51. *Journal des sçavans* (1677, 227–28); (1678, 441–43); (1679, 215–16); (1680, 21–24, 174–76, 275–76). Also see Deshayes (1685, 9); Du Torar (1688, 182–85); Clermont (1693, 112); and Ozanam (1693, 241–44). On the reception of the instrument, see Gerbino (2008).
52. On Bullet's collaboration with Blondel, see Gerbino (2010, 71–117). On his later involvement at Versailles, see Lemonnier (1911–1929, vol. 2, 71–91). For the aqueduct drawing, see Walton (1985, 52–53).

Photographic Credits

Archivio della Fabbriceria di San Petronio: Fig. 1
Author: Fig. 2
Beltrami (1887, 73): Fig. 3
Uffizi, Gabinetto dei Disegni e Stampe (by permission of the Ministero dei beni e delle attività culturali e del turismo): Figs. 4–9
Filippo Camerota: Fig. 10
Bibliothèque Nationale de France : Figs. 11–13

References

Ackerman, James. 1949. 'Ars Sine Scientia Nihil Est': Gothic theory of architecture at the Cathedral of Milan. *The Art Bulletin* 31(2): 84–111, reprinted in Ackerman (1991), 1211–1268.

Ackerman, James. 1954. Architectural practice in the Italian Renaissance. *Journal of the Society of Architectural Historians* 13(3): 3–41.

Ackerman, James. 1991. *Distance points: Essays in theory and Renaissance art and architecture*. Cambridge, MA/London: MIT Press.

Ackerman, James. 2002. *Origins, imitation, conventions: Representation in the visual arts*. Cambridge, MA: MIT Press.

Apostle, Hippocrates George. 1952. *Aristotle's philosophy of mathematics*. Chicago: University of Chicago Press.

Ascani, Valerio. 1991. I disegni architettonici attribuiti ad Antonio di Vincenzo. *Arte medievale* 1: 105–114.

Ascani, Valerio. 1997. *Il trecento disegnato: le basi progettuali dell'architettura gotica in Italia*. Rome: Viella.

Ash, Eric H. 2004. *Power, knowledge, and expertise in Elizabethan England*. Baltimore: Johns Hopkins University Press.

Ash, Eric H. (ed.). 2010. *Practical knowledge and the early modern state*. Chicago: University of Chicago Press.

Baldasso, Renzo. 2010. Portrait of Luca Pacioli and disciple: A new, mathematical look. *The Art Bulletin* 92(1–2): 83–102.

Barbaro, Daniele. 1567, repr. Milan: Edizioni il Polifilo, 1997. *I dieci libri dell'architettura di M. Vitruvio*, ed. Manfredo Tafuri. Venice: Francesco de Franceschi & Giovanni Chrieger.

Barbet, L.-A. 1907. *Les grands eaux de Versailles: Installations mécaniques et étangs artificiels*. Paris: H. Dunod, E. Pinat.

Barzman, Karen-edis. 2000. *The Florentine Academy and the early modern state: The discipline of Disegno*. Cambridge: Cambridge University Press.

Beaujouan, Guy. 1975. Réflexions sur les rapports entre théorie et pratique au Moyen âge. In *The cultural context of medieval learning*, ed. John Emery Murdoch and Edith Dudley Sylla, 437–484. Dordrecht/Boston: Reidel.

Becchi, Antonio. 2004. *Q.XVI. Leonardo, Galileo e il caso Baldi: Magonza, 26 marzo 1621*. Venice: Marsilio.

Beltrami, Luca. 1887. *Per la facciata del Duomo di Milano*. Milan: A. Colombo & A. Cordani.

Bennett, J.A. 1986. The mechanic's philosophy and the mechanical philosophy. *History of Science* 24(1): 1–28.

Bennett, J.A. 1993. Architecture and mathematical practice in England, 1550–1650. In *English architecture public and private: Essays for Kerry Downes*, ed. John Bold and Edward Chaney, 23–30. London: The Hambledon Press.

Bennett, J.A. 2006. The mechanical arts. In *The Cambridge history of science, vol. 3: Early modern science*, ed. Katherine Park and Lorraine Daston, 673–695. Cambridge: Cambridge University Press.

Betts, Richard J. 1993. Structural innovation and structural design in Renaissance architecture. *Journal of the Society of Architectural Historians* 52(1): 5–25.

Biagioli, Mario. 1989. The social status of Italian mathematicians, 1450–1600. *History of Science* 27(1): 41–95.

Bork, Robert. 2011. *The geometry of creation: Architectural drawing and the dynamics of gothic design*. Farnham: Ashgate.

Brioist, Pascal. 2009. 'Familiar Demonstrations in Geometry': French and Italian engineers and Euclid in the sixteenth century. *History of Science* 47(1): 1–26.

Brown, Gary I. 1991. The evolution of the term 'Mixed Mathematics'. *Journal of the History of Ideas* 52(1): 81–102.

Bruschi, Arnaldo. 1983. Cordini, Antonio. *Dizionario biografico degli Italiani*. http://www.treccani.it/biografie.

Bruti, Ottavio Revesi. 1627. *Archisesto per formar con facilita li cinque ordini d'architettura*. Vicenza: Amadio.

Bullet, Pierre. 1688. *Traité du nivellement, contenant la théorie et la pratique de cet art, avec la description d'un niveau nouvellement inventé*. Paris: Nicolas Langlois.

Camerota, Filippo. 2000. *Il compasso di Fabrizio Mordente: Per la storia del compasso di proporzione*. Florence: Oschki.

Camerota, Filippo. 2006. Teaching Euclid in a practical context: Linear perspective and practical geometry. *Science & Education* 15: 323–334.

Cantù, Cesare (ed.). 1877–85. *Annali della fabbrica del duomo di Milano*. Milan.

Carpo, Mario. 2003. Drawing with numbers: Geometry and numeracy in early modern architectural design. *Journal of the Society of Architectural Historians* 62(4): 448–469.

Carpo, Mario, and Francesco Furlan (eds.). 2007. *Leon Battista Alberti's Delineation of the city of Rome (Descriptio vrbis Romæ)*. Tempe: Arizona Center for Medieval and Renaissance Studies.

Cellauro, Louis. 2004. Daniele Barbaro and Vitruvius: The architectural theory of a Renaissance humanist and patron. *Papers of the British School at Rome* 72: 293–329.

Ciapponi, Lucia. 1984. Fra Giocondo da Verona and his edition of Vitruvius. *Journal of the Warburg and Courtauld Institutes* 47: 72–90.

Clermont, Sieur de. 1693. *La géométrie pratique de l'ingenieur*. Strasbourg: Schmuck.

Cocke, Thomas, and Peter Kidson. 1993. *Salisbury Cathedral: Perspectives on the architectural history*. London: H.M.S.O.

Cohen, Matthew. 2008. How much Brunelleschi? A late medieval proportional system in the Basilica of San Lorenzo in Florence. *Journal of the Society of Architectural Historians* 67(1): 18–57.

Cohen, Matthew. 2010. Quantification and the medieval mind: An imperfect proportional system in the Basilica of Santa Maria del Fiore in Florence. In *Some degree of happiness: Studi di storia dell'architettura in onore di Howard Burns*, ed. Maria Beltramini and Caroline Elam, 1–30. Pisa: Edizioni dell Normale.

Cohen, Matthew. 2013. *Beyond beauty: Reexamining architectural proportion through the Basilica of San Lorenzo in Florence*. Venice: Marsilio Editori.

Conforti, Claudia (ed.). 2002. *Architettura e tecnologia: Acque, tecniche e cantieri nell'architettura rinascimentale e barocca*. Rome: Nuova Argos.

Davis, Margaret Daly. 1977. *Piero della Francesca's mathematical treatises*. Ravenna: Longo.

Davis, Michael T., and Linda Elaine Neagley. 2000. Mechanics and meaning: Plan design at Saint-Urbain, Troyes and Saint-Ouen, Rouen. *Gesta* 39(2): 161–182.

Dear, Peter. 1995. *Discipline and experience: The mathematical way in the scientific revolution*. Chicago: University of Chicago Press.

Deshayes, le sieur. 1685. *La théorie et la pratique du nivellement*. Paris: chez l'autheur.

Dijksterhuis, Fokko Jan. 2007. Constructive thinking: A case for dioptrics. In *The mindful hand: Inquiry and invention from the late Renaissance to early industrialisation*, ed. Lissa Roberts, Simon Schaffer, and Peter Dear, 59–82. Amsterdam: Royal Netherlands Academy of Arts and Sciences.

Downey, Glanville. 1948. Pappus of Alexandria on architectural studies. *Isis* 38: 197–200.

Drake, Stillman. 1978. *Galileo Galilei: Operations of the geometric and military compass, 1606*. Washington, DC: Smithsonian Institution.

Du Torar, le sieur. 1688. *Leçons de géométrie pratique*. Paris: d'Houry.

Fernie, Eric. 1990. A beginner's guide to the study of architectural proportions and systems of length. In *Medieval architecture and its intellectual context*, ed. Eric Fernie and Paul Crossley, 229–237. London: Hambledon.

Fernie, Eric. 2002. Introduction. In *Ad Quadratum: The practical application of geometry in medieval architecture*, ed. Nancy Y. Wu, 1–9. Aldershot: Ashgate.

Field, J.V. 2005. *Piero della Francesca: A mathematician's art*. New Haven: Yale.

Frankl, Paul, and Erwin Panofsky. 1945. The secret of the mediaeval masons with an explanation of Stornaloco's formula. *The Art Bulletin* 27(1): 46–64.

Friedman, David. 1988. *Florentine new towns: Urban design in the late Middle Ages*. Cambridge, MA: MIT Press.

Frommel, Christoph L., and Nicholas Adams (eds.). 1994–2000. *The architectural drawings of Antonio da Sangallo the younger and his circle*. New York/Cambridge, MA/London: Architectural History Foundation/MIT Press.

Gabbey, Alan. 1993. Between *Ars* and *Philosophia naturalis*: Reflections on the historiography of early modern mechanics. In *Renaissance and revolution: Humanists, scholars, craftsmen and natural philosophers in early modern Europe*, ed. J.V. Field and Frank A.J.L. James, 133–145. Cambridge: Cambridge University Press.

Gagné, Jean. 1969. Du *quadrivium* au *scientia mediae*. In *Arts libéraux et philosophie au Moyen Âge: Acts du quatrième Congrès international de philosophie médiévale*, 975–986. Montreal: Institute d'études médiévales.

Gallois, Lucien. 1909. L'Académie des sciences et les origines de la Carte de Cassini. *Annales de géographie* 18(99): 193–204.

Gargiani, Roberto (ed.). 2008. *La colonne: nouvelle histoire de la construction*. Lausanne: Presses polytechniques universitaires romandes.

Gerbino, Anthony. 2008. The Académie royale des sciences and the gardens of Versailles. *SVEC [Studies on Voltaire and the Eighteenth Century]* 6: 69–95.

Gerbino, Anthony. 2010. *François Blondel: Architecture, erudition, and the scientific revolution*. London/New York: Routledge.

Gerbino, Anthony, and Stephen Johnston. 2009. *Compass and rule: Architecture as mathematical practice in England 1550–1750*. New Haven/London: Yale University Press.

Ghisalberti, Carla. 1994. Antonio di Vincenzo, plan and transverse section of Milan Cathedral. In *The Renaissance from Brunelleschi to Michelangelo: The representation of architecture*, ed. Henry A. Millon and Vittorio Magnago Lampugnani, 429–430. Milan: Bompiani.
Goldthwaite, Richard A. 1972. Schools and teachers of commercial arithmetic in Renaissance Florence. *Journal of European Economic History* 1(2): 418–433.
Goudeau, Jeroen. 2005. *Nicolaus Goldmann (1611–1665) en de wiskundige architectuurwetenschap*. Groningen: Philip Elchers.
Goudeau, Jeroen. 2006–07. A northern Scamozzi: Nicolaus Goldmann and the universal theory of architecture. *Annali di architettura* 18–19: 235–248.
Hadjitryphonos, Evangelia. 2009. Presentations and representations of architecture in Byzantium: The thought behind the image. In *Architecture as icon: Perception and representation of architecture in Byzantine art*, ed. Slobodan Ćurčić and Evangelia Hadjitryphonos, 113–154. Princeton: Princeton University Art Museum.
Howard, Deborah. 2011. *Venice disputed: Marc'Antonio Barbaro and Venetian architecture, 1550–1600*. New Haven/London: Yale University Press.
Jäger, Thomas. 2004. The art of orthogonal planning: Laparelli's trigonometric design of Valletta. *Journal of the Society of Architectural Historians* 63(1): 4–31.
Jayawardene, S.A. 2008. Pacioli, Luca. In *Complete Dictionary of scientific biography*, vol. 10, ed. Charles Coulston Gillispie, 269–272. New York: Scribner.
Juřen, Vladimir. 1974. Fra Giovanni Giocondo et le début des études vitruviennes en France. *Rinascimento* 14: 101–115.
Keller, A.G. 1976. Mathematics, mechanics and experimental machines in Northern Italy in the sixteenth century. In *The emergence of science in Western Europe*, ed. Maurice Crosland, 17–34. New York: Science History Publications.
Kidson, Peter. 1990. A metrological investigation. *Journal of the Warburg and Courtauld Institutes* 53: 71–97.
Kidson, Peter. 1999. Three footnotes to the Milan debates. In *Arte d'Occidente: Temi e metodi: Studi in onore di Angiola Maria Romanini*, ed. Antonio Cadei et al, 266–278. Rome: Sintesi informazione.
Kidson, Peter. 2008. Roriczer's iceberg. *Journal of the Warburg and Courtauld Institutes* 71: 1–20.
Laird, Walter Roy. 1983. The *Scientia Mediae* in Medieval Commentaries on Aristotle's *Posterior Analytics*. Ph.D. dissertation, University of Toronto.
Lefèvre, Wolfgang. 2000. Galileo engineer: Art and modern science. *Science in Context* 13(3–4): 281–297.
Lefèvre, Wolfgang. 2004. The emergence of combined orthographic projections. In *Picturing machines 1400–1700*, ed. Wolfgang Lefèvre, 209–244. Cambridge, MA: MIT Press.
Lemerle, Frédérique. 2000. *Les Annotations de Guillaume Philandrier sur le De Architectura de Vitruve, livres I à IV*. Paris: Picard.
Lemerle, Frédérique. 2011. *Les Annotations sur L'Architecture de Vitruve: livres V à VII*. Paris: Classiques Garnier.
Lemonnier, Henry (ed.). 1911–1929. *Procès-verbaux de l'Académie royale d'architecture, 1671–1793*. Paris: Jean Schemit.
Lenci, Liliana Guadagnino. 1975. Per Giovanni Poleni: Note e appunti per una revisione critica. *Atti dell'Istituto Veneto di Scienze, Lettere ed Arti: Classe di Scienze Morali, Lettere ed Arti* 134: 543–567.
Lennox, James G. 1986. Aristotle, Galileo, and "Mixed Sciences". In *Reinterpreting Galileo*, ed. William A. Wallace, 29–51. Washington, DC: Catholic University of America Press.
Levallois, Jean-Jacques. 1987. Picard géodésien. In *Jean Picard et les débuts de l'astronomie de précision au XVIIe siècle*, ed. Guy Picolet, 227–246. Paris: Éditions du CNRS.
Long, Pamela O. 2001. *Openness, secrecy, authorship: Technical arts and the culture of knowledge from antiquity to the Renaissance*. Baltimore: Johns Hopkins University Press.
Long, Pamela O. 2008. Hydraulic engineering and the study of antiquity: Rome: 1557–1570. *Renaissance Quarterly* 61: 1098–1138.

Long, Pamela O. 2011. *Artisan/practitioners and the rise of the new sciences, 1400–1600*. Corvallis: Oregon State University Press.

Lorber, Maurizio. 1989. I primi due libri di Sebastiano Serlio: Dalla struttura ipotetico-deduttiva alla struttura pragmatica. In *Sebastiano Serlio*, ed. Christof Thoenes, 114–125. Milan: Electa.

Loriferne, Hubert. 1987. L'Influence de Picard dans les travaux d'alimentation en eau du château de Versailles sous Louis XIV. In *Jean Picard et les débuts de l'astronomie de précision au XVIIe siècle*, ed. Guy Picolet, 275–311. Paris: Éditions du CNRS.

Losito, Maria. 1989a. Il IX libro del *De Architectura* di Vitruvio nei commentari di Daniele Barbaro (1556–1567). *Nuncius* 4: 3–42.

Losito, Maria. 1989b. La gnomonica, il IX libro dei *Commentari* vitruviani di Daniele Barbaro e gli studi analemmatici de Federico Commandino. *Studi Veneziani* n.s. 18: 177–237.

Lotz, Wolfgang. 1977. The rendering of the interior in architectural drawings of the Renaissance. In *Studies in Italian Renaissance architecture*, 1–65. Cambridge, MA/London: MIT Press.

Maccagni, Carlo. 1993. Leggere, scrivere e disegnare la 'scienza volgare' nel Rinascimento. *Annali della Scuola Normale Superiore di Pisa, Classe di Lettere e Filosofia, 3rd series* 23(2): 631–676.

Maccagni, Carlo. 1996. Cultura e sapere dei tecnini nel Rinascimento. In *Piero della Francesca tra arte e scienza*, ed. Marisa Dalai-Emiliani and Valter Curzi, 279–292. Venice: Marsilio.

Mack, Rosamond E. 2002. *Bazaar to Piazza: Islamic trade and Italian art, 1300–1600*. Berkeley: University of California Press.

Maffioli, Cesare S. 2010. *La via delle acque (1500–1700): Appropriazione delle arti e trasformazione delle matematiche*. Florence: Leo S. Olschki.

March, Lionel. 1998. *Architectonics of humanism: Essays on number in architecture*. Chichester: Academy Editions.

Marr, Alexander. 2011. *Between Raphael and Galileo: Mutio Oddi and the mathematical culture of late Renaissance Italy*. Chicago: University of Chicago Press.

Martini, Francesco di Giorgio. 1970. *La praticha di gieometria dal Codice Ashburnham 361 della Biblioteca Medicea Laurenziana di Firenze*. Milan: Giunti.

McKirahan Jr., Richard D. 1978. Aristotle's subordinate sciences. *The British Journal for the History of Science* 11(39): 197–220.

Meli, Domenico Bertoloni. 2006. *Thinking with objects: The transformation of mechanics in the seventeenth century*. Baltimore: Johns Hopkins University Press.

Mitrovic, Branko. 2004. *Learning from Palladio*. New York: W.W. Norton.

Nenci, Elio (ed.). 2005. *Bernardino Baldi (1553–1617) studioso rinascimentale: poesia, storia, linguistica, meccanica, architettura: atti del convegno di studi di Milano, 19–21 novembre 2003*. Milan: Franco Angeli.

Nussbaum, Norbert. 2011. Planning and building without writing: Questions of communication in Gothic Mason's Lodges. In *Architecture, liturgy and identity: Liber Amicorum Paul Crossley*, ed. Zoë Opačić and Achim Timmerman, 137–145. Turnhout: Brepols.

Ousterhout, Robert. 1999. *Master builders of Byzantium*. Princeton: Princeton University Press.

Ozanam, Jacques. 1693. *Méthode de lever les cartes de terre et de mer*. Paris: Michallet.

Özdural, Alpay. 1995. Omar Khayyam, mathematicians, and 'Conversazioni' with artisans. *Journal of the Society of Architectural Historians* 54(1): 54–71.

Pagliara, Pier Nicola. 1986. Vitruvio da Testo a Canone. In *Memoria dell'Antico nell'Arte Italiana*, vol. 3, ed. Salvatore Settis, 5–85. Turin: Einaudi.

Pagliara, Pier Nicola. 1988. Studi e pratica vitruviana di Antonio da Sangallo il Giovane e di suo fratello Giovanni Battista. In *Les Traités d'architecture de la Renaissance*, ed. Jean Guillaume, 179–206. Paris: Picard.

Pagliara, Pier Nicola. 2001. Giovanni Giocondo da Verona (Fra Giocondo). *Dizionario Biografico degli Italiani*. http://www.treccani.it/biografie.

Payne, Alina A. 1999. Architectural criticism, science, and visual eloquence: Teofilo Gallaccini in seventeenth-century Siena. *Journal of the Society of Architectural Historians* 58(2): 146–169.

Payne, Alina A. 2012. *The telescope and the compass: Teofilo Gallaccini and the dialogue between architecture and science in the age of Galileo*. Florence: Leo S. Olschki.

Pelletier, Monique. 2002. *Les cartes de Cassini. La science au service de l'État et des régions*. Paris: Éditions du CTHS.
Picard, Jean. 1671. *Mesure de la terre*. Paris: Imprimerie royale.
Ramelli, Antonio Cassi (ed.). 1964. *Luca Beltrami e il Duomo di Milano. Tutta gli scritti riguardanti la cattedrale pubblicati tra il 1881 e il 1914*. Milan: Ceschina.
Remmert, Volker R. 2009. Antiquity, nobility, and utility: Picturing the early modern mathematical sciences. In *The Oxford handbook of the history of mathematics*, ed. Eleanor Robson and Jacqueline Stedall, 537–563. Oxford: Oxford University Press.
Renn, Jürgen, and Matteo Valleriani. 2001. Galileo and the challenge of the Arsenal. *Nuncius* 16(2): 481–503.
Renn, Jurgen, Peter Damerow, and Simone Rieger. 2000. Hunting the white elephant: When and how did Galileo discover the Law of Fall? *Science in Context* 13(3–4): 299–419.
Roberts, Lissa, Simon Schaffer, and Peter Dear (eds.). 2007. *The mindful hand: Inquiry and invention from the late Renaissance to early industrialisation*. Amsterdam: Royal Netherlands Academy of Arts and Sciences.
Romanini, Angiola Maria. 1964. *L'architettura gotica in Lombardia*. Milan: Ceschina.
Romano, Elisa. 1987. *La capanna e il tempio: Vitruvio o Dell'architettura*. Palermo: Palumbo.
Rossi, Corinna. 2004. *Architecture and mathematics in ancient Egypt*. Cambridge: Cambridge University Press.
Rowland, Ingrid D. 2002. Vitruvius and technology. In *Architettura e tecnologia: Acque, tecniche e cantieri nell'architettura rinascimentale e barocca*, ed. Claudia Conforti, 244–258. Rome: Nuova Argos.
Rowland, Ingrid D. 2011. The Fra Giocondo Vitruvius at 500 (1511–2011). *Journal of the Society of Architectural Historians* 70(3): 285–289.
Saint-Victor, Hugh of. 1991. *Practical geometry [Practica Geometriae] attributed to Hugh of St. Victor*, ed. Frederick A. Homann. Milwaukee: Marquette University Press.
Sakarovitch, Joël. 1998. *Épures d'architecture: de la coupe des pierres à la géométrie descriptive, XVIe-XIXe siècles*. Basel: Birkhäuser.
Sanabria, Sergio Luis. 1982. The mechanization of design in the 16th century: The structural formulae of Rodrigo Gil de Hontañón. *Journal of the Society of Architectural Historians* 41(4): 281–293.
Sanabria, Sergio Luis. 1989. From Gothic to Renaissance stereotomy: The design methods of Philibert de l'Orme and Alonso de Vandelvira. *Technology and Culture* 30(2): 266–299.
Schlimme, Hermann (ed.). 2006. *Practice and science in early modern Italian building: Towards an epistemic history of architecture*. Milan: Electa.
Serlio, Sebastiano. 1996–2001. *On architecture*. New Haven: Yale University Press.
Serrai, Alfredo. 2002. *Bernardino Baldi: la vita, le opere, la biblioteca*. Milan: Edizioni Sylvestre Bonnard.
Shelby, Lon R. 1970. The education of medieval English master masons. *Mediaeval Studies* 32: 1–26.
Shelby, Lon R. 1972. The geometrical knowledge of medieval master masons. *Speculum* 47(3): 395–421.
Shelby, Lon R. (ed.). 1977. *Gothic design techniques: The fifteenth-century design booklets of Mathes Roriczer and Hanns Schmuttermayer*. Carbondale: Southern Illinois University Press.
Shirley, Rodney W. 2001. *The mapping of the world: Early printed world maps 1472–1700*, 4th ed. Riverside: Early World Press.
Smith, Pamela H. 2004. *The body of the artisan: Art and experience in the scientific revolution*. Chicago: University of Chicago Press.
Smith, Pamela H., and Benjamin Schmidt (eds.). 2007. *Making knowledge in early modern Europe: Practices, objects, and texts, 1400–1800*. Chicago: University of Chicago Press.
Smith, Pamela H., Amy Meyers, and Harold Cook (eds.). 2014. *Ways of making and knowing: The material culture of empirical knowledge*. Ann Arbor: University of Michigan Press.
Snyder, John P. 1993. *Flattening the earth: Two thousand years of map projections*. Chicago: University of Chicago Press.

Soppelsa, Maria Laura (ed.). 1988. *Giovanni Poleni idraulico, matematico, architetto, filologo (1683–1761): Atti della Giornata di studi: Padova, 15 marzo 1986*. Padua: Erredicì.
Swetz, Frank J. 2002. *Figura Mercantesco*: Merchants and the evolution of a number concept in the latter Middle Ages. In *Word, image, number: Communication in the middle ages*, ed. John J. Contreni and Santa Casciani, 391–412. Florence: SISMEL-Edizioni del Galluzzo.
Tafuri, Manfredo. 1987. Daniele Barbaro e la cultura scientifica veneziana del '500. In *Cultura, scienze e tecniche nella Venezia del Cinquecento: Atti del convegno internazionale di studio Giovan Battista Benedetti e il suo tempo*, 55–81. Venice: Istituto Veneto di Scienze, Lettere ed Arti.
Tafuri, Manfredo. 1989. *Venice and the Renaissance*. Trans. Jessica Levine. Cambridge, MA: MIT Press.
Taton, René. 1987. Picard et la *Mesure de la terre*. In *Jean Picard et les débuts de l'astronomie de précision au XVIIe siècle*, ed. Guy Picolet, 207–226. Paris: Éditions du CNRS.
Thoenes, Christof. 1990. Zur Frage des Massstabs in Architekturzeichnungen der Renaissance. In *Studien zur Künstlerzeichnung: Klaus Schwager zum 65. Geburtstag*, ed. Stefan Kummer and Georg Satzinger, 38–55. Stuttgart: Hatje.
Toker, Franklin. 1985. Gothic architecture by remote control: An illustrated building contract of 1340. *The Art Bulletin* 67(1): 67–95.
Tura, Adolfo. 2008. *Fra Giocondo & les textes français de géométrie pratique*. Genève: Droz.
Valleriani, Matteo. 2010. *Galileo engineer*. Dordrecht: Springer.
Van Liefferinge, Stefaan. 2010. The hemicycle of Notre-Dame of Paris: Gothic design and geometrical knowledge in the twelfth century. *Journal of the Society of Architectural Historians* 69(4): 490–507.
Vérin, Hélène. 1993. *La gloire des ingénieurs: l'intelligence technique du XVIe au XVIIIe siècle*. Paris: Albin Michel.
Victor, Stephen K. (ed.). 1979. Practical geometry in the high middle ages: *Artis cuiuslibet consummatio* and the *Pratike de geometrie*. Philadelphia: The American Philosophical Society.
Wallace, William E. 1995. Review of *The architectural drawings of Antonio da Sangallo the younger and his circle, Vol. 1: Fortifications, machines, and festival architecture by Christoph L. Frommel and Nicholas Adams*. *Journal of the Society of Architectural Historians* 54(1): 101–103.
Walton, Guy. 1985. *Versailles à Stockholm: dessins du Nationalmuseum*. Stockholm: Nationalmuseum.
Wassell, Stephen R., Kim Williams, and Lionel March (eds.). 2010. *The mathematical works of Leon Battista Alberti*. Basel: Springer Basel.
Weisheipl, J.A. 1965. Classification of the sciences in medieval thought. *Medieval Studies* 27: 54–90.
Wilkinson, Catherine. 1988. Renaissance treatises on military architecture and the science of mechanics. In *Les traités d'architecture de la Renaissance*, ed. Jean Guillaume, 467–474. Paris: Picard.
Wilson Jones, Mark. 2000. *Principles of Roman architecture*. New Haven/London: Yale University Press.
Wittkower, Rudolf. 1960. The changing concept of proportion. *Daedalus* 89(1): 199–215.
Wu, Nancy Y. (ed.). 2002. *Ad Quadratum: The practical application of geometry in medieval architecture*. Aldershot: Ashgate.
Yeomans, David. 2011. The geometry of a piece of string. *Architectural History* 54: 23–47.
Zenner, Marie-Thérèse. 2002. Imaging a building: Latin Euclid and practical geometry. In *Word, image, number: Communication in the middle ages*, ed. John J. Contreni and Santa Casciani, 219–246. Florence: SISMEL-Edizioni del Galluzzo.
Zervas, Diane Finiello. 1975. The Trattato dell'Abbaco and Andrea Pisano's design for the Florentine baptistery door. *Renaissance Quarterly* 28(4): 483–503.

Part I
Foundations

Our first section, on Vitruvius, may appear out-of-place in a book largely about early modernity, but its inclusion is calculated. On the level of practice, this section shows that the use of mathematics in architecture long predated the Renaissance. Indeed, the connection persisted in some form through the middle ages. What does seem to have disappeared in those intervening centuries was a self-consciously learned architectural practice, inflected by a textual and philosophical tradition of mathematics, an influence evident throughout *De Architectura*. It was this union of practice and erudition that Alberti largely resurrected, a fact that points to the other reason for beginning this volume with some discussion of Vitruvius. The study of this ancient author—along with the influence of scholarly humanism as a whole—was transformative for both the self-identity and intellectual culture of the profession. Any attempt to recuperate the meaning of this text was necessarily bound up with the whole apparatus of humanist mathematics.

The mathematical content of the *De Architectura* is diverse and diffuse, but the core surely lies in the notion of *symmetria*. Introduced at the head of Book 3, the concept underlies the design of temples and the proportional schema for what we now think of as the orders (Ionic in Book 3, Doric and Corinthian in Book 4), but, as Pierre Gros has pointed out, its import extends throughout the treatise, as a constant applicable to virtually all forms of architectural practice.[1] In essence, *symmetria* means commensurability, the application of a common measure, or module, to all the components of a complex work. For Vitruvius, the principle is more than merely operative, it is rooted in nature. The unity that it confers to the design of buildings finds an analogue in the human body, which exhibits a similar modular rigor. Human proportions, Vitruvius implies, using his own Polyclitus-inspired canon, evince the same thorough rationality that the architect must bring to the art building. The famous *homo bene figuratus* inscribed in a circle and square—overlapping images of perfection—confirms this "natural" harmony. These Platonic-Pythagorean overtones continue with a disquisition on the characteristics of the "perfect" numbers ten and six.

In addition to this metaphysical tradition, the *De Architectura* also reveals some knowledge of demonstrative geometry, at least in its more popular form. Book 9 on sundials begins with a celebration of Greek scientific heroes, whose "discoveries" are contrasted with and, to some extent, modeled on the exploits of Olympic athletes. It is here that Vitruvius trumpets the solution to several famous ancient mathematical problems, beginning with the "doubling of the square" ascribed to Plato, that is, how to find the side of a square twice the area of the original. The author continues, moving from the Pythagorean Theorem to Archimedes's crown problem and finally to Archytas and Eratosthenes's solutions to the "Delian" problem, that is, how to find the side of a cube twice the volume of the original (Book 9, Pref.1.14). Despite his enthusiasm, it is evident that Vitruvius's understanding is not that of a mathematician, but of a practitioner. Not only are the solutions relayed in anecdotal form, they are also couched in material, instrumental terms. Vitruvius explains the doubling of the square, for example, as a surveying problem, while the Pythagorean Theorem appears as no more than a method for making accurate set squares by means of joining a 3-, 4-, and 5-foot ruler. More telling is that the author does not appear to see the relationship between the two solutions in terms of the properties of right triangles. For him, the connection lies rather in their ease of application, in particular, the fact that they allow the practitioner to avoid dealing with irrational magnitudes. Some hint of this discomfort with incommensurable quantities emerges when Vitruvius explains that the side of a doubled square "is not discovered by means of numbers." As Pierre Gros has noted, geometrical constructions that involve rotating squares or diagonals are rare in the treatise, and when they do appear, Vitruvius is careful to subsume them in a larger net of simple whole-number relationships.[2]

De Architectura is clearly more of a manual than an academic treatise. Despite the learned references on the subject, it is evident that Vitruvius understands geometry as a means of execution, not a source for reflection. That is not to say that his design protocols are devoid of theoretical content, but that they tend to disguise that content in the form of operative techniques. In fact, historians have found evidence throughout the treatise of a complex and sophisticated intertwining of number and geometry, presented as a relatively simple and straightforward series of proportional instructions. As Louis Frey has argued, several of Vitruvius's recommendations for the design of colonnades, capitals, and entablatures reveal a carefully interlinked succession of "harmonic" and, in one case, "geometric" proportions. The former are related as terms of the triplets 5:7:10 and 12:17:24, while the latter are derived from the ratio 17:38.[3] One property of such proportions is that they provide whole-number approximations for the "irrational" geometry of the square and the double-square rectangle. Because they can be visualized by "rotating" sides and diagonals, these ratios may have served to reconcile geometrical form and arithmetical measure in a way that preserved the desirable—even mystical—characteristics of both. Whether Vitruvius understood the basis of these ratios is an open question. The text makes no mention of the theory of proportional means or of rational approximations as they were set out, for example, by Euclid. Given this silence, it is more likely that Vitruvius merely lifted the ratios from a Hellenistic building manual, a mathematical compilation, or some other intermediate source.

In his essay, Bernard Cache casts similar light on another part of *De Architectura*: Vitruvius's instructions for the layout of Roman houses. Although again presented under a practical veneer, these protocols differ in both kind and complexity from those found elsewhere in the book. Vitruvius regulates the elements of the courtyard not with a set of predetermined values, but with series of ratios that vary according to the courtyard's own dimensions. Instead of a fixed rule for all the parts of a design—otherwise the norm throughout the treatise—these instructions generate a series of *possible* proportions, under the rubric, "if x, then y". Moreover, as Cache shows, those potential outcomes are themselves linked by a proportional rule. Although this system does not conform to modern notions of a curvilinear function, the variation in series that it produces does appear to be mathematically continuous. This circumstance—first noticed by Auguste Choisy more than a century ago—raises the question of whether these instructions were derived from a learned mathematical tradition. Cache considers the possible origins of such a system, focusing especially on contemporary graphical practices in ancient drawing and construction.

Notes

1. Vitruvius (1999, 46–48, 188–91) and Vitruvius (1990, 3–11, 55–63). Also see Gros (2006 [1989]); Wilson Jones (2000, 40–43); Gros (2001); and more recently, Ilon and Goldstein (2008).
2. See Vitruvius (1999, 107–109, 281–82) and Vitruvius (1969, xix–xxi). Also see Gros (2006 [1976]).
3. Frey (1990). Admittedly, the terms harmonic and geometric are uncomfortably reminiscent of the intricate geometrical speculations that characterized older scholarship on proportions. The advantage of Frey's analysis is that it accounts for each of the examples mentioned above in terms of the numbers Vitruvius himself provides. Frey focuses on four cases: the diastyle temple with a four-column or tetrastyle portico, the eustyle temple with a six column or hexastyle portico, the Ionic capital, and Ionic entablature.

References

Frey, Louis. 1990. Médiétés et approximations chez Vitruve. *Revue Archéologique* (2): 285–330.

Gros, Pierre. 2001. La géometrie platonicienne de la notice vitruvienne sur l'homme parfait (De architectura, III, 1, 2–3). *Annali di architettura* 13: 15–24.

Gros, Pierre. 2006 [1976]. Nombres irrationnels et nombres parfaits chez Vitruve. In *Vitruve et la tradition des traités d'architecture, Fabrica et Racionatio: Recueil d'études*, 75–111. Rome: École française de Rome.

Gros, Pierre. 2006 [1989]. Les fondements philosophiques de l'harmonie architecturale selon Vitruve (*De architectura* III-IV). In *Vitruve et la tradition des traités d'architecture, Fabrica et Racionatio: Recueil d'études*, 271–280. Rome: École française de Rome.

Hon, Giora, and Bernard R. Goldstein. 2008. *From summetria to symmetry: The making of a revolutionary scientific concept*. Dordrecht: Springer.

Vitruvius. 1969. *De l'architecture: Livre IX*, ed. Jean Soubiran. Paris: Les Belles Lettres.

Vitruvius. 1990. *De l'architecture: Livre III*, ed. Pierre Gros. Paris: Les Belles Lettres.

Vitruvius. 1999. *Ten books on architecture*, ed. Ingrid D. Rowland and Thomas Noble Howe. Cambridge: Cambridge University Press.

Wilson Jones, Mark. 2000. *Principles of Roman architecture*. New Haven/London: Yale University Press.

Proportion and Continuous Variation in Vitruvius's *De Architectura*

Bernard Cache

It is important to balance Vitruvius's discussion of the architectural orders, centered on temples, with his sections on civil and, in particular, domestic architecture. It is in this domain, the subject of Book 6 (Chapters 3 and 4) of the *De Architectura*, that the relationships implied by the term *symmetria* appear explicitly, in both functional and aesthetic terms and without interference from the question of whether the recommended ratios are affected by the transformation of wooden temples to stone ones. Based on a review of his rules for designing *atria*, the Vitruvian conception of order as *genus* appears not as a fixed set of ideal relationships laid down once and for all, but as a series of variations in proportion. While certainly not obeying the concept of "function" as developed in the seventeenth century, these variations can nevertheless be shown to follow continuous curves interpolated from sets of derived values. In this respect, the Vitruvian project finds contemporary expression in today's CAD/CAM software.

The Atrium of the Country House

The instructions that Vitruvius gives for the plan of country houses begin with the atrium, the large central court around which the parts of the *domus* are distributed (Fig. 1). The *compluvium*, the unroofed space in the center of the atrium, owes its name to the fact that it allows rainwater to collect in the *impluvium*, or cistern,

---

Translated by Anthony Gerbino

B. Cache (✉)
School of Architecture, Civil, and Environmental Engineering (ENAC),
École Polytechnique Fédérale de Lausanne, BP 2129 (Bâtiment BP),
Station 16, CH-1015 Lausanne, Switzerland
e-mail: bernard.cache@epfl.ch

A. Gerbino (ed.), *Geometrical Objects: Architecture and the Mathematical Sciences 1400-1800*, Archimedes 38, DOI 10.1007/978-3-319-05998-3_2,

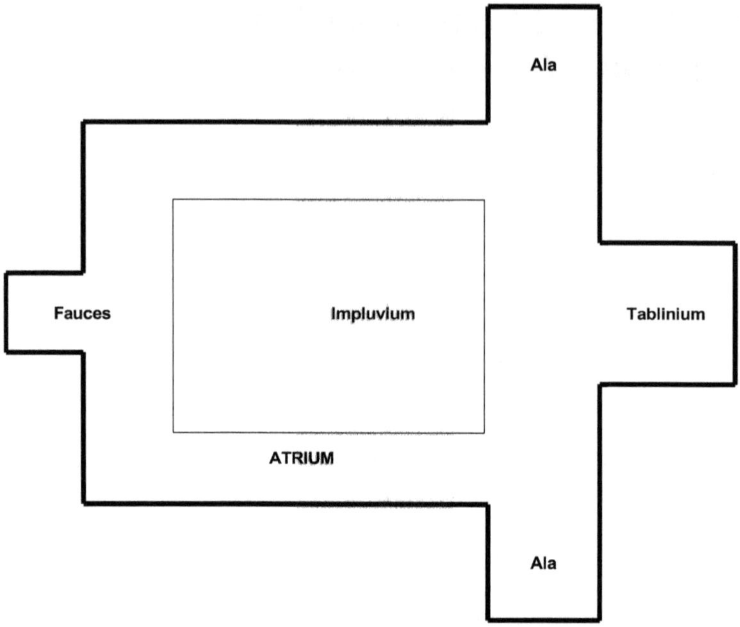

Fig. 1 Parts of the Roman Atrium, from *De Architectura*, Book 6, Chapter 3.4

below. In addition to the lateral wings, the *alae*, the other principal elements of Vitruvius's account are the *fauces*, a passage leading from the vestibule of the house, and the *tablinium*, at the other end of the atrium. As its name suggests, this room may have served to house the *tabulae*, or wax covered tablets inscribed with the accounts of the house, but it may also have served other purposes. Varro, for example, relates that it was used to host meals in summer.[1]

Vitruvius's rules for designing *atria* consist essentially of a series of instructions, in which the principal dimensions of the component spaces depend on each other according to the following sequence, with the preceding value determining the subsequent:

- length of the atrium
- width of the atrium
- width of the wings (*alae*)
- width of the *tablinium*
- width of the *fauces*

However, instead of the preceding value being linked to the next by a fixed proportion, Vitruvius subjects the four relationships between these five elements to what we would call dependant variables. We will look at these case by case.

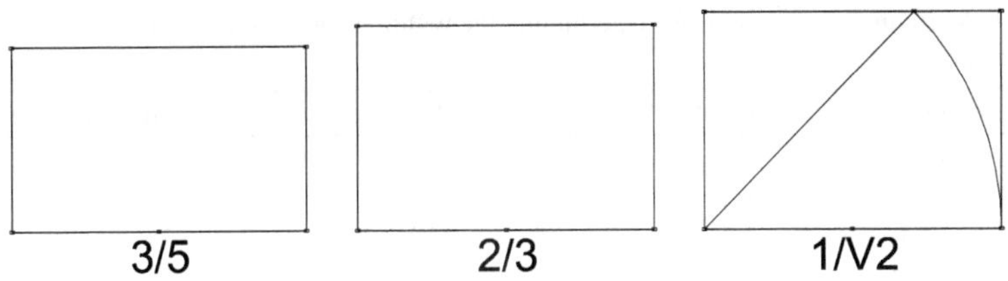

Fig. 2 Variations of the Roman Atrium (ratio *A*/*a*), from *De Architectura*, Book 6, Chapter 3

The Length-to-Width Ratio of the Atrium, *A:a*

The first relation is described as depending on a typological choice between three genres of atrium of increasing width. Given a length *A*, the width *a* of the atrium is calculated by choosing one of three proportions, formulated in the following manner:

- dividing the length in 5 parts, 3 will be given to the width
- dividing the length in 3 parts, 2 will be given to the width
- a square being constructed on the width, the length will be equivalent to the diagonal.

The series poses a problem of consistency: in the first two formulations, it is the length *A* that determines the width *a*, while in the third it is the width that determines the length. The third formulation, moreover is different in kind, as it is geometrical, while the first two are numerical. In any event, the instructions seem to correspond to a choice between one of three types of rectangle (Fig. 2).

Aisle Width to Atrium Length, *L:A*

Next, Vitruvius moves on to the rules for calculating the width of the aisles *L*. These are determined by the length of the atrium *A*. More precisely, the aisle widths are given in terms of a proportion, aisle width to atrium length (*L:A*), which itself varies as a function of the actual length of the atrium. Vitruvius's instructions are given in Table 1. We might call this a "second order" variation, $L = f(L{:}A)$, where the ratio *L:A* itself depends on *A*. Another way of expressing this relationship is by the formula $L = f(g(A))$.

Auguste Choisy was the first to note, in his 1909 translation of Vitruvius, that the proportions of the *alae*, as they diminish with respect to the length of the atrium, seem to imply a continuous variation. If the mean points of the five atrium lengths *L* (35, 45, 55, 70, and 90 ft) are plotted on a graph against the corresponding ratios of atrium length to aisle width *L:A*, the resulting points very closely approximate a curve, which Choisy identified as a hyperbola (Fig. 3).[2] He also found evidence of

Table 1 Vitruvius's instructions for determining the width of the aisle *L* in relation to the length of the atrium *A*

Atrium length	Ratio *L:A as recommended by Vitruvius*	Equivalent fractions
From 30 to 40 ft	1:3	2/6 (0.333)
From 40 to 50 ft	1:3.5	2/7 (0.285)
From 50 to 60 ft	1:4	2/8 (0.250)
From 60 to 80 ft	1:4.5	2/9 (0.222)
From 80 to 100 ft	1:5	2/10 (0.200)

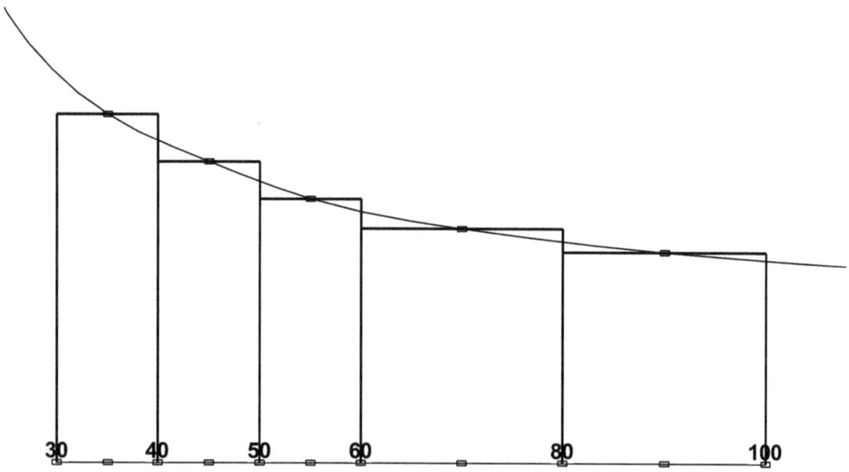

Fig. 3 Variations of the ratio *L/A*, from *De Architectura*, Book 6, Chapter 3.4

an attempt to approximate continuous variation in two other sequences that Vitruvius had recommended: the optical corrections for the width of columns (Book 3, Chapter 3) and for the height of architraves (Book 3, Chapter 6), implying a parabola and hyperbola respectively.[3]

That Vitruvius's instructions-in-series translate into a continuous variation is likely. As Choisy had suggested, the recommendations in these cases may be simplified rules-of-thumb derived from a learned mathematical tradition. But which tradition? Where did these recommendations originate? In one interesting analogy, Choisy related the curve implied by the rules for *atria* to the *scamilli impares*, the "unequal benches" mentioned by Vitruvius in Book 3 (Chapter 4). The *scamilli*, understood today either as small leveling blocks or as the ordinates of a full-scale construction drawing, are described in the text as the means of producing the subtly rising curve of the stylobate, or temple platform. It is by reference to the process of interpolating a curve, in this case that of a light chain hanging from the edges of the stylobate (inverted to produce a shallow mound), that the technique can be linked to the description of the atrium.[4]

In a very stimulating article, Gilbert Hallier has picked up this question, adducing other examples of this phenomenon.[5] Hallier also refers to the design of sundials, such as the monumental one drawn on the pavement of the Campus Martius, near the Ara Pacis. The curves of the dial—some 150 m long—would have been traced by the tip of the shadow of the obelisk through the day at different times of the year. Here we are indeed dealing with curves plotted point by point. Moreover, those curves are hyperbolas, formed by the intersection of the horizontal dial plane with the cone of solar rays passing by the tip of the gnomon.

This suggestion, however, probably reaches too far. Although the properties of hyperbolas were known at least since the time of Menaechmus in the fourth century BCE, we have no evidence that ancient astronomers had conceived the lines of sundials in this way. Book 9 of *De Architectura*, the sole surviving ancient treatise on sundials, makes no mention of the kinds of curves produced by the moving shadow of the gnomon. Nor does the word *hyperbola* appear in the second-hand references that we have of the treatise by the astronomer Diodorus of Alexandria concerning a method for drawing meridian lines.[6]

Hallier probably also goes too far in the other direction, attributing the curve of variation implied in the ratio $(L{:}A){:}A$ to the geometrical tradition stemming from Apollonius of Perga and Pappus of Alexandria. Apollonius had no doubt gathered most of the elements for solving the problem of constructing a conic through five points, but, as Heath explains, such constructions are not found in his *Treatise on Conic Sections*.[7] Much later, some three centuries after Vitruvius, Pappus would produce a method for constructing an ellipse from five given points, working from a problem that involved finding the diameter of a column from a fragment. Pappus's solution, however, is not general and supposes that four of the five points are found on two parallel lines.[8] In fact, the construction of a conic section from five arbitrary points derives from a theorem of projective geometry that was not formulated explicitly until the nineteenth century.[9] Despite its color of practical usefulness, the study of conics does not seem to have elicited any direct application, either in perspective or gnomonics. To take one striking example, the concept of the visual cone formed by rays from the eye or from a specific object is well attested in ancient times, but its consequences—for a system of representation consisting in the intersection of the cone by a "picture plane"—are nowhere picked up. Euclid himself, who is reported to have written his own treatise on conic sections, describes the image of chariot wheels viewed obliquely as oblong, not as ellipses.[10]

These considerations must necessarily invalidate Hallier's conjectures. The historical problem posed by Vitruvius's text involves not the construction of a curve from given points or lines, but rather the determination of fractional values in series in a way that happens to approximate a certain curve. It follows, too, that these ratios cannot have originated as a hyperbola. Despite the seeming accuracy of Choisy's formulas, Greek mathematical thought did not provide the techniques necessary to model such complex curves arithmetically. These objections, however, do not fundamentally alter the fact that we are dealing with a second order variation, that is to say, a variation in proportional relationships where the coefficient $L{:}A$ is itself depending on the variable A, expressed in increments and interpolable,

Table 2 Vitruvius's proportional series *L:A* (columns 1 and 2), as interpreted by Herman Geertman (columns 3 and 4) (to be read with Fig. 4)

Atrium length	Ratio *L:A as recommended by Vitruvius*	Geometrical series	Approximation used
From 30 to 40 ft	1:3 = 2/6 (0.333)	$1:2\sqrt{2}$ (0.354)	$\sqrt{2} \approx 15/10$
From 40 to 50 ft	1:3.5 = 2/7 (0.289)	$1:2\sqrt{3}$ (0.289)	$\sqrt{3} \approx 17.5/10$
From 50 to 60 ft	1:4 = 2/8 (0.250)	$1:2\sqrt{4}$ (0.250)	$\sqrt{4} \approx 20/10$
From 60 to 80 ft	1:4.5 = 2/9 (0.222)	$1:2\sqrt{5}$ (0.224)	$\sqrt{5} \approx 22.5/10$
From 80 to 100 ft	1:5 = 2/10 (0.200)	$1:2\sqrt{6}$ (0.204)	$\sqrt{6} \approx 55/10$

moreover, in a continuous form. More importantly, as Hallier shows, the ratios seem to correspond to archaeological reality, falling within a cluster of points produced by the analysis of the remains of roughly 100 Roman villas.[11]

This formulation of variation clearly goes well beyond the simple concept of proportion, strictly speaking. That is not to say that Vitruvius somehow anticipates the modern concept of "function", which would only appear in the seventeenth century. For this, Vitruvius would have had to overcome a deeply rooted epistemological obstacle to the concept of a change *in* change. To Aristotle, for example, change was an irreducible category belonging to the order of the pure event. It is worth noting that Galileo himself did not go so far as to elaborate a concept of acceleration.[12] This Vitruvian variation is, instead, best seen as one of many incremental steps necessary for the formation of the concept of the continuous mathematical function. One of the interests of the *De Architectura* lies precisely in this and other such contributions to the archeology of the modern sciences.

For his part, Herman Geertman has developed a competing interpretation of the Vitruvian ratios of the atrium. Geertman sees the ratios as an attempt to simplify and approximate not a curve, but a diminishing geometric series defined by the ratio $1:2\sqrt{n}$. This interpretation has a very different orientation in that it focuses not on an implied continuity, but on the discontinuity resulting from the approximate roots of a series of consecutive integers.[13] Vitruvius's instructions, as interpreted by Geertman, appear in Table 2.

This interpretation has a number of strengths. In the first place, it is based on a geometric pattern conceivably rooted in an ancient design technique, namely lengthening a given rectangle by means of its diagonal. At full-scale, such a procedure would have made use of stakes and string (Fig. 4). Moreover, similar ratios appear in other passages of the text. Geertman notes that Vitruvius recommends apparent approximations of $1:\sqrt{5}$ for the width and height of doors in Doric temples and of $1:\sqrt{6}$ for Ionic temples (Book 4, Chapter 6). Finally, Geertman's interpretation rests on methods of approximating square roots that would conceivably have been codified at least in the fourth century BCE. As Geertman and others have argued, Vitruvius may have inherited standard approximations for such values from Hellenistic mathematical texts.[14]

The main weakness of the hypothesis, however, relates to this last point: Geertman's series relies on at least one rather imprecise approximation, in

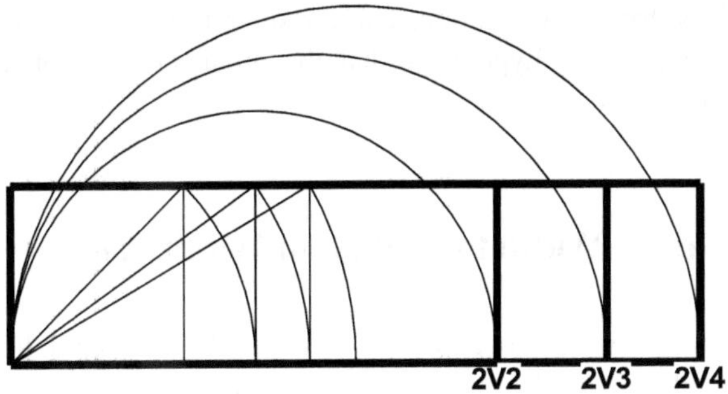

Fig. 4 Generation of rectangles of the ratio $2\sqrt{n}$

particular, that of $\sqrt{2}$ to 15/10 or 3/2. This would have been among the least accurate of the available approximations for this value, differing from the next closest (7/5) by more than 6 %. It also requires explaining why Vitruvius, a few lines above, where he lists three types of atrium by length to width, would have distinguished the ratio 3/2 from the geometrical process leading to the ratio $\sqrt{2}/1$. This divergence may spring from an inaccuracy in the manuals or graphical constructions that the author relied on, but it is nevertheless jarring, given his earlier instructions. Perhaps the most incongruous aspect of this hypothesis is that it ignores the straightforward and consistent series that Vitruvius himself provides, to replace it with a conjectural and more complicated one.

Tablinium Width to Atrium Width, *T:a*

In this regard, the case of the atria is certainly exemplary. For if we continue the examination of the other elements, namely the *tablinium* and *fauces*, we find the same characteristic approach. Regarding the *tablinium*, Vitruvius says explicitly:

> For smaller atria cannot have the same principles of symmetry that larger ones do. It we use the proportions of larger atria in the design of smaller ones, the *tablinum* and the *alae* will be too small to be functional. If, on the other hand, we use the proportional systems of smaller atria to design the larger ones, the dependent rooms will seem vacant and oversized. Therefore I thought that the principles for the dimensions of *atria* should be recorded precisely in the interests of function and appearance (Book 6, Chapter 3.5).[15]

On this basis, the architect explains that the ratio *T:a*, which determines the width of *tablinium* as a function of the width of the atrium, will be 2/3 for *atria* 20 ft wide, 1/2 for *atria* 30–40 ft wide, and 2/5 for those between 40 and 60 ft wide. Note here that the author provides three increments rather than five. This reduction in the number of variables reflects a different approach to dealing with the subsidiary spaces of the atrium, also evident in the rules for dealing with entryways, or *fauces*, below. For the moment, it is worth noting the mathematical consequences of this change. Although three increments might still plausibly correspond to points on a continuous curve, they alone cannot provide the construction of the curve itself, at least for a conic section. This consideration, in case any more were needed, further weakens the hypothesis that the architect had conceived of these points as a hyperbola.

Fauces Width to Tablinium Width, *f:T*

The same reasoning that characterizes the discussion of the *alae* and *tablinium* also applies to the *fauces*, but with a still further reduction in the number of increments. For these spaces, Vitruvius declares simply: "The entryways for smaller atria should be determined by the width of the *tablinum*, minus one-third; those of the larger *atria* should be one-half (Book 6, Chapter 3.6)"[16] This formulation, reduced now to only two values, indicates that Vitruvius intended his readers to adopt a different approach in determining the dimensions of this room. Rather than moving abruptly between only two ratios, it is more likely that he expected practitioners to gradually interpolate the proportions for *atria* of intermediate size, even according to a linear variation, as suggested by the two extreme values of the ratio of the *fauces* to *tablinium f/T*. In the absence of explicit rules, Vitruvius seems to be recommending a trial-and-error process of interpolation, reminiscent of the notion of "correction". This idea, mentioned throughout *De Architectura*, is always described with a combination of two words, *adiectio/detractio*, as though to suggest that the method proceeds by estimation, sometimes by adding, sometimes by taking away. In such cases, Vitruvius implicitly calls on the architect to exercise his own qualities of *ingenium* and *acumen*, talent and skill.

Although often discussed in relation to the use of optical refinements, the dual concept *adiectio/detractio* is not confined to that field. The terms appear, in fact, in the introduction to the chapters on the atrium, in a general formulation that relates only partially to the visual appearance of a building. Here, *adjectio/detractio* appear as an *ad hoc* method of fine-tuning a given proportional system:

> Thus, once the principle of the symmetries has been established and the dimensions have been developed by reasoning, then it is the special skill [*acuminis*] of a gifted architect to provide for the nature of the site, or the building's appearance, or its function, and make adjustments by subtractions or additions, should something need to be subtracted from or added to the proportional system, so that it will seem to have been designed correctly with nothing wanting in its appearance (Book 6, Chapter 2.1).[17]

This explanation for correcting a set of "symmetries" seems to point to a visual, or more specifically, a graphical method of interpolation. To determine the correct ratio *f/T* between two extreme values requires that it be visually calibrated according to the length of the atrium, which is itself situated between the larger and the smaller *atria*. In addition, the coefficients of proportionality governing the relationships of the *tablinium* to the atrium and of the *alae* to the atrium themselves vary depending on the length of the atrium. These intermediate cases, defined only by a limited set of values, would be difficult to determine without the aid of an elementary diagram.

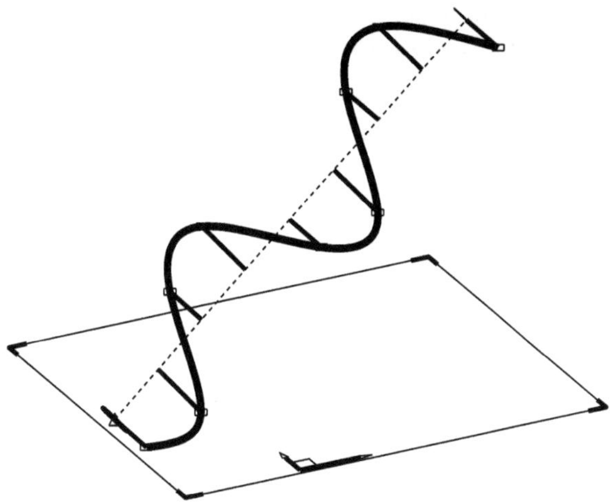

Fig. 5 Interpolation of an Archimedean screw using splines, from *De Architectura*, Book 10, Chapter 6.2

We know of similar graphical procedures in ancient design and construction. Aristotle—an unusual source in this context—speaks of a flexible, leaden rule used to replicate molding profiles.[18] The *De Architectura* itself provides other examples. Like the passage on the *scamili impares*, they appear to relate to the point-by-point construction of curves. In the chapter on baths, Vitruvius describes how to hang a plaster ceiling from metal arcs suspended from rafters in order to mimic a curved vault (Book 5, Chapter 10.3).[19] In explaining the construction of the water screw, the author gives explicit instructions for wrapping strips of willow or chasteberry around a beam so as to build up a helicoid (Fig. 5). This lattice of lateral and longitudinal strips forms a cylindrical graph, on which one literally plots the path of the spiral: "Where the lines have been drawn along the length, the transverse scorings create intersections, and these intersections determine specific points (Book 10, Chapter 6.1)." These supple branches, coated with pitch, constitute the physical equivalent of our contemporary curve-approximating software for Beziers, splines or NURBS. The word "spline" derives, in fact, from a craft context of just the sort Vitruvius describes, to designate flexible strips forced to pass through specified points. We can imagine an analogous attempt to regulate the proportions of the atrium by virtue of drawn plans. In some respects, these would follow a pre-established proportional or schematic logic, but in others, they would have to be estimated more-or-less faithfully by the eye. Indeed, Vitruvius emphasizes the role of visual judgment in this process, "so that [the whole] will seem [*videatur*] to have been designed correctly with nothing wanting in its appearance [*in aspectu*]."[20]

Whatever the tools used to achieve it, it is evident that Vitruvius's conception of the atrium possesses a high degree of elaboration. Taken as a whole, his instructions clearly form a system or, more precisely, a variational one. The consistency of the system is not always easy to achieve, but it is described well enough that we can construct an organizational diagram for it—the kind required, incidentally, in computer-aided design and manufacturing (Fig. 6). We see, in this case, that two of the interrelated variables—the length of the atrium, and the ratio of its length to its

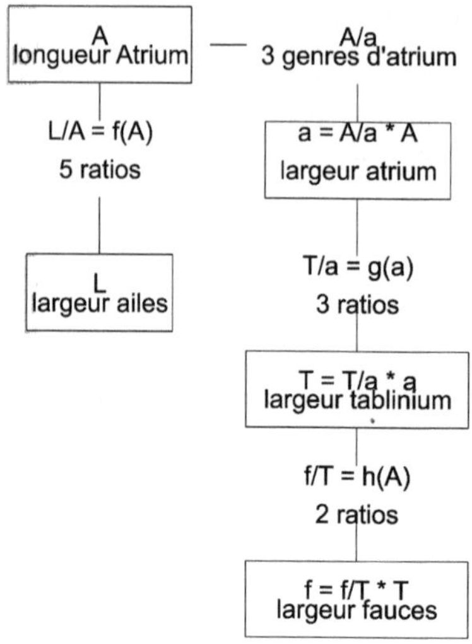

Fig. 6 Organizational diagram for the composition of the atrium, based on *De Architectura*, Book 6, Chapter 3.3 [The *boxes* denote entities to be dimensioned, while the *unframed terms* refer to the variable ratios that govern them.]

width—launch the two deductive chains that determine the dimensioning of the wings as well as that of the *fauces* and *tablinium*. Everything therefore depends on the first two decisions regarding the length and type of the atrium.

That Vitruvius's ratios for the atrium cannot be related to modern continuous functions, such as a hyperbola, should come as no surprise. Yet, it is also clear that the proportional series varies in a way that seems to imply some sort of interpolated continuity. This is what I have termed—for want of a better expression—a "second order" variation. To be sure, modern mathematicians would have a totally different notion of continuity, but it is enough only to open an up-to-date CAD-CAM package to see that Vitruvius's methods are in other ways not far from our own. To describe a continuous variation, all that is necessary is to input a set of values and let the software interpolate the resulting curve.

Notes

1. The explanation given by Vitruvius here closely reflects the definitions that Varro gives for the words *domus*, *aedes*, *cavum*, *aedium*, *impluvium*, and *atrium*. See Varro (1977–1979, 1st ed. 1938, 151–53) (Book V, 160–161). Further on the *tablinium*, see Riposati (1939, Book I, 29).
2. The curve corresponds to the equation $L{:}A = 1/9 + 70/9\ (1/A)$. His values for all five points come within three decimal places of Vitruvius's fractions. Solving for the atrium width L, reduces this to the linear equation $L = 1/9\mathrm{A} + 7.77$ ft. Choisy (1909, vol. 1, 230–36; vol. 4, pl. 62, Fig. 3).
3. Choisy (1909, vol. 1, 149–156; vol. 2, pl. 30, 31).
4. See the very detailed commentary in Vitruvius (1990, 139–145), which provides several interpretations for controlling the curvature of the stylobate.

For the current state of the question, including recent archaeological discoveries, see Bankel (1999) and Haselberger (1999).

5. Hallier (1989).
6. This method, known from Arabic sources and the surviving writings of the agrimensor Hyginus, derives the meridian from any three shadows made during the day. See Neugebauer (1975, vol. 2, 840–43). My thanks to Bernard Vitrac for bringing this important work to my attention.
7. See Heath (1896, cli–clvi).
8. See Pappus of Alexandria (1982, Book VIII, chapter 16). Also see Heath (1921, vol. 2, 434–437).
9. The theorem was discovered independently by William Braikenridge and Colin Maclaurin c. 1733. See Coxeter (1964, 85).
10. See, for example, Euclid, *Optics*, see Definition 2 and Proposition 46. On the concept of the visual cone, with reference to Roman sources, see Haselberger (1999, 57–58).
11. Hallier (1989, 199).
12. Panza (1989, Chapter 2).
13. Geertman (1984).
14. See Heath (1921, vol. 1, 60–63; vol. 2, 323–24). Also see Gros (2006 [1976]).
15. Vitruvius (1999, 79).
16. Vitruvius (1999, 79).
17. Vitruvius (1999, 78).
18. Aristotle describes this building tool in terms of a metaphor for laws that are applicable only to particular situations. "In fact this is the reason why all things are not determined by law, that about some things it is impossible to lay down a law, so that a decree is needed. For when the thing is indefinite the rule also is indefinite, like the leaden rule used in making the Lesbian moulding; the rule adapts itself to the shape of the stone and is not rigid, and so too the decree is adapted to the facts." Aristotle (1925, 1137b).
19. Vitruvius (1999, 72).
20. "…uti id videatur recte esse formatum in aspectuque nihil desideretur." Vitruvius (1999, 78).

Photographic Credits

All figures by author

References

Aristotle. 1925. *Nicomachean Ethics*. Trans. W.D. Ross. Oxford: Clarendon Press.

Bankel, Hansgeorg. 1999. *Scamilli inpares* at an early Hellenistic Ionic propylon at Knidos: New evidence for the construction of a curvature. In *Appearance and essence, refinements of classical architecture: Curvature*, ed. Lothar Haselberger, 127–138. Philadelphia: The University Museum, University of Pennsylvania.

Choisy, Auguste. 1909. *Vitruve*. Paris: Lahure.

Coxeter, H.S.M. 1964. *Projective geometry*. New York: Springer.

Geertman, Herman. 1984. Vitruvio e i rapporti numerici. *Bulletin Antieke Beschaving* 59(1): 53–62.

Gros, Pierre. 2006 [1976]. Nombres irrationnels et nombres parfaits chez Vitruve. In *Vitruve et la tradition des traités d'architecture, Fabrica et Racionatio: Recueil d'études*, 75–111. Rome: École française de Rome.

Hallier, Gilbert. 1989. Entre les règles de Vitruve et la réalité archéologique: *l'atrium toscan*. In *Munus non Ingratum: Proceedings of the international symposium on Vitruvius' De Architectura and the Hellenistic and Republican Architecture*, Leiden 20–23 January 1987, ed. H. Geertman and J.J. de Jong, 194–211. Leiden: Babesch.

Haselberger, Lothar. 1999. Old issues, new research, latest discoveries: Curvature and other classical refinements. In *Appearance and essence, refinements of classical architecture: Curvature*, ed. Lothar Haselberger, 1–68. Philadelphia: The University Museum, University of Pennsylvania.

Heath, Thomas (ed.). 1896. *Apollonius of Perga: Treatise on conic sections*. Cambridge: Cambridge University Press.

Heath, Thomas. 1921. *A history of Greek mathematics*. Oxford: Clarendon.

Neugebauer, Otto. 1975. *A history of ancient mathematical astronomy*. Berlin/Heidelberg/New York: Springer.

Panza, Marco. 1989. *La Statua di Fidia: Analisi filosofica di una teoria matematica, Il calcolo delle flussioni*. Milan: Unicopli.

Pappus of Alexandria. 1982 [1st ed. 1933]. *La Collection Mathématique*, ed. Paul Ver Eecke. Paris: Blanchard.

Riposati, Benedetto (ed.). 1939. *M. Terenti Varronis De vita populi Romani*. Milan: Società Editrice "Vita e Pensiero".

Varro. 1977–1979, 1st ed. 1938. *On the Latin Language*. Trans. Roland G. Kent. Loeb Classical Library. Cambridge, MA: Harvard University Press.

Vitruvius. 1990. *De l'architecture: Livre III*, ed. Pierre Gros. Paris: Les Belles Lettres.

Vitruvius. 1999. *Ten books on architecture*, ed. Ingrid D. Rowland and Thomas Noble Howe. Cambridge: Cambridge University Press.

Part II
Mathematics and Material Culture in Italian Renaissance Architecture

Part II, on fifteenth- and sixteenth-century Italy, also considers the use of mathematics in design, particularly as it is embedded in the profession's material culture. The following three essays concentrate on Renaissance architectural drawings, instruments, and, not least, buildings themselves. This shift in emphasis allows our contributors to focus on the physical character of the design and construction process, an aspect of architectural practice that is often obscured by idealizing, literary form of the treatise. One of the overarching themes of this section has to do with the way mathematical techniques of design were both stymied by—and ultimately adapted to—the constraints posed by the site itself. Measurement and scale drawing, in particular, were crucial in translating the design into built form, but these techniques were themselves subject to broader changes in mathematical knowledge during the period concerned.

Francesco Benelli looks at the way in which the architect and builders of the Palazzo del Podestà in Bologna (begun 1472) sought to translate the building's ideal proportions into actual dimensions, a task made difficult by an irregular site and the need to utilize the walls of a pre-existing building. The attempt to preserve the image of perfect geometrical regularity in the façade involved a complex mix of arithmetic manipulation, optical fudging, and basic concerns over cost. The emphasis on consistency of measure was no doubt partly carried over from medieval practice. What does appear novel in the Palazzo del Podestà is the "optical" character of its proportions, that is, the dimensional fine-tuning intended either to emphasize or to retain the appearance of modular rigor, particularly in elevation. These adjustments—which can be distinguished from normal constructional "tolerance"—were probably determined by the architect, Aristotele Fioravanti. They suggest that he was familiar with the architectural values of early humanism. The application of whole-number ratios in the elevation, for example, may derive from a reading of Vitruvius, but Alberti is another possible source. Benelli points to the proportions of the window surrounds, similar to prescriptions in *De Re Aedificatoria* as well as in Alberti's built work. More provocative is the idea that an "optical correction" has been used to hide an irregularity in one of the lateral bays. Vitruvius's speculations about the negative effect of void spaces may have inspired the architect to think of the arched bays as compositional elements in themselves.

Ann Huppert is also interested in how architects negotiated the material and physical restrictions imposed by their commissions. Her contribution focuses on the way Renaissance design practice was shaped by the use of algorism, that is, calculations using Hindu-Arabic numerals. A comparison between Antonio da Sangallo the Younger and his contemporary Baldassarre Peruzzi forms the basis of her study. Drawings from the hands of these two practitioners survive in great quantity. Scaled to on-site dimensions and copiously annotated, they afford some of the earliest and most varied evidence for the use of paper calculations in architectural practice. This new and powerful tool offered architects several benefits: for converting often-incompatible local dimensions, for estimating the cost of materials, and for communicating with workers and patrons—all directly on the drawing. More important was the way in which Peruzzi, in particular, was able to integrate calculations directly into the design process, using them to generate alternative projects and compare them to each other. This ability points to a crucial difference between the two architects. Although both men were masters of the technique, Huppert shows that Peruzzi found more creative, intuitive, and practical ways of incorporating numerical calculations into the design and building process. The difference is striking, especially given Sangallo's greater theoretical interests in mathematical and humanist scholarship.

The third essay of this section, by David Friedman, shifts our attention from architectural to urban design. Following his groundbreaking work on medieval town plans, the author began several years ago to look at the adoption of modern mapping and planning techniques during the sixteenth and seventeenth centuries. This was by no means a story of straightforward technical progress. As Friedman argues, the limitations of available instruments and the complexity of urban environments made the principal methods of survey—particularly the compass traverse—unpredictable and often unreliable. His focus in this chapter is a single drawing, a proposal from the late 1550s for a new convent in the center of Rome. The proposal is unusual in that it incorporates the results of a geometrical survey in a densely built area, where such techniques were difficult to carry out. Friedman's microscopic attention to the plan's details reveals the practical and conceptual obstacles to the adoption of geometrical survey for urban planning. Needle holes, erased lines, and compass pricks show the project's halting and rather imprecise construction. With no accurate printed map of the city to go on, all the data about the neighboring blocks had to be generated anew. Yet, despite its provisional quality, this document portended another very powerful visual tool. The large-scale survey plan would come into its own in the seventeenth century, allowing new and unimagined levels of control and definition over urban space.

The Palazzo del Podestà in Bologna: Precision and Tolerance in a Building *all'Antica*

Francesco Benelli

Mario Carpo has shown how to map out a parallel history of Renaissance architecture and numeracy through the study of treaties and theory.[1] He also suggests that such results are much more difficult to obtain when examining built works. Existing architectural drawings, especially those from fifteenth-century Italy, are scarce, as are accurate surveys and written construction documents.

In light of these obstacles, historians may usefully consider the evidence of the buildings themselves. The Palazzo del Podestà in Bologna offers an important case study for some of the quantitative and numeric features of built architecture of the last quarter of the fifteenth century. It shows, in particular, how imperfections in construction and difficult site conditions could hinder the much-desired ideal of geometrical, mathematical, and proportional exactitude that was already well diffused in both the theory and practice of Renaissance architecture.[2] The Palazzo, a project of the early 1470s, can serve as a model for understanding how the idea of a building conceived on the model of the geometric grid—with precisely calculated, exact, and whole measurements—was the strongest prerogative of the well-educated Renaissance architect (Fig. 1).[3] Such characteristics imply a knowledge of precise geometrical and mathematical rules, the ability to render meticulous and accurate drawings, and to execute them in built form. It also reflects the capabilities of stonemasons to create architectural elements of great precision.

The Palazzo del Podestà illuminates how architects reconciled a desire for geometrical accuracy and modularity—an ideal condition of simplicity but also a necessity for building a portico *all'antica*—with the irregularities inherent in a pre-existing medieval site and the need to fully exploit the existing foundations. The building reveals, in particular, how careful attention to minute differences between architectural elements helped the designer to form an analyzable, modular, and geometrically

F. Benelli (✉)
Department of Art History and Archaeology, Columbia University, 826 Schermerhorn Hall, 1190 Amsterdam Avenue, New York, NY 10027, USA
e-mail: fb2013@columbia.edu

A. Gerbino (ed.), *Geometrical Objects: Architecture and the Mathematical Sciences 1400-1800*, Archimedes 38, DOI 10.1007/978-3-319-05998-3_3,

Fig. 1 Palazzo del Podestà, Bologna

quantifiable "exact" space. The use of such optical refinements, in particular, was precocious. The technique was in principle known only through the obscure and still poorly known writings of Vitruvius and the treatise of Leon Battista Alberti, both of which were at that time still available only in manuscript form.[4] Finally, the Palazzo suggests that dimensional exactitude and repeatability were important not only for the composition of a building but also for reasons of cost and efficiency.

Correcting Irregularities

On 9 November 1472, the *Comune* of Bologna commissioned a model that would determine the basic volumetric features and architectural elements of the new *Palazzo Comunale*. The project was intended to utilize the existing foundations and walls of a previous medieval building constructed, along with the related piazza, in 1200–1203.[5]

While the document does not mention the architect in charge, we can attribute the work, for various reasons, to the illustrious Bolognese engineer Aristotele Fioravanti.[6] Aristotele—his name itself expresses, not by chance, a connection with geometry—was already known both throughout the Italian peninsula and abroad for his engineering endeavors. These began in 1451–1452 with the excavation and transportation of giant monolithic roman columns from the area of the church of Santa Maria sopra Minerva in Rome to the Vatican, to be employed in the choir of the new Basilica.[7] In 1455, he successfully transported Bologna's tower of the Magione 13 m

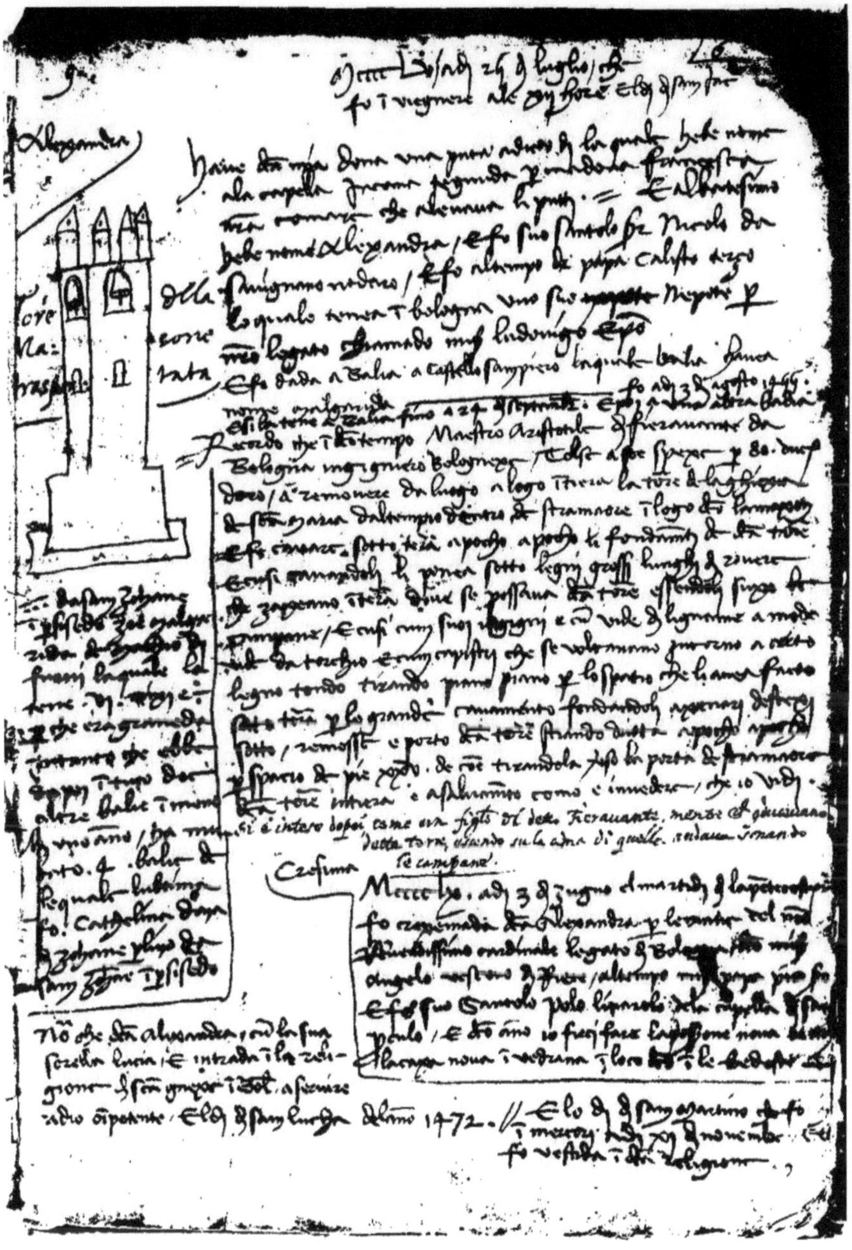

Fig. 2 Tower of the Magione, Bologna, Biblioteca Archiginnasio

to the opposite side of Strada maggiore (Fig. 2), a feat for which he gained considerable fame and enough work to keep him busy under three papacies and at several Italian and foreign courts, some as far away as Hungary and Russia.[8] Fioravanti's renown must also have brought him the commission for the Palazzo del Podestà, but

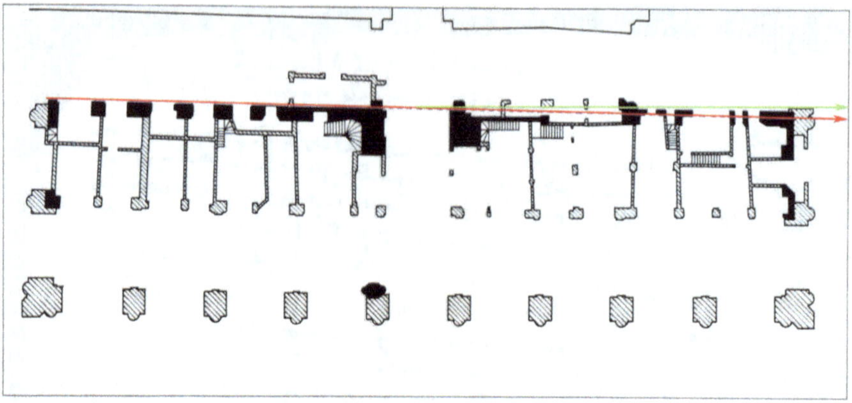

Fig. 3 Measured ground plan of the Palazzo del Podestà

he did not oversee the reconstruction. By 1484, when work began on the site, he had already died, probably during a trip to Siberia.[9]

The primary goal of the project was to obtain a series of arches around the pre-existing medieval envelope of the building that were as similar as possible to one another and that presented the greatest rhythmic, proportional, and modular coherence.[10] The end result is visible in a modern, measured survey of the building, which also makes clear the relationship between the earlier building and later additions (Fig. 3).[11] The footprint of the Renaissance arcade running around the perimeter is indicated with hatching, while the earlier medieval structure is marked in black on the plan. This structure consisted principally of a massive rear wall running east-west and divided into narrow shops. This long wall was capped on its extremities by two perpendicular end walls, and it was pierced by a passageway on axis with the central arch. The passageway, centered on the intersection of a minor Roman road to the rear of the Palazzo, contained the foundations of the bell tower.[12] Another given established by the earlier building—which is not visible on this plan—was the height of the floor of the great hall—the *sala magna*—on the second story.[13]

These pre-existing elements—particularly the perpendicular sections on either end of the medieval wall—determined the building's basic proportional schema. The two terminal walls established both the depth of the internal shops and the inner boundary line of the arcaded loggia. These elements had to remain fixed in order to respect the cross axis of the central passage and the piazza on the exterior. Those medieval end walls would also determine the general width of the bays to be used throughout the façade.

Considering the desire for precise regularity in the arcade, the pre-existing condition of the site presented several problems. The first was that the bays defined by the terminal walls of the medieval structure were not compatible with the length of the façade, being slightly too broad to provide even multiples that could be contained within its length. More importantly, the two ends were themselves of different width. As is shown on the plan, the line of the Palazzo's rear medieval wall followed the slightly crooked outline of the roman road to the rear of the structure. This road deviated 1.5° to the north from the base of the bell tower at the center of our building, creating an extra 60 cm on the eastern side. The deviation is shown by the red and green arrows on the plan. The result of these slight discrepancies was that the perimeter of the building, far from being regular in form, embodied an axial incongruity that made the site incompatible with the demands of regular bays *all'antica* with pillars, engaged columns and arches.

A survey of the building's elevations reveals the intriguing solution to this problem (Table 1, Fig. 4). In confronting it, Fioravanti—or his builder—created subtle differences in the width of the front and side bays that are invisible to the naked eye. Discrepancies are normal to certain extent—especially in porticoes. In a Renaissance building like this one, a normal construction "tolerance"—that is, the inaccuracy inherent in contemporary building methods—would be within 5 or 6 cm for every 700. The difference of 5 cm between the two corner bays of the front façade, for example, is relatively negligible and therefore we can consider them to be equal. However, in other parts of the building, the measurements diverge by as much as 61 cm. Bay 8—the narrowest in the building—differs from bay 1 by this amount, or just under two Bolognese feet.[14] Such an easily measured and evenly quantified irregularity which corresponds incidentally to the length of two Bolognese bricks—rules out the possibility that the discrepancy was simply within the normal "tolerance" of early Renaissance builders.[15]

To solve the first problem, the lateral arcades were made to fit a subtly different dimensional scheme that that of the façade. The side bays, in other words, are on average 20 cm wider than those on the long façade. It is a very slight difference of less that 3 %, but nonetheless an important one. The second problem—the long wall section caused by the crooked road—was solved in a similar way. To compensate for the additional length, the architect widened the corresponding bay by 33 cm more than its counterpart on the opposite end. Of the lateral bays, only this one is perceptibly bigger. The other three vary within a difference of at most 6 cm. The module chosen for the short ends of the building is therefore based on the opposite bay (13) on the west. This solution, however, necessitated another irregularity. Because they were restricted by the railed architrave above, the lateral arches, particularly those of the longer bay (1) are actually squashed. The height of the ground story could not be greatly altered in order to keep the hall on the same level with the *iter in voltis*, a terraced path leading from the rear to other buildings in the complex (Figs. 5 and 6).[16] This deformation, however, is too slight to be visually perceptible. It has only come to light through our measured survey.

Table 1 Dimensions in the facade of the Palazzo del Podestà, Bologna (to be read with Fig. 4)

	1	2	3	4	5	6	7	8	9	10	11	12	13
A	734	707	681	681	680	680	673.5	673	682	680	682	702	701
B	525	514	480	483	480	486	476	475	482	481	481	507	495
C	320.5	304	282.5	295	300	304	292.5	299.5	300	300	307	307	312
D	736	706	671	681	698.25	684	668.5	677.5	677.25	680	687.5	703.5	728.5
s	115	114	104.5	99	102.5	102	97.5	93	98	94	101	105.5	111
d	114.5	101	97	100.5	101	91	93	99	94	99.5	94.25	105.5	119

Measures expressed in cm

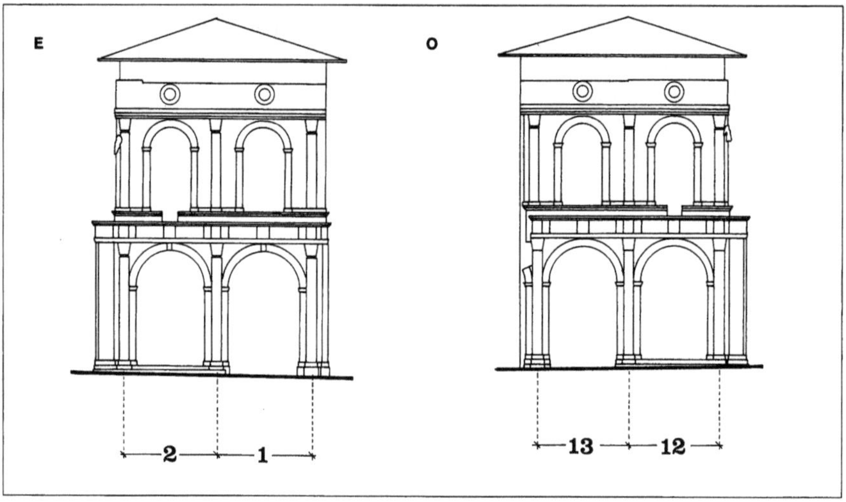

19

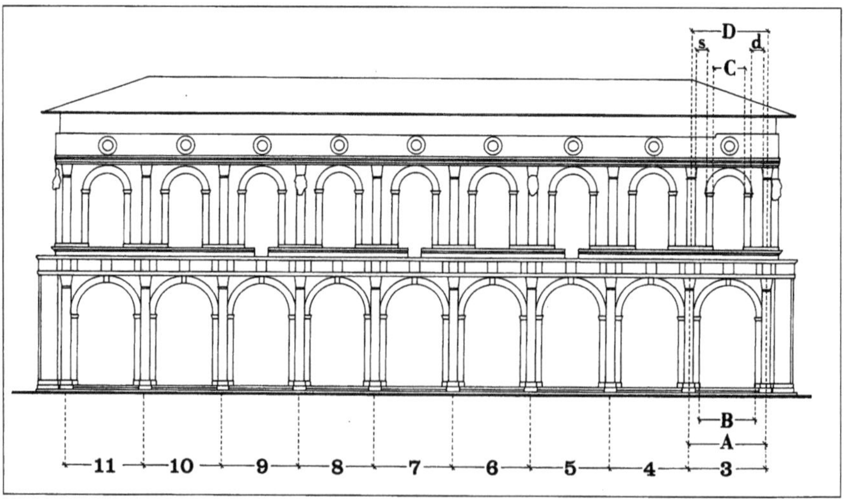

20

Fig. 4 Side and front elevations of the Palazzo del Podestà (to be read with Table 1)

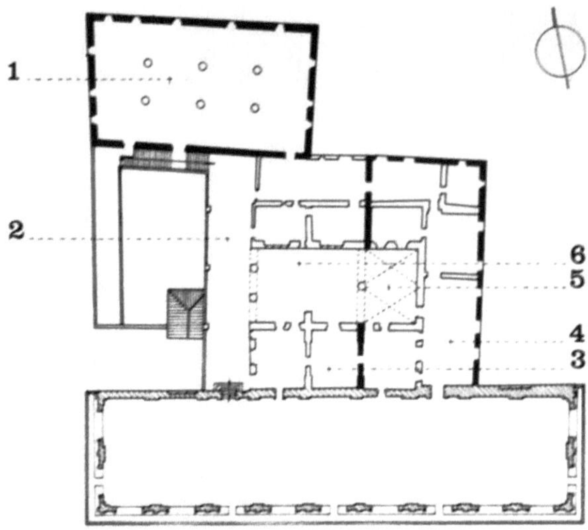

Fig. 5 Upper story plan of the Palazzo del Podestà, with building complex to the rear

This solution of adjusting the width of the bays is economical and logical, as it allows for corrections to be made through the manipulation of the "empty spaces", while leaving the masonry elements among them as consistently equal as possible. The strategy also serves a more subtle visual purpose, particularly on the east side, where the difference in width between the two lateral arches is most perceptible. By concentrating the variations in the arches, rather than in the piers, the narrow bay (2) at the head of the portico benefits from the optical effect of the voids surrounding it. The problem here is reminiscent of one of the several passages on optical corrections that Vitruvius describes in regard to the Ionic temple. In Book 3 of the *De Architectura*, Vitruvius states that a corner column appears thinner to the eyes, as if it were consumed by the air all around (Book 3, Chapter 8).[17] It can therefore be built bigger in order to correct the negative effect of the space around it, thereby making it look like all the others. The architect of the Podestà may have been inspired by this idea, but in a roundabout way, for he has used it to amplify the perceived width of a void rather than a solid. By using the empty spaces of the bays as an element of composition, the incompatibility of the two shorter sides is more readily concealed.[18]

Composition and Planning

The composition of the principal façade took a more straightforward path. The front elevation was established using the bay as the basic module, the width obtained by dividing the entire façade into nine intervals. That module measures 18 Bolognese feet—a whole number that is also easily divisible by 2 and 3. This module provided

Fig. 6 Rear facade of the Palazzo del Podestà

in turn whole-number dimensions for the long side of the pedestal at the base of the half-columns, which measure 3 ft, and the diameter of the base of the half-column, which is reduced to 2 ft (Fig. 7). In other words, the half-column is proportioned according to the measurement of the bay, with its diameter being a whole number so to facilitate the process of construction. The result is a ratio of exactly 1:9. At the same time, the height of the column works out to be 10 times its diameter. This ratio is rather slender for the Corinthian order, but is understandable in view of the unchangeable height of the ground floor.[19] The rest of the measurements are not multiples of the Bolognese metric system, but are calibrated according to the proportions established by the half-columns, giving the impression, in this case, that the secondary dimensions have been determined by geometrical constructions rather than arithmetical calculations.

The depth of the façade pillars, on the other hand, was not determined by the same proportional logic. That measurement was based on the depth of the angle

Fig. 7 Palazzo del Podestà, column base

piers, which was derived, in turn, by the length of the pre-existing medieval wall sections. There also appear to have been structural considerations at work here. Although lightened by great arched windows, the walls of the great hall on the upper story are further weighed down by the massive roof. As was known empirically in Gothic practice, the outward thrust of the roof truss required an increase in the depth of the pillar to counteract it. As a result, the pillars are deeper than is structurally necessary. Ultimately, the formal and structural logic of the building is based on the development of measurements in two different directions (along with two different rationales). On the surface of the façade, the architectural elements are determined according to "formal" issues of rhythm and proportion, while in the depth of the façade they are determined by the dimensions of the pre-existing structure and by structural considerations, with the pillars understood as having an important weight-bearing function.

The second level of the façade follows the same logic of the ground story (Fig. 8). The ratio between base and height of the pilaster is 1:7.5, while the width and height of the bays are equal, creating a square. This ratio is replicated in the widows, where the height of the small pilasters is equal to the total width of the window, a relationship further emphasized by the squared moldings of the small pilasters. Furthermore, the wall space between the larger and smaller pilasters is a quarter of the height of the window pilasters. Given the lack of obstacles posed by a pre-existing structure, the simple proportional relationships in the upper level stand out. Indeed, the prevalence here of the ratios 1:1 and 1:4 suggests some familiarity with Alberti's theory and practice.[20] It is also evident that some of the bays on this level are laid out carelessly. Indeed, the imprecise placement of certain pilasters is readily apparent to the

Fig. 8 Palazzo del Podestà, upper story window

naked eye, but these dimensional blunders appear quite random. That is, they do not contradict the sense of geometry in the building, appearing instead to be the result of sloppy construction.

Despite these discrepancies, the upper story also reveals some surprising subtleties. These are found on the eastern side, where the size and spacing of the windows have been slightly shifted to counteract visually the irregular widths of the bays below. While the window over bay 1 is positioned symmetrically over its corresponding archway, the window over bay 2 has been made 16.5 cm narrower, to correlate it, presumably, with the narrower archway below. That window, moreover, has been distinctly slipped toward the center of the façade. Rather than equalizing the space on either side of the window, the architect pushed the irregularities to one side, so that at least one of the dimensions—the distance of the window to the corner pier—matched those on the right bay.

Given this analysis, it seems that Fioravanti was conscious of the fact that applying an ideal geometrical system to a project, especially in the form of a grid, was not enough to reach acceptable formal results. In order to overcome the limitations imposed by the site, he considered the modular grid as a system not of geometric points but rather of clusters or bounded areas, within which vertical elements could be flexibly arranged. This operation was undertaken with a view to regularizing the proportional and rhythmical effects of the façade composition. Indeed, he appears to have often positioned irregularities in those areas of the elevation where they could be most easily hidden. That this was sometimes done through optical corrections suggests that Fioravanti worked with precise plans and elevations in order to articulate the relationships between elements, in particular the axes of the bays. Such a practice would also fit with Alberti's prescriptions.[21]

Construction

By keeping the dimensions of the masonry elements as consistent as possible, the builders obtained a second important advantage, namely the reduction of costs. The building records held in the Bologna State Archive reveal some interesting facts regarding this process.[22] In particular, a payment made to the stonemason Antonio Frangipani on 19 March 1492 for the masonry of the western side tells us not only the date of completion of the body of the building, but also lists the pieces paid for, moving from the lowest to the highest. From the prices in the document, we can see that individual elements such as pedestals, capitals, brackets, and *roxuni* (the rose motifs on the rusticated bosses), but also rounded forms such as the two great arches of the portico and the circular windows of the attic story, were all priced at consistent rates (Fig. 9). The Corinthian capitals of the ground story, for example, were much more expensive (16 *lire*) than the flat pilaster capitals on the upper level (6 *lire*) (Fig. 10). By the same token, varying quantities of different elements that required roughly the same amount of work—such as the rose motifs of the piers, the half-columns, and the architraves and cornice moldings of the large order—all cost the same: 8 *scudi* and 6 *denari.*

One curious feature of the price list suggests that the masons did not know how to calculate the length of curves. The sums paid for the masonry of the first floor arches are about 25 % more expensive than those paid for their corresponding second story cornices, which have identical moldings and are set in bays of the same width. We can surmise therefore that the 25 % increase is therefore based solely on the workmanship of non-rectilinear elements. This surcharge, however, is inadequate. The circumference, for example, of the ground-floor arch in bay 2, measured on the intrados, is 8.07 m (21.6 Bolognese feet). That is 36 % longer than the width of the bay. Although masons' rules-of-thumb are by nature rough and approximate, this discrepancy does seem surprisingly large.

Fig. 9 Palazzo del Podestà, side facade

The masons may very well have charged a different amount for each arch, according to its characteristic size and form. As we have seen, almost all of the arches vary in size according to the different widths of the bays, particularly those on the short sides, which are considerably larger than those of the principal façade. But that variation would not have affected the consistent and standardized means of pricing per piece of masonry. That mechanism points to the use of exact and repeated dimensions—one might say serial or even standard—to optimize the production and cost of individual elements. By standardizing the size of the parts while manipulating the voids, not only are the irregularities of the façade better hidden, the masonry elements themselves can be produced offsite, thereby reducing costs and simplifying the construction process. The construction of the Palazzo del Podestà shows how a clear and regular geometry not only facilitates the composition of a building but also serves as a cost-efficient parameter of its construction, providing a powerful tool for patron, architect, and builders alike.

Fig. 10 Palazzo del Podestà, capitals from the half-columns of the ground story, to be compared with those from the pilasters of the upper story (Fig. 8)

Notes

1. Carpo (2003). For a recent account of the role of quantification and measurement in early Renaissance architecture, see Cohen (2013).
2. The notion of precision appears at the beginning of Leon Battista Alberti's *De Re Aedificatoria,* in relation to the role of the architectural drawings. See Alberti (1988,7). The bibliography on the issue of proportion in Renaissance Architecture has become vast and unmanageable. See Wittkower (1988). For a recent overview on the topic see Curti (2006, 65–138).
3. The use of the geometrical grid in order to reproduce the human body in painting was already known during the Gothic era. Villard de Honnecourt, for example, represents a woman's head proportioned in this way. See Barnes (2009, fol. 38). Lorenzo Ghiberti describes a human figure using the same method in the

first *folio* of his third commentary. See Ghiberti (1998). This tool was also widely diffused among architects of that time as a tool for composition and design. Antonio Averlino (Filarete) begins the plan of Sforzinda's Cathedral from a square grid of 15 modules on each side, though its function is only for proportioning and dimensioning, less for composing. Another grid appears in the book for the design of the *Casa Regia*. See Averlino [Filarete] (1972, vol. 1, 182–183, 207–208; vol. 2, Fig. 24). For another example, see Martini (1967, Fig. 236). A famous architectural drawing that clearly shows the grid as a tool of design is Bramante's plan of Saint Peter's, Uffizi 20. See Thoenes (2006) with bibliography on the drawing. Further analysis of fifteenth- and sixteenth-century plans and drawings would likely turn up other examples.

4. For the history of Vitruvius' treatise, see Pagliara (1986). For the history of Alberti's treatise, see Orlandi (1994). Also see Burns (1998, 120).
5. Libri Mandatorum, reg. 17, fol. 30r, Archivio di Stato di Bologna. Published in Sighinolfi (1909, 57–58, 147).
6. For the attribution, see Benelli (2001, 47–68 and 2005a, 100–103) with complete bibliography on Aristotele Fioravanti. For a brief but detailed biography of Fioravanti, see Ghisetti Giavarina (1997).
7. Müntz (1878, vol. 1, 83, 108). Bertolotti (1886, 2).
8. The transport was accomplished not by dismantling the tower but by splitting it from its foundations and carrying it vertically on a wooden cart over a track at the bottom of a trench. See Pattaro (1976).
9. For the phases of construction, see Benelli (2005a, 73–87). For Fioravanti in Russia, the bibliography is quite extensive, but see the complete list in Ghisetti Giavarina (1997).
10. Earlier attempts to reconstruct the proportional basis of the façade have been suggested by De Angelis and Nannelli (1976) and Licciardello (1991). Neither of these reconstructions, however, considers the irregularity of the site.
11. The survey, made with traditional methods, was organized and executed by the author with the help of Anna Maria Moro, Lucia Bacchiani, and Vittorio Pizzigoni.
12. This road served to link the Via Emilia—connecting Rimini with Milan across the Po Valley—with the new *Platea Comunis,* a rectangular piazza created along with the first communal palace from 1200 to 1203. Within the city wall, the Via Emilia was the *decumanus maior* of Roman Bologna, to become one of the busiest streets of the city during the Medieval period and beyond.
13. Part of the medieval structure was revealed during heavy restorations by Alfonso Rubbiani beginning in 1905. For a brief synthesis of the restoration, see Mazzei (1979).
14. One Bolognese foot is 0.380098 cm. See Martini (1883, vol. 1, 92). A Bolognese brick during the Renaissance was 28.50 cm long, 12.66 wide, and 6.33 high, or 9 by 4 by 2 Bolognese inches. See Benelli (2005b). Reprinted in Ricci (2007, 75–94).
15. This sort of "elastic system" is actually common among medieval Bolognese porticoes, which were normally adjusted according the preexisting façade to which they were attached. For a similar topic, see Hubert (2001, 33–34).

16. This elevated terrace was built from 1438 to connect the medieval palaces of the compound, the so-called Palazzo di Re Enzo and the palace of the Capitano del Popolo, with the Palazzo del Podestà, also creating a uniformed loggia façade facing west. The original *iter in voltis* was destroyed in 1572 for the construction of a building to house the Auditori della Sacra Rota, employed in the local Vatican Court. It was rebuilt in the same shape by Alfonso Rubbiani in the beginning of the twentieth century. On this feature, see Benelli (2005a, 74, 108).
17. "Etiamque angulares columnae crassiores faciendae sunt ex suo diametro quinquagesima parte, quod eae ab aere circumciduntur et graciliores videntur esse aspicientibus." Vitruvius (1997, vol. 2, 247). The same optical refinement was underlined by Alberti (1988, 215–216). Vitruvius was known in medieval Bologna. A manuscript copy of the text belonged to the university lecturer and canon Giovanni Calderini as early as the fourteenth century. See Ibanez (1998, 62). Further copies of the treatise are accounted between 1426 and 1455. They belonged to Carlo Ghisilieri and Cardinal Bessarion, the latter appointed *Legato apostolico* in Bologna from 1450 to 1455. See Hubert (2001, 35).
18. The dimensional consistency of the pillars also reflects an appreciation of ancient practice, as in, for example, the Colosseum, the Tabularium, and in general all Roman theaters. As Christoph Thoenes has pointed out, such consistency is rare in Renaissance loggias. The Bolognese case is best seen therefore not a superficial imitation of Roman classical architecture, but rather as an attempt to recreate its "structural substance". Thoenes (1998, 59–65).
19. Both Alberti (Book VII, Chapter 6) and Filarete (Book VIII) recommend the height of a Corinthian column to be eight times its diameter. See Alberti (1988, 201) and Averlino [Filarete] (1972, vol. 1, 218). Vitruvius' description of the Corinthian order is more complex (Book III, chapt. 5, 1–9; Book IV, chapt. 1, 11). Though generally broader, it includes in some cases proportions of 1:10. See Vitruvius (1997, vol. 1, 255–259, 369–375). Alberti faced same problem in designing the façade of Santa Maria Novella in Florence, where he was also forced to stretch his columns beyond their theoretically ideal proportions. See Benelli (2005a, 93).
20. Fioravanti may have known Alberti from 1451 to 1452, when they are both documented at the Vatican working under Nicholas V. It was precisely at this time, in fact, between December 1451 and January 1452 that Alberti presented his manuscript of the *De Re Aedificatoria* to the Pope. See Burns (1998, 120). On Alberti's work for Nicholas V, see Tafuri (1992, 33–88) and Frommel (2005). Fioravanti may also have been familiar with Alberti's project for the Tempio Malatestiano in Rimini, which also involved "wrapping" a classical envelope around a pre-existing medieval building. For a chronology of the Tempio, see Hope (1992).
21. Examples of optical refinements in Alberti's buildings are found, for example, in the portal of Santa Maria Novella in Florence and in the façade of Sant'Andrea in Mantua. See Bulgarelli (2007) and Curti (2007).
22. "Massarolo dei Lavori, Spese relative al Palazzo del Podestà," 13r–15r, Archivio di Stato di Bologna. Partially published in Valeri (1895, 251). Also see Valeri (1896, 78) and Zucchini (1912, 14).

Photographic Credits

Author: Fig. 1, 3–10
Biblioteca Archiginnasio, Bologna: Fig. 2

References

Alberti, Leon Battista. 1988. *On the art of building in ten books*. Trans. Joseph Rykwert, Neil Leach, and Robert Tavernor. Cambridge, MA/London: The MIT Press.

Averlino [Filarete], Antonio. 1972. *Trattato di architettura*. Milan: Edizioni il Polifilo.

Barnes, Carl F. (ed.). 2009. *The portfolio of Villard de Honnecourt (Paris, Bibliothèque nationale de France, MS Fr 19093): A new critical edition and color facsimile*. Aldershot: Ashgate.

Benelli, Francesco. 2001. Il Palazzo del Podestà di Bologna fra tradizione ed innovazione. In *L'Architettura a Bologna nel Rinascimento (1460–1550): Centro o periferia?* ed. Maurizio Ricci, 47–68. Bologna: Minerva Edizioni.

Benelli, Francesco. 2005a. Il Palazzo del Podestà di Bologna nel quattrocento. Storia e architettura. In *Nuovi Antichi*, ed. Richard Schofield, 66–119. Milan: Electa.

Benelli, Francesco. 2005b. Pedre cotte e pedre vive: Note sull'uso del mattone e della pietra a Bologna fra Medioevo e Rinascimento. In *Aspetti dell'abitare e del costruire a Roma e in Lombardia tra XV e XVI Secolo*, ed. Augusto Rossari, 71–91. Milan: Edizioni Unicopli.

Bertolotti, Antonino. 1886. *Artisti bolognesi, ferraresi ed alcuni altri del già Stato Pontificio in Roma nei secoli XV, XVI e XVII: Studi e ricerche tratte dagli archivi romani*. Bologna: Regia Tipografia.

Bulgarelli, Massimo. 2007. Rappresentazione, natura e artificio nell'opera di Alberti. In *Leon Battista Alberti: Teorico delle arti e gli impegni civili de "De Re Aedificatoria"*, ed. Arturo Calzona et al., 571–603. Florence: Leo S. Olschki.

Burns, Howard. 1998. Leon Battista Alberti. In *Storia dell'architettura italiana*, ed. Francesco P. Fiore, 114–165. Milan: Electa.

Carpo, Mario. 2003. Drawing with numbers: Geometry and numeracy in early modern architectural design. *Journal of the Society of Architectural Historians* 62(4): 448–469.

Cohen, Matthew. 2013. *Beyond beauty: Reexamining architectural proportion through the Basilica of San Lorenzo in Florence*. Venice: Marsilio Editori.

Curti, Mario. 2006. *La Proporzione: Storia di un'idea da Pitagora a Le Corbusier*. Rome: Gangemi Editore.

Curti, Mario. 2007. Valori ottici e idea di progetto nel *De Re Aedificatoria*. In *Leon Battista Alberti: Teorico delle arti e gli impegni civili de "De Re Aedificatoria"*, ed. Arturo Calzona et al., 667–682. Florence: Leo S. Olschki.

De Angelis, Carlo, and Paolo Nannelli. 1976. La facciata del Palazzo del Podestà a Bologna. *Arte Lombarda (special issue: Aristotele Fioravanti a Mosca: Convegno sugli architetti italiani del Rinascimento, Bologna 1975)* 44–45: 79–82.

Frommel, Christoph L. 2005. Il San Pietro di Nicolò V. In *La Roma di Leon Battista Alberti*, ed. Francesco P. Fiore, 102–111. Milan: Skira.

Ghiberti, Lorenzo. 1998. *I commentarii: Biblioteca nazionale centrale di Firenze, II, I, 333*. Florence: Giunti.

Ghisetti Giavarina, Adriano. 1997. Fioravanti (Fieravanti), Aristotele. *Dizionario biografico degli Italiani*. http://www.treccani.it/biografie.

Hope, Charles. 1992. The early history of the Tempio Malatestiano. *Journal of the Warburg and Courtauld Institutes* 55: 51–154.

Hubert, Hans W. 2001. L'Architettura bolognese del primo Rinascimento. Osservazioni e problemi. In *L'Architettura a Bologna nel Rinascimento (1460–1550): Centro o periferia?* ed. Maurizio Ricci, 29–46. Bologna: Minerva Edizioni.

Ibanez, Ignacio Gonzales-Varas. 1998. *Dietro il muro del Collegio di Spagna*. Bologna: Clueb.
Licciardello, G. Billings. 1991. Notes on the architectural patronage in Bologna of the Bentivoglio. Ph.D. dissertation, Columbia University.
Martini, Angelo. 1883. *Manuale di metrologia*. Torino: Ermanno Loescher.
Martini, Francesco di Giorgio. 1967. *Trattati*. Milan: Edizioni il Polifilo.
Mazzei, Otello. 1979. *Alfonso Rubbiani. La maschera e il volto della città*. Bologna: Cappelli.
Müntz, Eugène. 1878. *Les Arts à la cour des Papes pendant le XVe et le XVIe siècle*. Paris: E. Thorin.
Orlandi, Giovanni. 1994. Le prime fasi nella diffusione del trattato architettonico Albertiano. In *Leon Battista Alberti*, ed. Joseph Rykwert and Anne Engel, 96–105. Milano: Electa.
Pagliara, Pier Nicola. 1986. Vitruvio da testo a canone. In *Memoria dell'antico nell'arte Italiana*, vol. 3, ed. Salvatore Settis, 5–85. Turin: Einaudi.
Pattaro, Sandra Tugnoli. 1976. Le Opere Bolognesi di Aristotele Fioravanti Architetto e Ingegnere del Secolo Quindicesimo. *Arte Lombarda (special issue: Aristotele Fioravanti a Mosca: Convegno sugli architetti italiani del Rinascimento, Bologna 1975)* 44–45: 35–70.
Ricci, Maurizio (ed.). 2007. *Storia dell'architettura come storia delle tecniche costruttive*. Venice: Marsilio.
Sighinolfi, Lino. 1909. *L'Architettura Bentivolesca in Bologna e il Palazzo del Podestà*. Bologna: Libreria Beltrami.
Tafuri, Manfredo. 1992. *Ricerca del Rinascimento. Principi, città, architetti*. Turin: Einaudi.
Thoenes, Christof. 1998. Bramante e la 'Bella Maniera degli Antichi'. *Sostegno e Adornamento*, 59–65. Milan: Electa.
Thoenes, Christof. 2006. 'Templi Petri Instauracio': Giulio II, Bramante e l'antica basilica. In *Colloqui d'architettura*, ed. Augusto Roca De Amicis, 60–85. Rome: Nuova Argos.
Valeri, Francesco Malaguzzi. 1895. Der Palast der Podestà in Bologna. *Repertorium fur Kunstwissenschaft* 13(4): 251.
Valeri, Francesco Malaguzzi. 1896. La chiesa della Santa a Bologna. *Archivio Storico dell'Arte* 1–2 (2nd series): 78.
Vitruvius. 1997. *De Architectura libri decem*. Trans. Antonio Corso and Elisa Romano, ed. Pierre Gros. Turin: Einaudi.
Wittkower, Rudolf. 1988, 1st ed. London: The Warburg Institute 1949. *Architectural principles in the age of humanism*, 4th ed. London/New York: Academy Editions/St. Martin's Press.
Zucchini, Guido. 1912. *La facciata del Palazzo del Podestà, dal sec. XV al XIX*. Bologna: Luigi Beltrami.

Practical Mathematics in the Drawings of Baldassarre Peruzzi and Antonio da Sangallo the Younger

Ann C. Huppert

Combining technical practice with aesthetic intent, Renaissance architecture was by nature a mathematical art. Although the limitations of surviving documents hinder efforts to discern what Italian Renaissance architects knew of mathematics, where they learned it, and how they applied this knowledge, extant drawings from the period offer one means of addressing these questions.[1] Inscribed numerals and calculations, in particular, abound in the drawings by two leading architects of early sixteenth-century Italy, Baldassarre Peruzzi and Antonio da Sangallo the Younger, suggesting that both attained a high degree of numeracy.

Comparing these contemporaries is also revealing since, while each incorporated mathematics as a central element in their architectural practice, their approaches diverge in ways that point to and illuminate significant differences in their background and design methods. Sangallo, who was born in Florence, lived from 1484 to 1546. His initial training was in carpentry but once in Rome, where he was based for most of his career, he turned exclusively to architecture. Sangallo was involved in designing the new basilica of St Peter at the Vatican for much of his professional life, having succeeded Raphael in 1520 as head of the project. As papal architect, Sangallo's projects were wide ranging, including fortifications throughout the Italian peninsula.[2] Peruzzi, born in Siena in 1481, instead trained as a painter. His career also focused on architectural work, especially in Rome, where he worked at St Peter's under Bramante and Raphael and then as second architect to Sangallo. After the Sack of the city in 1527, he returned to Siena, where he was named head of the Cathedral works and architect to the Republic, responsible for the refortification of the city. He died in Rome in 1536.[3]

A.C. Huppert (✉)
Department of Architecture, University of Washington,
P.O. Box 355720, Seattle, WA 98195-5720, USA
e-mail: ahuppert@uw.edu

A. Gerbino (ed.), *Geometrical Objects: Architecture and the Mathematical Sciences 1400-1800*, Archimedes 38, DOI 10.1007/978-3-319-05998-3_4,

Of surviving Italian architectural drawings from the early sixteenth century, those by Peruzzi and Sangallo form the largest quantity. Finished drawings for major projects, such as the basilica of St Peter and the Palazzo Farnese in Rome, emphasize their respective graphic abilities, but as a rule the corpuses of both men contain a greater quantity of preliminary sketches and studies.[4] These more typical sheets, inscribed with dimensions and calculations, demonstrate that Peruzzi and Sangallo were extremely numerate and skilled in basic mathematics. Indeed, both architects littered their drawings with figures, showing basic arithmetic, correctly calculated and using the latest methods.[5] In general, Peruzzi's figures appear on drawings that relate to concrete projects, while Sangallo's mathematics also reflect more abstract and theoretical pursuits, and these prove to be key distinctions. The appearance and use of numbers in these sources reveal the ways that mathematical knowledge contributed to the professional requirements and design processes of the Renaissance architect. Both Peruzzi and Sangallo demonstrated skills that reveal their earliest general education, yet the contrast in how they applied their knowledge reflects differences in their subsequent professional training.

"New Math" Education: Numbers, Algorism, and Measurement

Sangallo and Peruzzi were fully at ease using Hindu-Arabic numerals and employing them in basic arithmetical calculations. These are skills that one might safely assume today but that could not be taken for granted in the Renaissance, a period in which even literacy was far from universal.[6] While the mathematician Leonardo da Pisa, known as Fibonacci, had promulgated this new notation system to Italy in the thirteenth century, it in fact supplanted Roman numerals only gradually.[7] Fibonacci also taught the application of these figures in "algorism", or basic arithmetic, as appears on Peruzzi and Sangallo's drawings.[8] This "new" number system was increasingly taught in Italian *abaco* schools by the fourteenth and fifteenth century.[9] *Abaco* here refers not to the instrument called an abacus but to the algorithmic methods introduced from the east, which were used in particular by merchants.

Since we lack documentation about the specific early education that either Sangallo or Peruzzi received, the mathematical aptitude displayed in their drawings provides valuable evidence that each received standard *abaco* instruction in their native cities.[10] By the fifteenth century, communal schools providing instruction in mercantile mathematics existed throughout Tuscany. Such general education was offered to boys of about age 10 or 11 and therefore preceded the specific training that Peruzzi and Sangallo received in their respective crafts.[11] The *abaco* treatises began with explanations of Hindu-Arabic numbers and the basic operations of mathematics (addition, subtraction, multiplication and division), then progressed to the application of these operations in increasingly more complex problems or

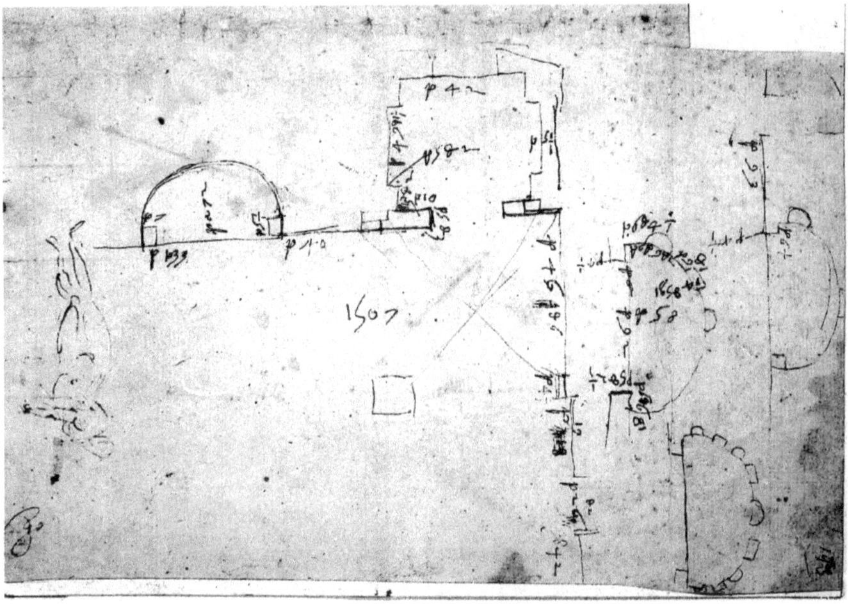

Fig. 1 Baldassarre Peruzzi, plan of the Baths of Diocletian (UA 528r detail)

exercises, involving fractions, translations between monetary systems, and simple geometry.[12] These exercises refer specifically to architecture only on occasion but, as Nicholas Adams has noted, it is in general easy to see their applicability for an architect.[13] Moreover, many of the *abaco* masters, those who taught and wrote treatises on practical mathematics, were also closely involved in the engineering and building trades in their own and nearby cities.[14]

Measurement was one fundamental architectural application for *abaco* skills, and measuring ancient monuments also served as essential training for Renaissance architects. Evidence of this practice appears in some of the earliest drawings we have by our two architects. On his measured plan of the Baths of Diocletian, dated 1507, Peruzzi's abbreviated notations identify the measurement units as contemporary Roman *canne*, *palmi*, and *digiti* (Fig. 1). Sangallo's comparable study does not indicate the units, but notes on the other side of the sheet show that he was using the Florentine *braccio*, as he tended to do on most of his early drawings (Fig. 2). Both architects also made drawings of antique details. In these, Peruzzi consistently identified the units of measure, and frequently he used the ancient Roman *piede* (Fig. 3).

Such studies were not new by this date. It is noteworthy, though, that Sangallo and Peruzzi's surveys surpass those of their predecessors in both comprehensiveness and level of detail, as a comparison of drawings from a generation earlier demonstrates. Francesco di Giorgio, for example, with whom Peruzzi trained, tended to limit the measurements on his drawings of ancient structures, which often took the form of overall plans rather than details (Fig. 4). In contrast, Sangallo's master and

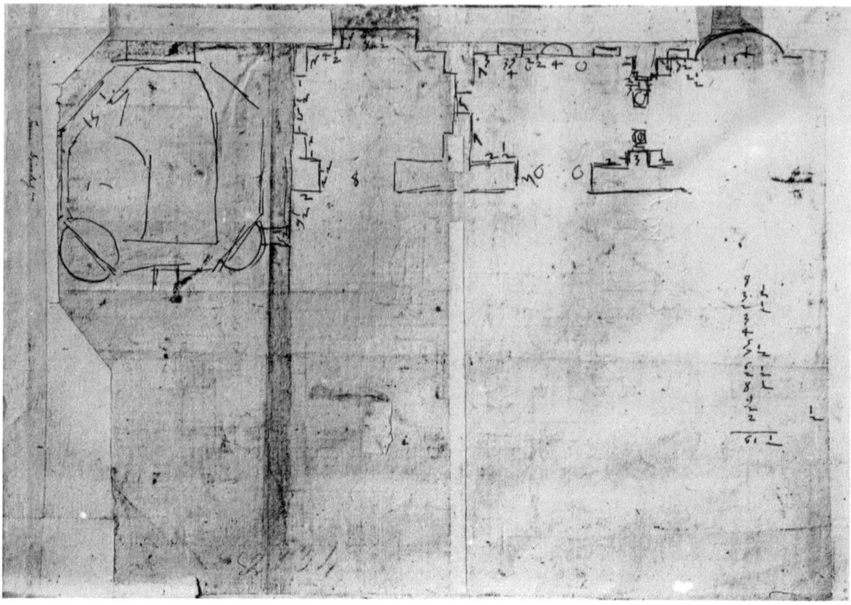

Fig. 2 Antonio da Sangallo the Younger, plan of the Baths of Diocletian (UA 2134v detail)

uncle, Giuliano da Sangallo, produced many measured detail studies, such as those in the Codex Barberini sketchbook. Even so, the work of Antonio exhibits far greater attention to discrete elements and individual measurements (Fig. 5).

The very process of measuring could engender complexity, since changes of location required adopting different local dimensional units. The measure most familiar to Sangallo and Peruzzi was the *braccio*, equivalent to an arm's length or about three-fifths of a meter. Different city-states, however, had their own version: the *braccio* of Sangallo's native Florence was marginally shorter than the Sienese *braccio* familiar to Peruzzi. Upon transferring to Rome in the early 1500s both architects encountered the smaller local unit, the *palmo*, roughly one-quarter of a meter in length. Drawings by both suggest that they could readily adapt to new contexts. They each employed the local *trabucco* and *braccio*, for example, in their respective fortification designs for Piacenza. Peruzzi measured his project for a new bastion using the local units, introducing on one of his drawings a scale line of 48 *trabucchi*, alongside which he noted the relationship of 6 *braccia* to each *trabucco* (Fig. 6). The scale on a later project drawing by Sangallo for the same city shows a simple ratio of 1 to 2 between the local *braccio* and the Roman *palmo* (Fig. 7).[15]

Such convenient whole-number ratios were the exception; more often, converting between local measurements would produce complex fractions. Peruzzi's drawings in particular suggest that he did not shun such complexity. On a measured sketch of a column base in Rome, he noted that one Roman *palmo* was equal to 22-4/5 *minuti* of one Florentine *braccio*. Perhaps more surprisingly, he also included such precise

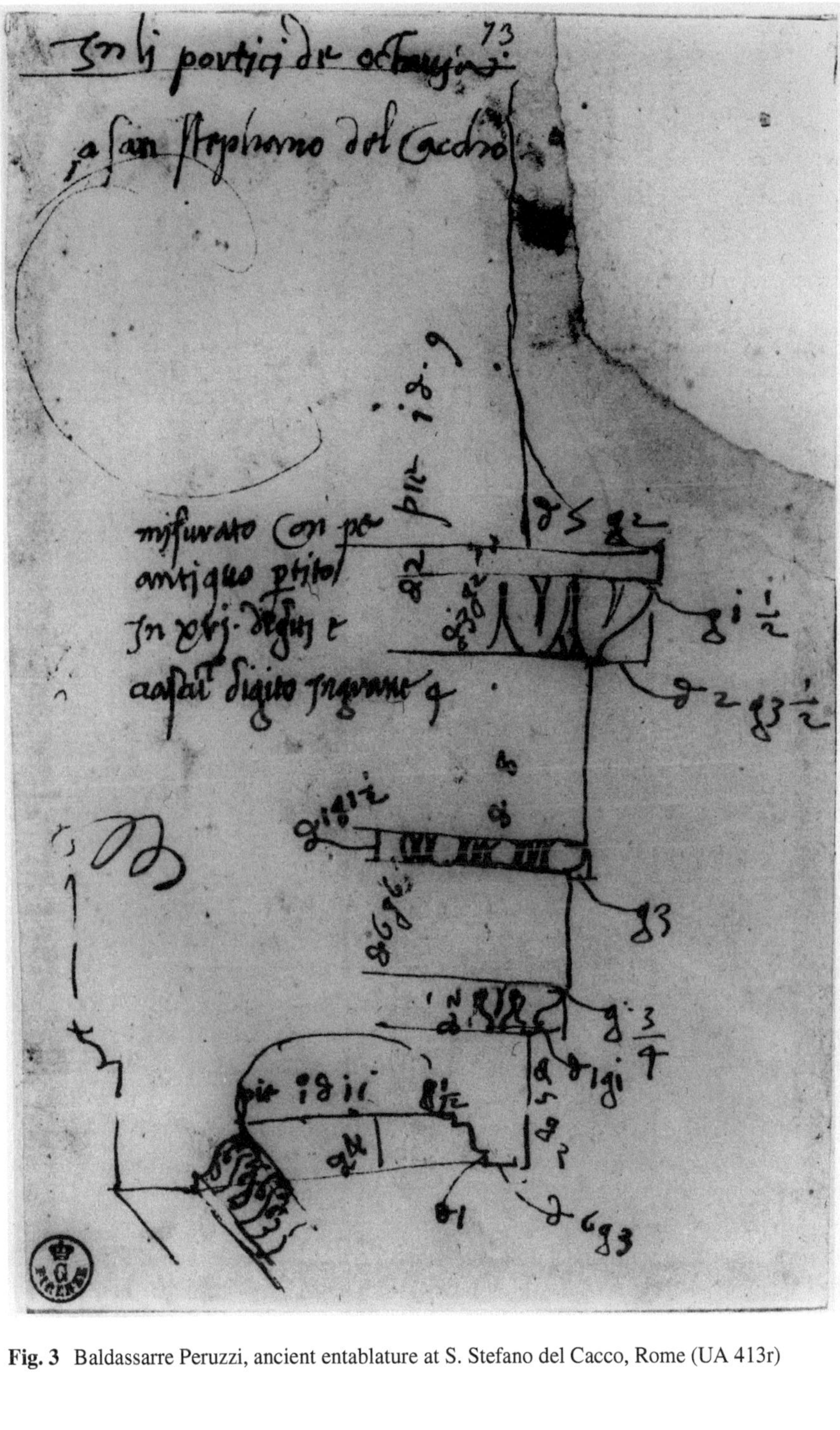

Fig. 3 Baldassarre Peruzzi, ancient entablature at S. Stefano del Cacco, Rome (UA 413r)

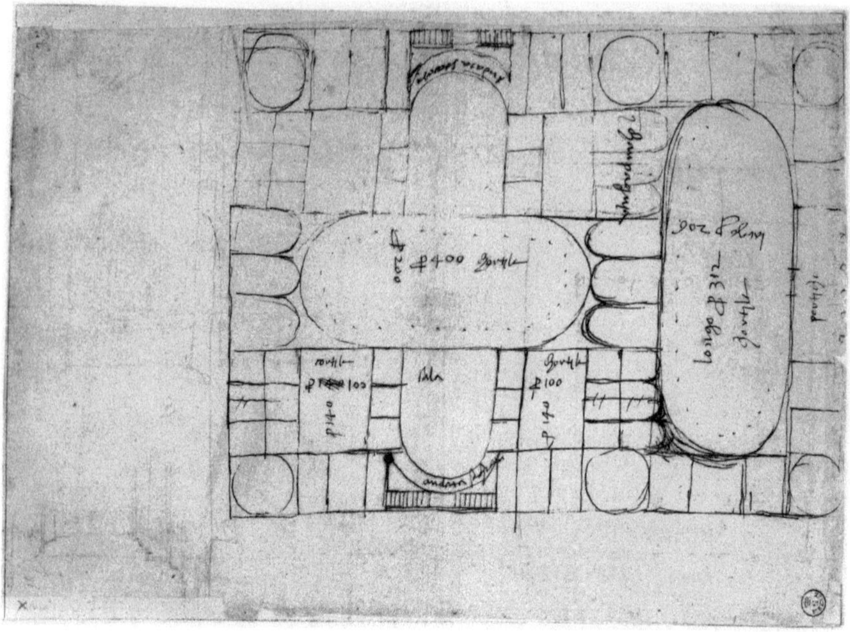

Fig. 4 Francesco di Giorgio Martini, plan of the Flavian palace on the Palatine, Rome (UA 328r)

relationships on clean presentation drawings. On a plan for the Ricci Palace in Montepulciano, Peruzzi recorded the ratio of the Roman *palmo* to the local *braccio* as 1 to 2-16/25, which he represented by the fraction 8 over 12-1/2.[16] Such translations between units of measure directly parallel the problem faced by merchants needing to convert between different local currencies, and in this respect, Peruzzi's comfort with complex fractions may represent an element of his Sienese education.[17] A different kind of source, an accounting ledger, offers more direct evidence of Peruzzi's knowledge of merchant accounting. As *camerlengo* or treasurer from 1515 to 1516, Peruzzi maintained the cash book for the Confraternity of San Rocco in Rome.[18] This ledger might seem to have an obvious architectural corollary in the account books for building projects. However, among the records that exist for large projects of the period, it is unusual to find project accounts in the architect's own hand.[19]

In general, Peruzzi appears somewhat more comfortable than Sangallo with arithmetic, but he differs also in his attitude toward numbers themselves. His facility with difficult fractions provides one point of comparison, the use of recorded dimensions and graphic scales offers another. Peruzzi was not consistent in his use of scale lines, which sometimes appear without dimensional units, and instead tended to rely much more on inscribed measurements. This contrasts with the practice of Sangallo, who varied his use depending on the purpose of the drawing. As Mario Carpo has noted, Sangallo typically reserved inscribed measures for sketches and working drawings, and included scale lines for final project or presentation

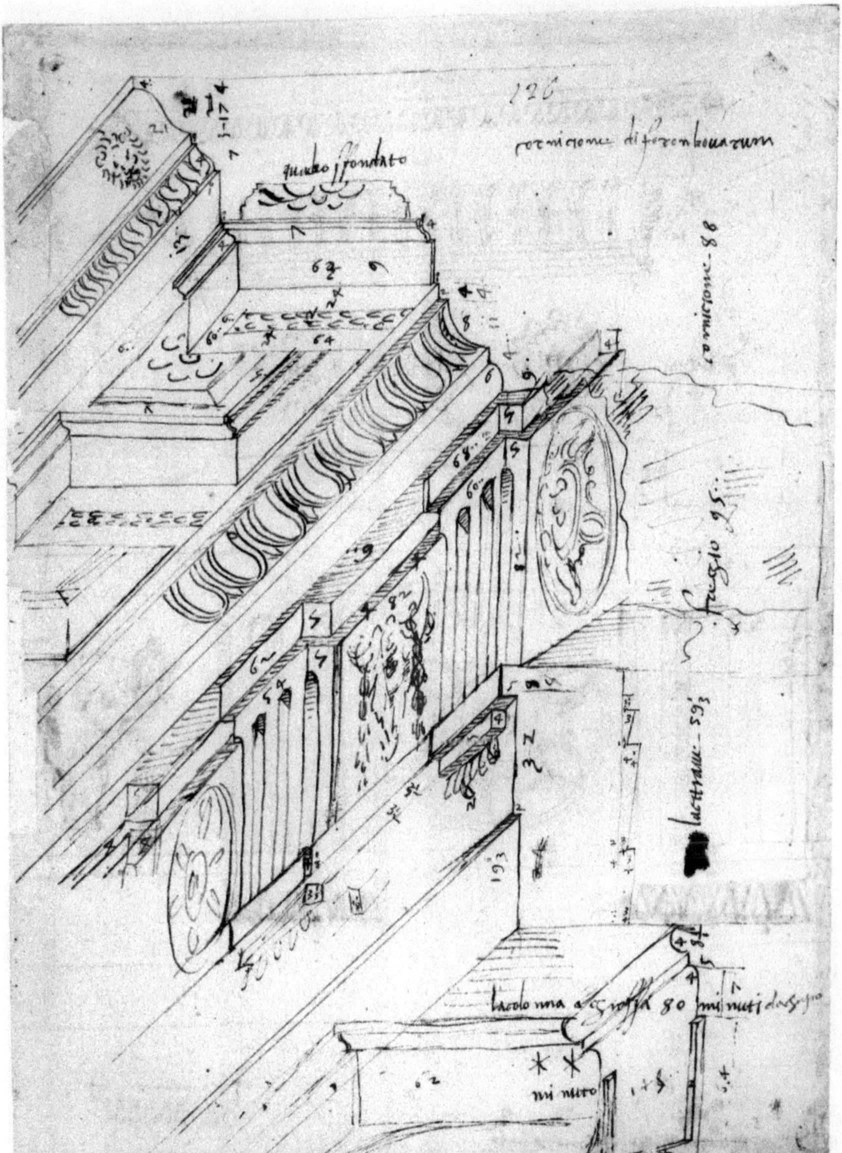

Fig. 5 Antonio da Sangallo the Younger, entablature from the Basilica Aemilia, Rome (UA 1413v)

drawings.[20] Two plans for an unrealized fortified palace in Casigliano show this transition clearly. The copious measurements on an initial sketched plan for the palace give way, in Sangallo's more finished version, to a discreet scale line at the bottom of the sheet (Figs. 8 and 9).[21] Peruzzi, by comparison, seems to have

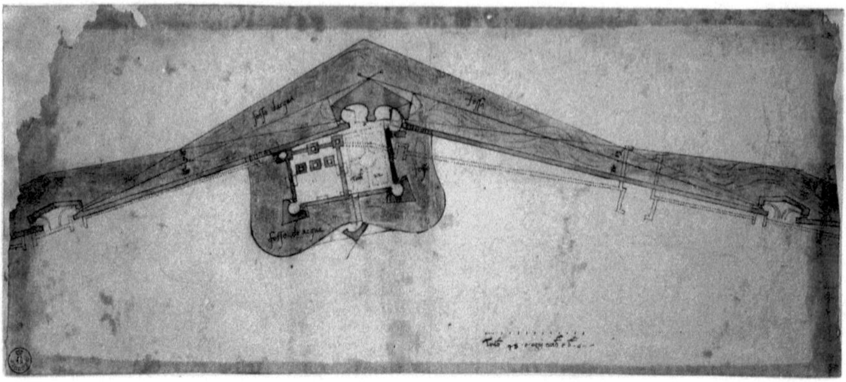

Fig. 6 Baldassarre Peruzzi, fortification design, Piacenza, (UA 459r)

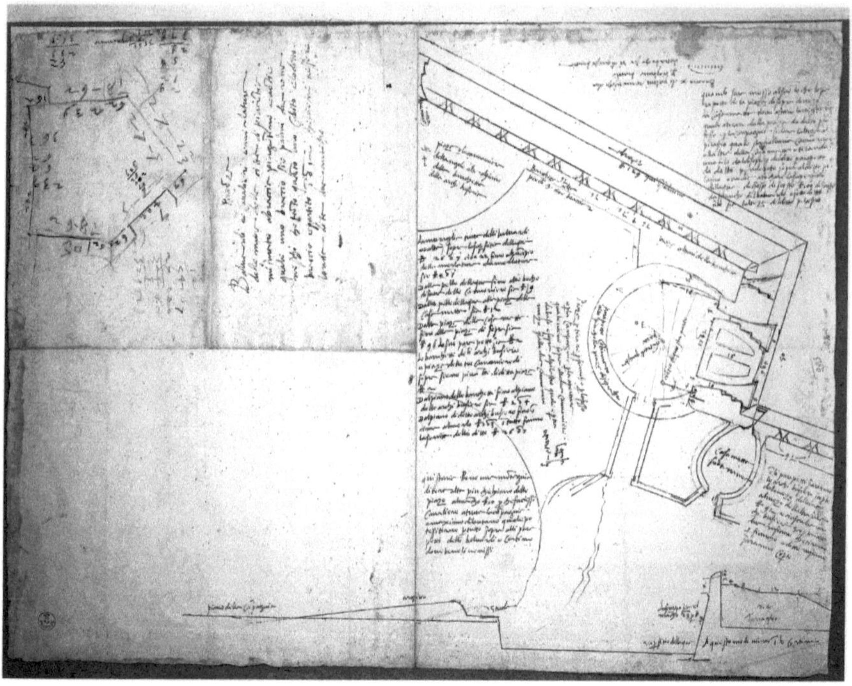

Fig. 7 Antonio da Sangallo the Younger, fortification design, Piacenza, (UA 808r)

remained fully wedded to the actual dimensions, even on carefully finished drawings. His plan for the renovations to the Rocca Sinabaldi completed around 1530, is just one of many examples (Fig. 10).[22] Here the extensive measurements and careful recording of dimensions for all the parts suggest that he wanted to exert a degree of precision and control over the realization of his designs, and that he retained a

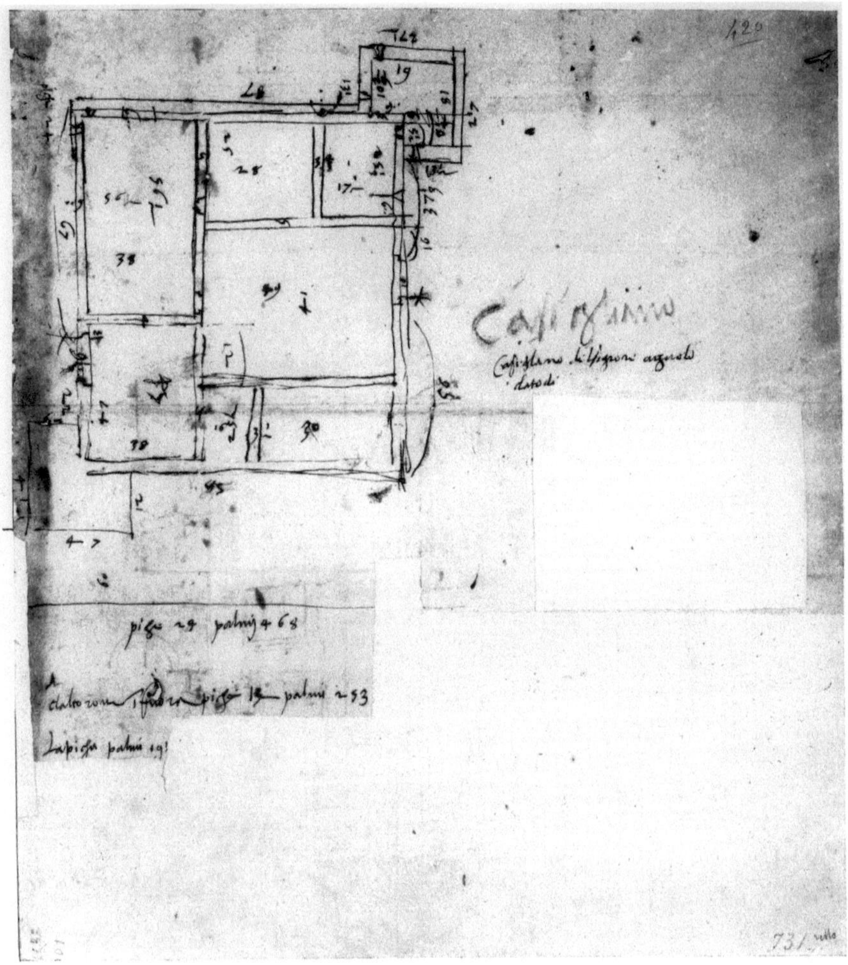

Fig. 8 Antonio da Sangallo the Younger, fortified palace project, Casigliano (UA 731r)

greater faith in the numbers themselves than in proportional relationships alone. This almost-obsessive attachment to numbers distinguishes Peruzzi, not only from Sangallo but from other contemporaries as well.

As this brief overview indicates, the mathematics found in Sangallo and Peruzzi's drawings are very practically grounded. Although well removed from the realm of "high" or theoretical mathematics, their work is nonetheless representative of the most advanced state of the field. Along with the recovery of classical mathematical texts, it was precisely in the area of practical mathematics that historians of the discipline have identified the major contributions of the Italian Renaissance. According to Warren Van Egmond, an inherent shift towards abstraction associated with the writing, calculating and conceiving of numbers in the new Hindu-Arabic system was decisive for the period.[23]

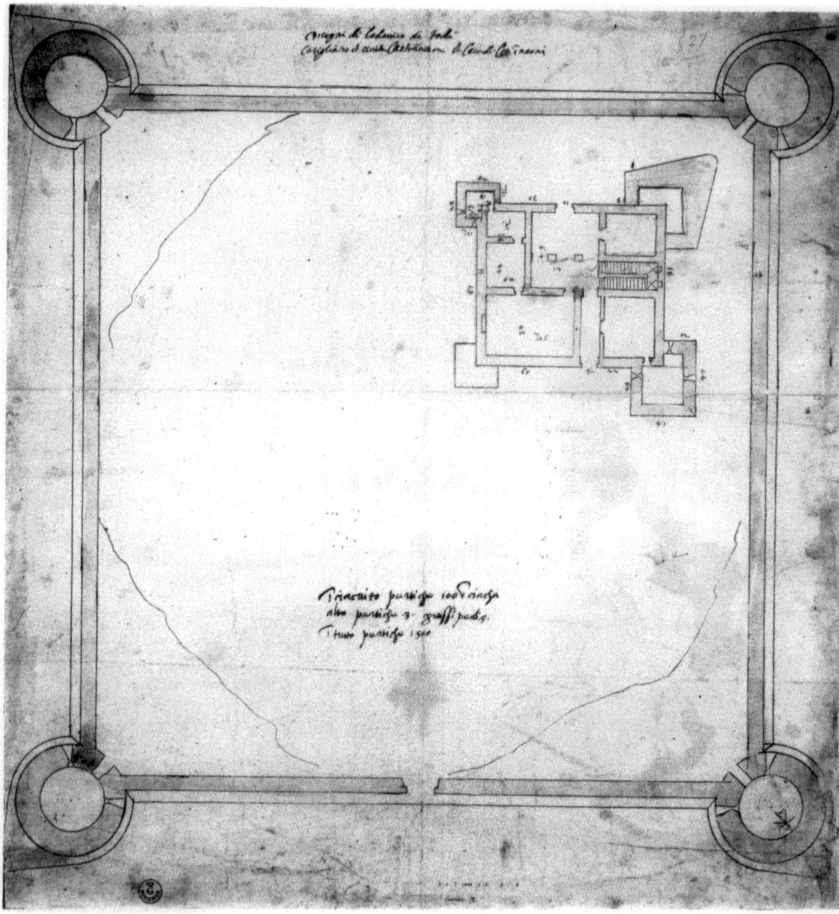

Fig. 9 Antonio da Sangallo the Younger, fortified palace project, Casigliano (UA 839r)

Practical Mathematics and Design

Measurement and unit conversion routinely called on an architect's mathematical abilities, but these procedures formed only one aspect of Renaissance architectural practice. How were calculations actually involved in the design process? How might different levels of mathematical ability have led to different design choices? Sangallo and Peruzzi's drawings reveal answers to these questions in two important, often overlapping areas: cost estimates and structural design.

Estimates of costs and materials lay within the purview of the Renaissance architect, just as they do today. On a preliminary study by Sangallo for unrealized fortifications in Amelia, for example, columns of figures adjacent to the plan reveal his effort to calculate the overall dimensions with reference to the total cost, estimated in the

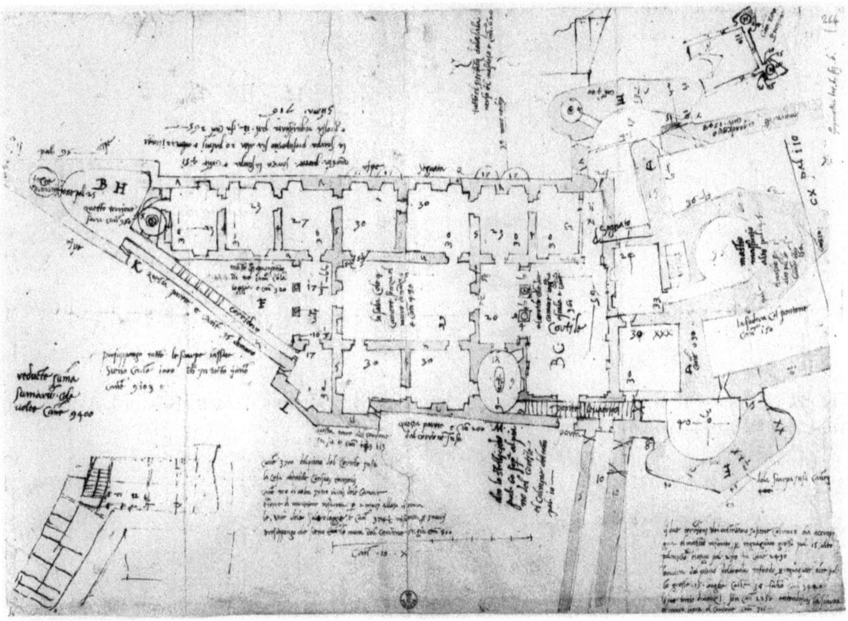

Fig. 10 Baldassarre Peruzzi, plan of the Rocca Sinabaldi (UA 579r)

Roman coinage of *giuli*.[24] It is worth observing, however, that notes of this sort only rarely appear on surviving drawings from this period, including Sangallo's. His many surviving drawings for St Peter's demonstrate this point. Over the course of forty years, twenty-six as chief architect, Sangallo left more than one hundred pages of designs and studies for this project. Yet, none of these include notes on cost or material estimates.

In this regard, too, Peruzzi was distinctive. A number of his drawings show evidence of cost calculations alongside the designs, including several for new St Peter's. This juxtaposition is striking, for it suggests that his numeracy allowed him to integrate considerations of economy within the design process, that is, to weigh the cost and effectiveness of different solutions directly on the drawing. We can see this process at work in two Peruzzi drawings of St Peter's. That these seem to date from the early 1530s is significant since, during this period in which construction resumed following the Sack, the pope may have been particularly concerned about the price of the new building.[25]

Another group of Peruzzi's drawings from Siena shows this phenomenon more clearly. After a fire destroyed the roof of the church of San Domenico in late 1531, Peruzzi proposed five different schemes—in eighteen surviving autograph sheets—for restoring and modernizing its interior while re-using its surviving medieval walls. Although his proposals eventually were rejected in favor of rebuilding the original trussed, wooden ceiling and simple hall nave, Peruzzi's designs attest to both the grandeur and fecundity of his imagination. Far from visionary, however, the designs were firmly grounded in practical considerations of measurement, structure, and materials.[26]

Fig. 11 Baldassarre Peruzzi, project for the church of San Domenico, Siena (Oxford, Ashmolean Museum, WA 1944.102.40)

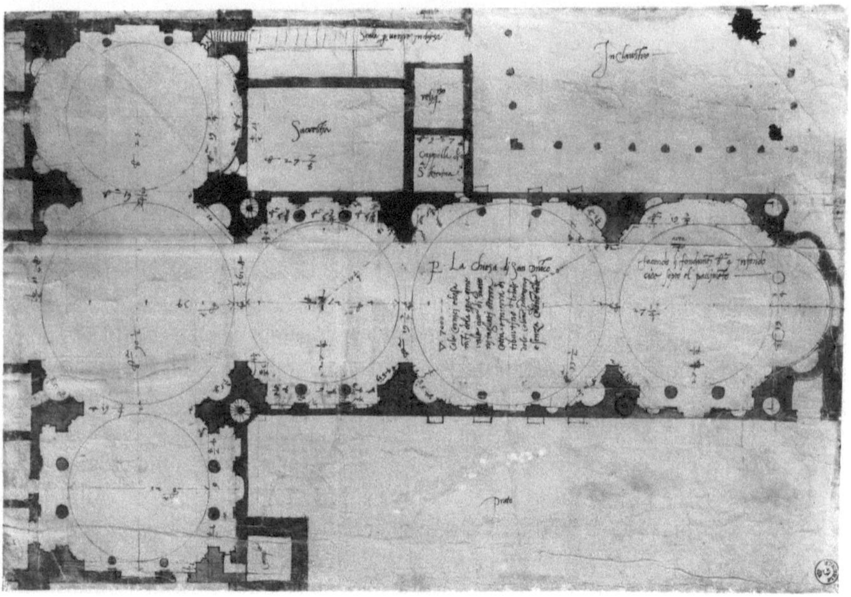

Fig. 12 Baldassarre Peruzzi, project for San Domenico, Siena (UA 339r)

A magisterial drawing, showing a sectional view through the nave, reflects one of the more ambitious of the proposals, all of which began in plan with the T-shaped outline of the existing structure (Fig. 11). Within this shell, Peruzzi proposed over the nave a series of three sail vaults of differing diameters, supported on angled piers that project into the central space. The corresponding plan shows the nave intersected by a transept at its east end (Fig. 12). The circles of alternating sizes indicate the proposed vaults over each bay, and Peruzzi also provided extensive measurements,

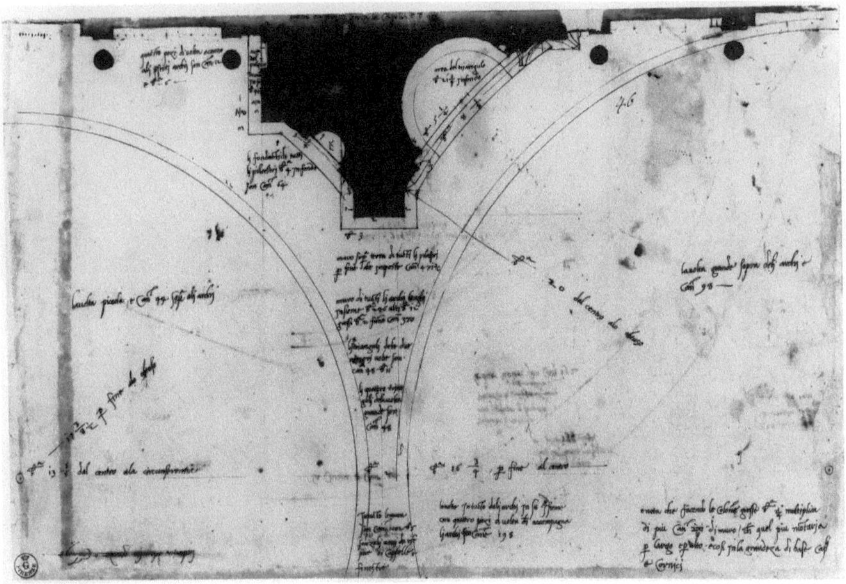

Fig. 13 Baldassarre Peruzzi, project for San Domenico, Siena (UA 344r)

including the overall dimensions for the length of the nave and transept, diameters for the individual vaults, dimensions of smaller elements, such as the piers and niches along the perimeter. He also included a summary cost estimate.[27]

The complexity of the vaults led Peruzzi to study both the structural elements and building mass in more depth on separate drawings, including one that shows a single main pier of the nave (Fig. 13). His annotations distinguish between the differently sized vaults to either side of the projecting pier. For each dome, he included two different radius dimensions, one for the circles as drawn and a second for the distance to the face of the pier. Whereas the section shows a seamless sail vault, the drawing and notes make it clear that Peruzzi conceived the structure and calculated the wall dimensions as though it were a saucer dome placed atop four pendentives. The extensive notes around the pier and down the center of the sheet refer to the total wall masses calculated for various elements, including the piers, arches, and the pendentives, or "*trianguli*." A revealing note appears at the lower right: it registers the change to the overall dimensions entailed in increasing the width of the nave pier, serving, in effect, as a variant proposal expressed in numerical terms. Graphically he explored this alternative on a separate sheet.[28]

Additional studies on the reverse of this sheet represent Peruzzi's further analysis of the vault dimensions (Fig. 14). Amid arcs that relate to the lines of the vaults, Peruzzi provided dimensions for "*la cherica*," referring to the crown of the vault. This crowning section, the dome segment that rises from above the transverse arches that span the nave, functioned structurally as a true dome.[29] In other notes, he referred to obtaining the overall area ("*superficie*") of different sections by slicing

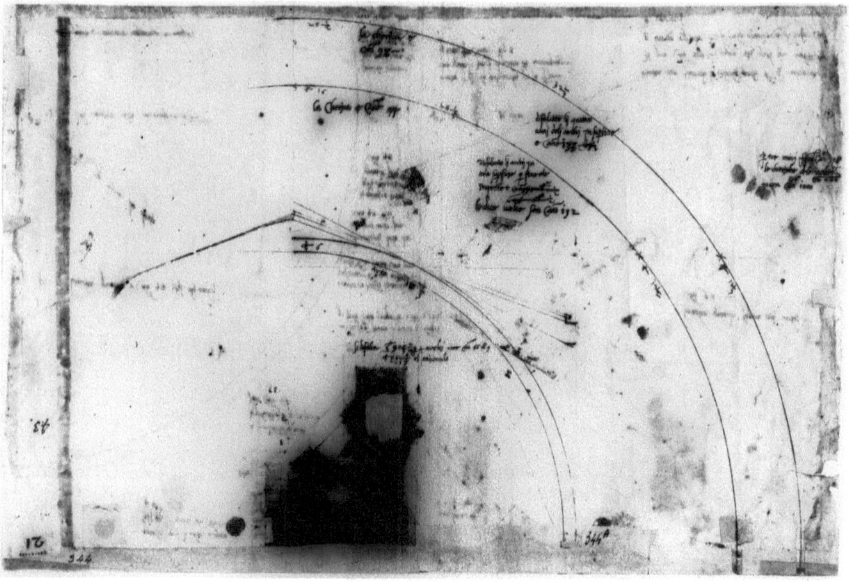

Fig. 14 Baldassarre Peruzzi, project for San Domenico, Siena (UA 344v)

through ("*falcato*") at various points.[30] Although the fragmentary and hermetic nature of the annotations hinders a full analysis of Peruzzi's numbers, the written notes demonstrate that he was thinking in terms of overall surface areas. The numbers therefore represent quantities that would have financial and structural implications for constructing the vaults.

Peruzzi's care to fix exact dimensions reflects a newfound measure of control over the building process. Such control was necessary partly for reasons of economy, given that exact measurements enabled a precise materials survey, but also for communicating ideas effectively to builders as well as to patrons. As Richard Betts has argued, Renaissance architectural practice was characterized by complex new structural forms that challenged traditional methods of construction.[31] We find Sangallo similarly exploiting drawing as a communicative tool in his work at St Peter's, in particular, for the great wooden model of the basilica. More than the new building itself, it was the model, constructed between 1539 and 1546, that most occupied Sangallo in the final years of his career. The extensive graphic record for this project includes a number of drawings for the dome as it was to be realized in the model. As Christof Thoenes has shown in his detailed analyses, these drawings demonstrate a distinctive approach to design problems.[32] For Sangallo, complex arithmetic could at times prove an insoluble challenge. In these and other instances, he often fell back on empirical methods of design and calculation that reflect his practical training, rather than relying on the abstract power of numbers.

The sequence begins with a preliminary study, in which Sangallo shifted away from Bramante's initial proposal for a hemispherical dome modeled on the Pantheon

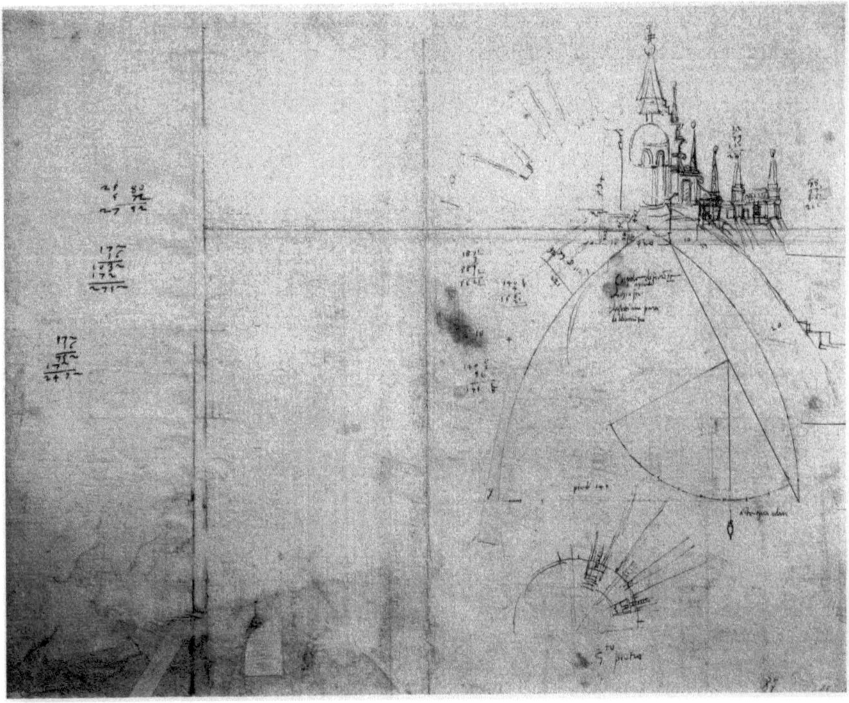

Fig. 15 Antonio da Sangallo the Younger, design for the dome of St. Peter's, Rome (UA 87r)

in favor of a pointed profile inspired by that of Florence cathedral (Fig. 15). In part, the drawing studies the proportions of the dome, which is narrower and more steeply raked than Brunelleschi's original. Yet it also illustrates a concern with structure, since a more steeply inclined vault would produce less outward thrust.[33] To examine this effect, Sangallo ruled a line from the base of the dome to that of the lantern, juxtaposing this with a literal depiction of the tools he used to measure the angle of inclination. Notably, this quadrant and plumb line are instruments of construction, not of design.

On this sheet and its reverse, Sangallo explored alternatives for a series of radiating piers to support the dome at its base, and it was in his attempt to determine the dimensions of these elements that the architect encountered a considerable mathematical obstacle. A partial plan at the lower right corner of the verso shows him investigating the necessary dimensions of these supporting piers (Fig. 16). The arithmetic surrounding the plan reflects his calculations of the overall circumference of the dome at its base and his effort to subdivide this into 48 equal bay divisions, the "*vani* 48" he noted below. However, with the pier elements of this chosen size, the sum of the dimensions did not add up to the given circumference.[34] At this point, as Thoenes has proposed, Sangallo switched methods. For his subsequent drawing, Sangallo resorted to a geometric rather than arithmetic solution: using a compass and straight edge he established the scale of 150 *palmi*, at the left, and from that, determined the dimensions of the radiating piers at the lower edge (Fig. 17).

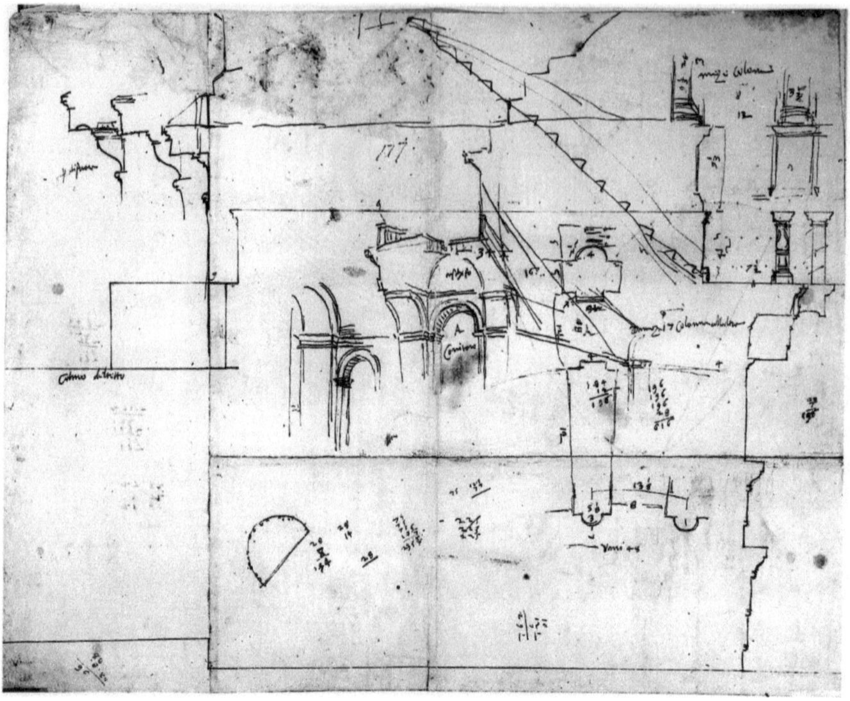

Fig. 16 Antonio da Sangallo the Younger, design for the dome of St. Peter's, Rome (UA 87v)

The last drawing of the sequence outlined the definitive form of the buttresses and other elements (Fig. 18). This final drawing, made at the scale of the model, is nearly 2 m in height and served as a guide for the dome as executed in wood.[35] In addition to the drawing, showing a pointed profile, Sangallo also offered a written counter-proposal for the dome. The extended note in the center of the sheet, which gives an elaborate geometrical procedure for constructing an alternate dome profile in the shape of an ellipse, offers further insight into Sangallo's empirical, somewhat non-intuitive approach to design.[36] A small diagram above the text shows a compass over a square inscribed within a circle, illustrating the instruments he used to derive the requisite profile. Thoenes, an architectural historian, worked with Wolfgang Böhm a mathematician, to analyze the drawing and the complex method Sangallo described. The first step would be to curve a sheet of paper over a carefully dimensioned wooden barrel-vault form. Then, using a fixed compass, an arc is traced from one mid-point of the barrel to the other. Next the paper is laid on a flat table, producing a half oval, which then is completed. Rotating the sheet 90° and bisecting the oval along its second, shorter axis produces the desired arch of the dome.

Why such a complicated form of construction? As Thoenes concludes, although sound in terms of the arithmetic and geometry involved, Sangallo's method for deriving the curvature was cumbersome and impractical, as well as nearly

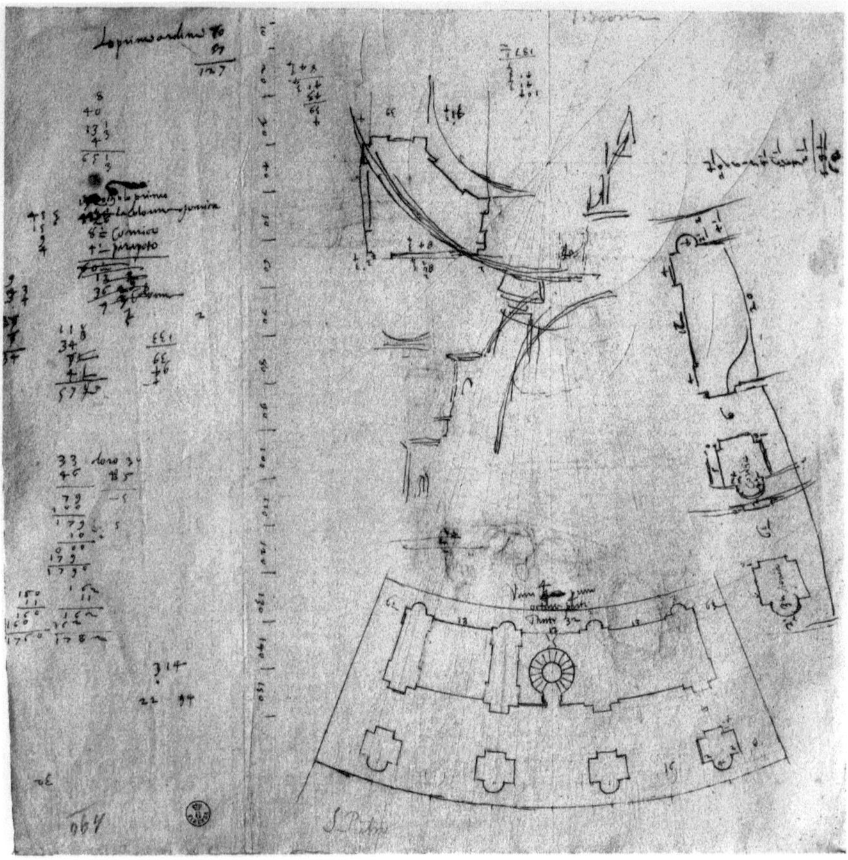

Fig. 17 Antonio da Sangallo the Younger, design for the dome of St. Peter's, Rome (UA 798r)

impossible to replicate at the full scale of the building.[37] Both aesthetic and structural considerations seem to have driven the choice. Sangallo recognized that, potentially, the form could combine measurements that are proportionally related.[38] Second and more importantly, the ellipse combines the advantages of the two basic models that Sangallo was contemplating. Its narrower profile and raised crown offered a compromise between the stability of the Gothic pointed form and the aesthetically-preferable, but less stable, classical shape of the hemisphere.[39]

Pragmatism and Theory

The oval also interested Peruzzi, one of the first Italian architects to use this form in plan proposals.[40] His design method, however, stands in stark contrast to that of Sangallo. Where the latter pursued an empirical and indirect solution for the sake of

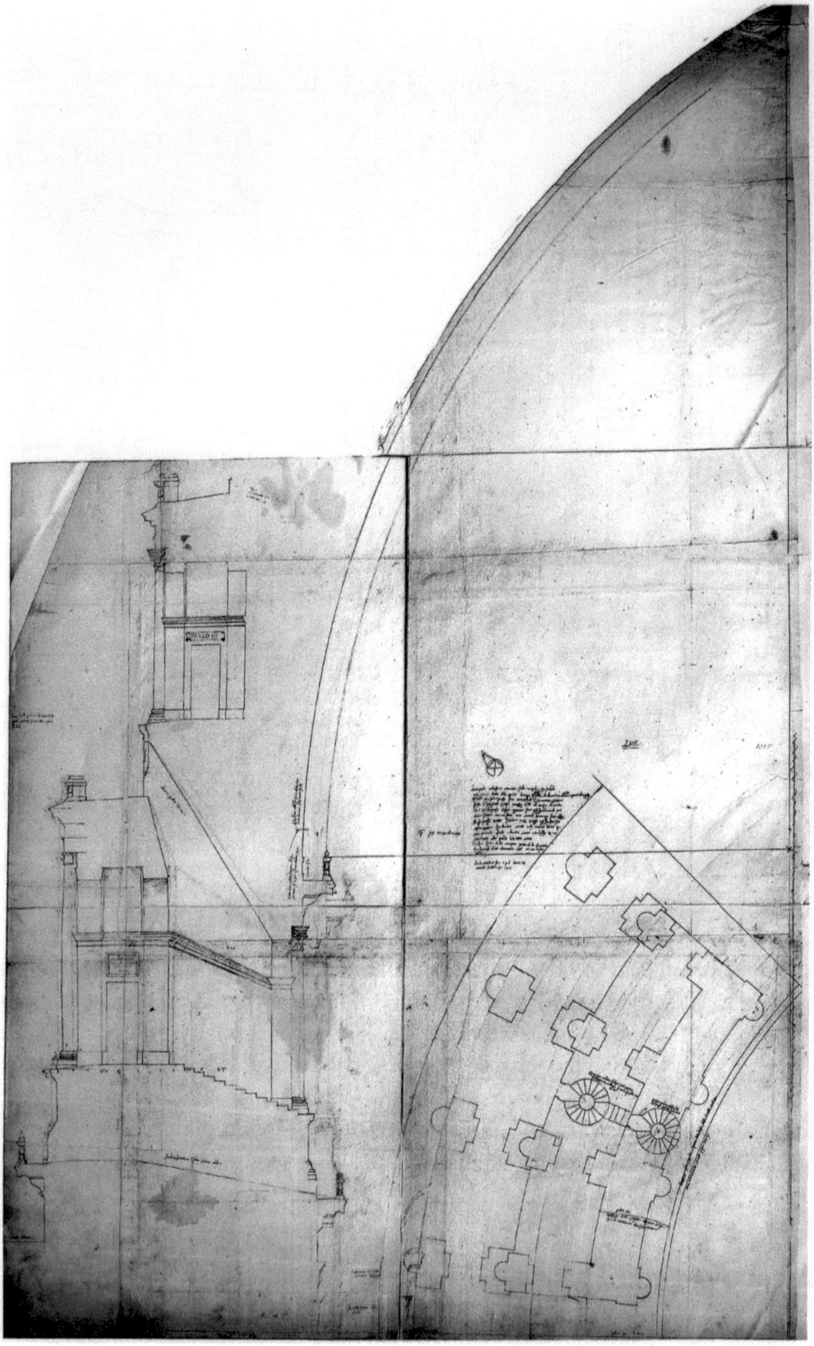

Fig. 18 Antonio da Sangallo the Younger, design for the dome of St. Peter's, Rome (UA 267r)

Fig. 19 Baldassarre Peruzzi, design for the Palazzo Massimo alle Colonne, Rome, with oval construction diagrams (UA 531r)

a desired outcome, Peruzzi's approach was to leverage his mathematical abilities to produce simpler and more practical results. His exploration of the construction of ovals provides a clear basis for comparison (Fig. 19). Sebastiano Serlio, who published Peruzzi's diagrams, praised one in particular for its beauty and ease of use. This method, which appears at the right edge of Peruzzi's sheet, is based on intersecting circles. Following Serlio's recommendation, it became the standard for approximating an ellipse in Renaissance architectural practice.[41] Whereas both designers used practical mathematics for their geometric constructions, when compared with Sangallo's complicated technique for deriving an oval, Peruzzi's method points to a distinctive pragmatism in his work.

Peruzzi displayed the same mathematical intuition in his multiple studies of the scrolling volute of the Ionic capital, and on one drawing in particular (Fig. 20). Typically this spiral form is constructed by joining progressively smaller circular arcs, according to proportional rules that dictate the step-by-step diminution of the compass and the placement of the compass point on the volute's center.[42] In his study, Peruzzi began with a foreshortened perspectival grid of the kind he constructed for scenographic designs, using the diminishing spans between the orthogonals to provide the measures for the decreasing radii of the volute below. Peruzzi himself labeled the source of the construction with the inscription "*per piramide*." Peruzzi's drawing demonstrates an impressive insight: that the constant diminution within a perspectival construction represented a proportional sequence that could be transferred to an area conceptually far removed from it.[43] However, he made no reference to one of the key sources for the form of the volute, the instructions provided by the ancient author Vitruvius.[44]

This omission points to another difference that separates Peruzzi and Sangallo, namely their approach to learning and erudition. Sangallo clearly sought authority in the works of both ancient and contemporary authors. Several of his mathematical and geometrical studies attest, for example, to the influence of the Veronese architect and engineer, Fra Giovanni Giocondo. A scholar of wide-ranging interests, Giocondo worked alongside both Sangallo and Peruzzi at St Peter's until his death in 1515. Sangallo, on the verso of a geometrical study, a diagram of a 12-pointed figure inscribed within a circle, included the credit "*Geometrio di fra Iochundo*" (Fig. 21).[45] This explicit attribution is significant, for it parallels Sangallo's practice of noting and comparing literary and archaeological sources in his studies of ancient architecture.[46]

Peruzzi's approach was different. While he was aware of contemporary developments in practical mathematics, purely theoretical studies are completely absent in his work. Nor do his drawings or designs carry the kind of scholarly references that litter Sangallo's. Peruzzi's archaeological studies provide a case in point, particularly in his frequent use of the Roman *piede* as a measurement unit (Fig. 3). The exact length of the *piede* as used in antiquity had been rediscovered by the humanist Angelo Colocci in the early sixteenth century. Like Bramante, who used the measure to establish the dimensions of the Belvedere Courtyard in 1505, it is characteristic of Peruzzi to exploit erudite antiquarian knowledge for practical use.[47] Such an application demonstrates that Peruzzi recognized the instrumental value of mathematics, yet he also appreciated its social import. One scholar has seen Sangallo's quest for a theoretical basis for design as an avenue towards legitimacy, but a parallel argument might be made regarding Peruzzi's practical approach towards with numbers and quantities.[48]

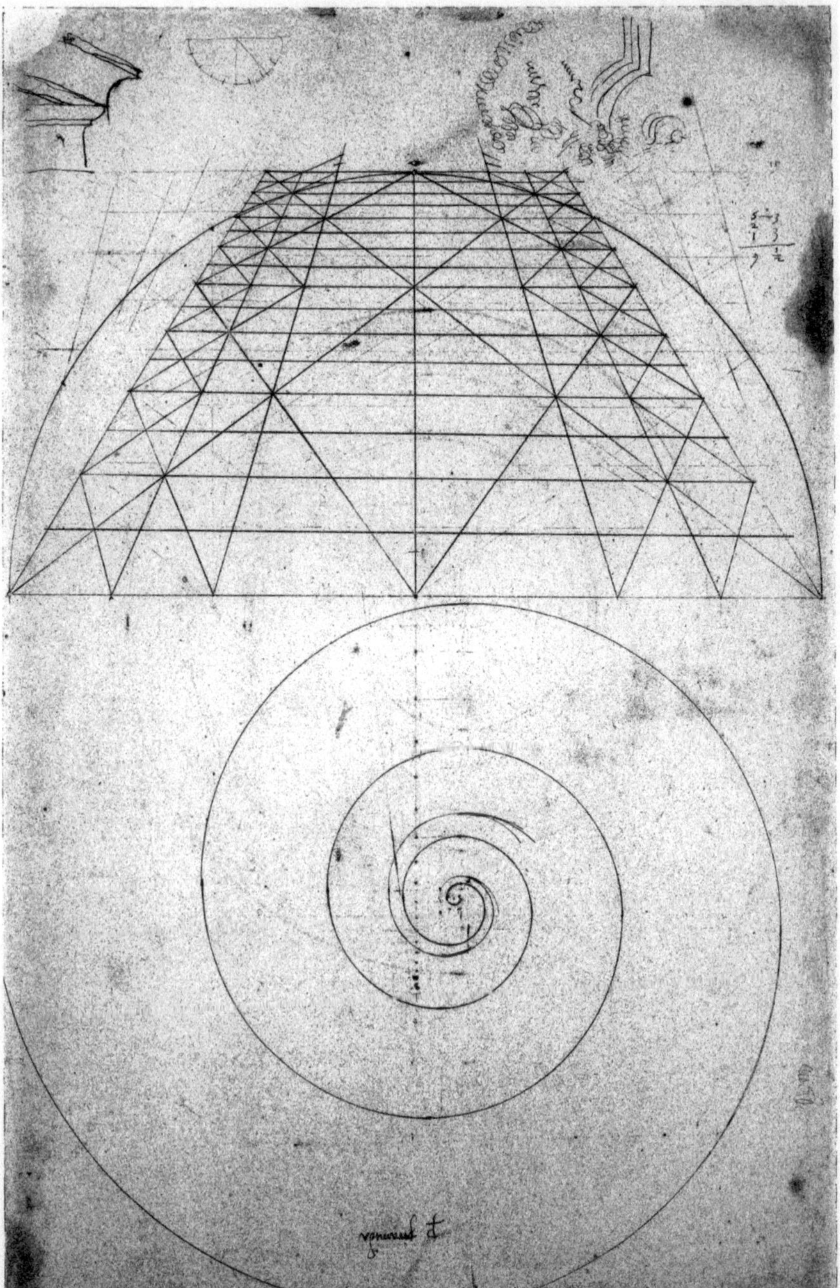

Fig. 20 Baldassarre Peruzzi, Ionic volute study (UA 465v)

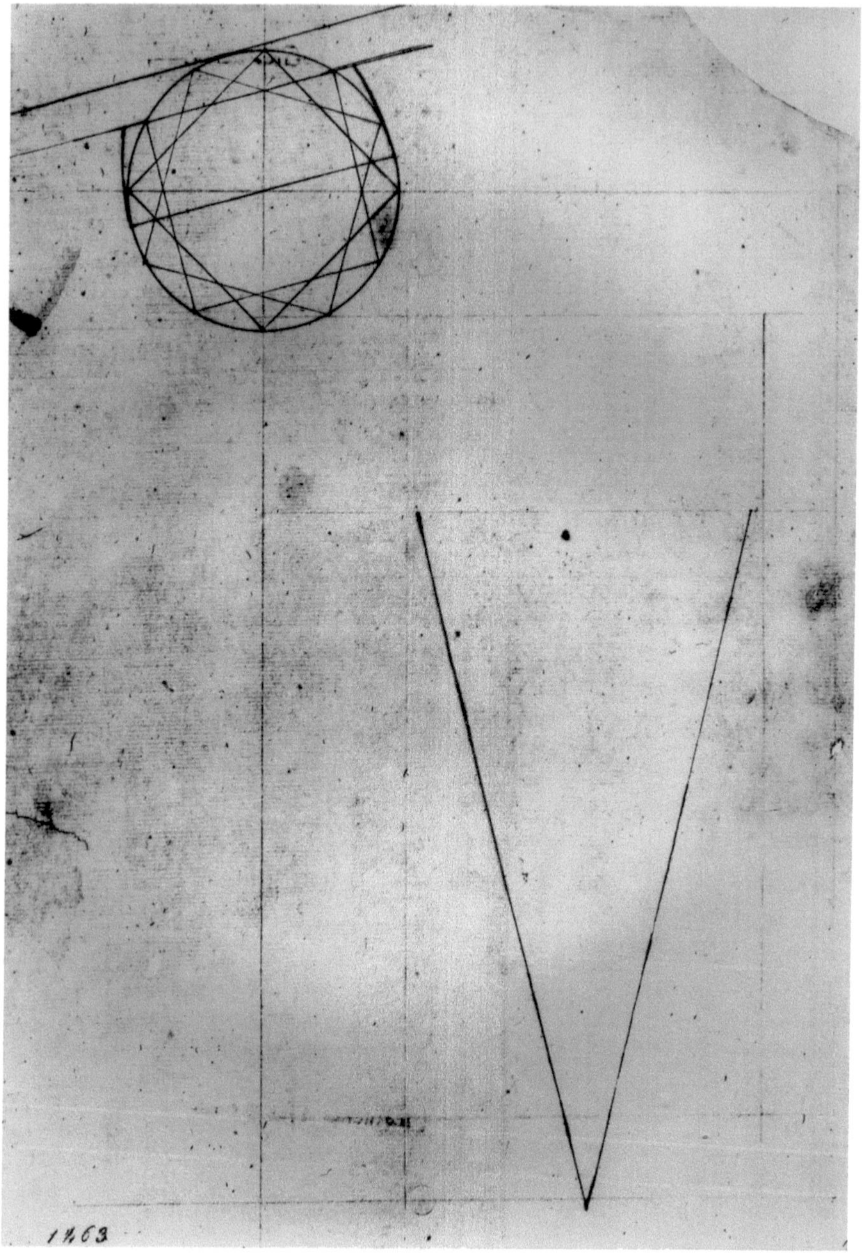

Fig. 21 Antonio da Sangallo the Younger, geometrical study after Fra Giocondo (UA 1463r)

Both Peruzzi and Sangallo developed a knowledge of mathematics from their earliest education. The differences in their application of this knowledge, however, reflect their professional training, and several of the drawings examined above epitomize these distinctions. Peruzzi's apprenticeship as a painter, and his mastery of perspective in particular, are clearly evident in his great longitudinal section for San Domenico and his Ionic volute study (Figs. 11 and 20). By contrast, the appearance of a plumb line in Sangallo's St Peter's dome study and the great technical drawing of the wooden model exemplify his background in construction (Figs. 15 and 18). These drawings manifested the particular skills that each brought to the field of architecture and with which each was most comfortable. But many others among the surviving corpuses point to the skills each focused on specifically, because these had not formed part of their particular training. Seen in this light, then, each sought through their particular use of mathematics to transcend the limitations of their respective backgrounds. Peruzzi's preoccupation with numbers reflects his legitimation through the practicalities of measure and quantity, while Sangallo's investment in the abstraction of theory was his comparable search for validation.

Notes

This essay developed out of a paper first presented at the Annual Meeting of the Society of Architectural Historians, Vancouver, April 2005, and benefitted from discussions at the Italian Renaissance Seminar, University of Oxford, and the Bibliotheca Hertziana, Rome, in February and March of 2005. Francesco Benelli, Jim Bennett, Filippo Camerota, Mario Carpo, Rob Corser, Anthony Gerbino, Richard Goldthwaite, Stephen Johnston, Martin Kemp, Pamela Long, Gervase Rosser, and Christof Thoenes all graciously shared expertise. I am also indebted to the Scott Opler Foundation and Worcester College, Oxford for the fellowship that supported this research.

1. Architectural treatises offer another avenue for such exploration, as Mario Carpo recently has demonstrated in Carpo (2003).
2. Bruschi (1983).
3. Adams (1996).
4. Ackerman (1954) in Ackerman (1991, esp. 370–73).
5. See for example Peruzzi's calculations on a plan for San Domenico in Siena, Gabinetto dei Disegni e Stampe degli Uffizi, Florence, Architettura (hereafter UA) 545r, or Sangallo's page of calculations on UA 858r. Wurm (1984, 237); Frommel and Adams (1994–2000, vol. 1, 343).
6. Grendler's conservative estimate of the literacy rate for Florentine males in 1480 is between 30 and 33 %; *abaco* training presupposed the ability to read and write. Grendler (1989, 77–78).
7. Wide-scale adoption of the new, Hindu-Arabic number system seems to have varied by field but in general occurred slowly. Roman numerals continued to

predominate in account books, for example, through the fifteenth century, despite the institution of communal *abaco* schools in central and northern Italy during the fourteenth century. See Berggren (2002, 361); Franci and Rigatelli (1982, 22); Arlinghaus (2004).

8. Swetz (1987, xiv, 12–14). Raffaella Franci and Laura Toti Rigatelli have traced the influence of Fibonacci on Italian abaco education in numerous publications that build upon their fundamental study of 1982, *Introduzione all'aritmetica mercantile* (as in note 7). See especially their recent contribution, Franci and Rigatelli (2002, 45–66).
9. Grendler (1989, chapter 3).
10. Christoph L. Frommel discusses Sangallo's education and training in Frommel (1994, 1–22). Frommel also summarizes the little that is known of Peruzzi's earliest years in Frommel (2005, 4–7). For further discussion of Peruzzi's education, see Huppert (2008) and Huppert (2015).
11. Professional apprenticeships typically occurred between ages 12 and 15. Grendler (1989, 22 and 306–11); Grendler (1995, 167–68); and Franci and Rigatelli (1982, 32–33).
12. Van Egmond (1981, 21–26) and Moscadelli (1991, 209).
13. Adams (1985, 386).
14. Pietro dell'Abaco, active in Siena later in the fifteenth century, was one such instructor. Another example is Giovanni d'Abacco, who consulted on the construction of the dome of Florence's cathedral in the early fifteenth century. See Scolari (1994, 586).
15. On the projects for Piacenza see Adorni (1986).
16. See Peruzzi's drawings, UA 632r and 355r. A comparably complex fraction of 17 and 3 over 8-1/2 occurs on a study for San Domenico on UA 344r (Fig. 16). Wurm (1984, 463, 325).
17. Franci and Toti Rigatelli point out that, as a result of its location on the via Francigena, Siena in particular was a center for currency exchange. Franci and Rigatelli (1982, 29).
18. Arch. Arcisped. S. Rocco, vol. 233, Archivio di Stato di Roma. Peruzzi served as *camerlengo* for the 12 month period beginning September 1515.
19. For example, extensive records survive for Michelangelo's work at San Lorenzo in Florence, but the expenses and accounting were overseen by others, in this case bankers, leaving the architect more freedom to concentrate on design. See Wallace (1994, 87, 138).
20. Carpo (2003, 466–67 and n. 54).
21. See Fabiano T. Fagliari Zeni Buchicchio, "U 731 A *recto*," in Frommel and Adams (1994–2000, vol. 2, 147).
22. Ognaretto (1998, 49–68).
23. Van Egmond (1986, 53).
24. Fabiano T. Fagliari Zeni Buchicchio, "U 724 A *recto*," in Frommel and Adams (1994–2000, vol. 1, 111 and 284).
25. Total costs in *scudi* appear along the edges of Peruzzi's drawings, UA 16r and 18r, together with calculations of overall dimensions. Wurm (1984, 499, 505).

26. See Wurm (1992) and Huppert (2008).
27. The extensive note within the central nave bay on UA 339r includes an overall cost of 7,000 *scudi*.
28. Peruzzi study on UA 343r shows the same pier form but with the increased dimensions. Wurm (1984, 231).
29. The same measurements for "*la cherica*" that he labeled on the verso, 44 and 98 *canne*, appear on the recto of the sheet. Here Peruzzi identified these as the size of the vaults above the arches, without specifying what aspect of the vault he referred to ("*la volta piccola e can[ne] 44 sop[r]a ali archi*", "*la volta grande sopra deli archi e can[ne] 98*"). Peruzzi did not provide sufficient information to explain the relationship of these *canne* measurements to other dimensions on the drawings, such as the radii of the vaults.
30. The two notes towards the center of the sheet read: "*di falcato li quattro vani deli archi in superficie e can[n]e 133 b 19-1/2,*" and "*di falcato li archi in una superficie per fine ale imposte e le due volte son can[ne] 132*". Below this he wrote "*si di falca b[raccia] 905 2/14 p[er] 4 archi cioe can[ne] 56 b [raccia] 9... [*illegible number*] el semicirculo.*"
31. Betts (1993, 5–6).
32. For his analysis of Sangallo's sequence of drawings see Thoenes (1997) and his catalog entries in Frommel and Adams (1994–2000, vol. 2, 101–3, 129–31, and 153).
33. Thoenes (1997, 192), calls this study "a graphic investigation of statics."
34. Working with the interaxial dimension for the piers of 13-1/8 *palmi* would give him a total circumference of 630 *palmi*. However, this did not match the circumference of 616 *palmi* that he correctly converted from the diameter of the dome, 196 *palmi*, using the Archimedean approximation of 3-1/7 for π.
35. Sangallo identified the scale for the sheet as "*colli palmi del modello.*"
36. Thoenes identified the alternative arch described in the note as the one used in the model. See Thoenes (1997, 194–97).
37. Thoenes critiques this non-algebraic curve on practical terms in Thoenes (1994, 641).
38. He concluded his written note with a proposal for an elliptical form with a ratio of base diameter (196 *palmi*) to dome height of 3:2, which he realized in the built model. Thoenes (1997, 196–97).
39. Sangallo identified the profile variations as "*todescho*" and "*antico buono,*" the latter having "*piu gratia.*" For analysis of Sangallo's design as a solution that surpasses both Gothic and classical forms, see Benedetti (2009, 65–77).
40. Lotz (1955, 7–99).
41. Serlio (1996–2001, vol. 1, 27). Also see Rosin (2001) and Kitao (1974, 31–35).
42. For Renaissance interest in the Ionic volute, see Günther (1988, 221–25) and Losito (1993).
43. Poggi (2005, 450, 455).
44. Vitruvius provides instructions for constructing the volute in Book III, chapter 5. See Vitruvius (1999, 52).

45. See Scaglia (1994, 87); Frommel and Adams (1994–2000, vol. 1, 233–34) and Ciapponi (1984).
46. On the differing degrees of reliance on Vitruvius between Sangallo and Peruzzi, see Huppert (2015), chapter 2.
47. On the rediscovery of the Roman *piede*, see Günther (1988, 227–29) and Rowland (1998, 133–34). Günther also discusses a wide-spread interest in ancient weights and measures among humanists in the early sixteenth century. Among architects, use of the ancient *piede* became prevalent, and the unit appears frequently in drawings by Antonio da Sangallo.
48. Pagliara (1986, 52–54).

Photographic Credits

Uffizi, Gabinetto dei Disegni e Stampe (by permission of the Ministero dei beni e delle attività culturali e del turismo): Figs. 1–10, 12–21
Biblioteca Apostolica Vaticana: Fig. 7
Ashmolean Museum, Oxford: Fig. 11

References

Ackerman, James. 1954. Architectural practice in the Italian Renaissance. *Journal of the Society of Architectural Historians* 13(3): 3–41.

Ackerman, James. 1991. *Distance points: Essays in theory and Renaissance art and architecture*. Cambridge, MA/London: MIT Press.

Adams, Nicholas. 1977. Baldassare Peruzzi: Architect to the Republic of Siena 1527–1535. Ph.D. dissertation, New York University.

Adams, Nicholas. 1984. Architecture for fish: The sienese dam on the Bruna river – Structures and designs, 1468-C.1530. *Technology and Culture* 25: 768–797.

Adams, Nicholas. 1985. The life and times of Pietro dell'Abaco, A Renaissance estimator from Siena, active 1457–1486. *Zeitschrift für Kunstgeschichte* 48(3): 384–395.

Adams, Nicholas. 1996. Peruzzi, Baldassare. Oxford Art Online. www.oxfordartonline.com.

Adorni, Bruno. 1986. Progetti e interventi di Pier Francesco da Viterbo, Antonio da Sangallo il giovane e Baldassarre Peruzzi per le fortificazioni di Piacenza e di Parma. In *Antonio da Sangallo il giovane: la vita e l'opera*, ed. Gianfranco Spagnesi, 349–372. Rome: Centro di Studi per la Storia dell'Architettura.

Arlinghaus, Franz-Josef. 2004. Bookkeeping, double-entry bookkeeping. In *Medieval Italy: An encyclopedia*, vol. 1, ed. Christopher Kleinhenz, 147–150. New York: Routledge.

Benedetti, Sandro. 2009. *Il grande modello per il San Pietro in Vaticano: Antonio da Sangallo il Giovane*. Rome: Gangemi.

Berggren, J. Lennart. 2002. Medieval arithmetic: Arabic texts and European motivations. In *Word, image, number: Communication in the middle ages*, ed. John J. Contreni and Santa Casciani, 351–365. Florence: SISMEL-Edizioni del Galluzzo.

Betts, Richard J. 1993. Structural innovation and structural design in Renaissance architecture. *Journal of the Society of Architectural Historians* 52(1): 5–25.

Bruschi, Arnaldo. 1983. Cordini, Antonio. *Dizionario biografico degli Italiani*. http://www.treccani.it/biografie.

Carpo, Mario. 2003. Drawing with numbers: Geometry and numeracy in early modern architectural design. *Journal of the Society of Architectural Historians* 62(4): 448–469.

Ciapponi, Lucia. 1984. Disegni ed appunti di matematica in un codice di Fra Giocondo da Verona (Laur. 29, 43). In *Vestigia: Studi in onore di Giuseppe Billanovich*, vol. 1, ed. Rino Avesani et al., 181–196. Rome: Edizioni di storia e letteratura.

Franci, Raffaella, and Laura Toti Rigatelli. 1982. *Introduzione all'aritmetica mercantile del Medioevo e del Rinascimento. Realizzata attraverso un'antologia degli scritti di Dionigi Gori (sec. XVI).* Urbino: Quattro Venti.

Franci, Raffaella, and Laura Toti Rigatelli. 2002. L'eredita di Leonardo Fibonacci. In *Fibonacci tra arte e scienze*, ed. Luigi A. Radicati di Brozolo, 45–66. Cinisello Balsamo: Silvana.

Frommel, Christoph L. 1994. Introduction: The drawings of Antonio da Sangllo the Younger: History, evolution, method, function. In *The architectural drawings of Antonio da Sangallo the Younger and his circle*, vol. 1, ed. Christoph L. Frommel and Nicholas Adams, 1–60. New York/Cambridge, MA/London: Architectural History Foundation/MIT Press.

Frommel, Christoph L. 2005. 'Ala maniera e uso delj bonj antiquj': Baldassarre Peruzzi e la sua quarantennale ricerca dell'antico. In *Baldassarre Peruzzi (1481–1536)*, ed. Christoph L. Frommel et al., 3–83. Venice: Marsilio.

Frommel, Christoph L., and Nicholas Adams (eds.). 1994–2000. *The architectural drawings of Antonio da Sangallo the Younger and his circle.* New York/Cambridge, MA/London: Architectural History Foundation/MIT Press.

Grendler, Paul F. 1989. *Schooling in Renaissance Italy: Literacy and learning, 1300–1600.* Baltimore/London: Johns Hopkins University Press.

Grendler, Paul F. 1995. What Piero learned in school: Fifteenth-century vernacular education. In *Piero della Francesca and his legacy*, ed. Marilyn A. Lavin, 161–174. Washington, DC: National Gallery of Art.

Günther, Hubertus. 1988. *Das Studium der Antiken Architektur in den Zeichnungen der Hochrenaissance*. Tübingen: Ernst Wasmuth.

Huppert, Ann C. 2001. The archaeology of Baldassarre Peruzzi's architectural drawings. Ph.D. dissertation, University of Virginia.

Huppert, Ann C. 2008. Roman models and Sienese methods: Baldassarre Peruzzi's designs for San Domenico. In *L'ultimo secolo della Repubblica di Siena: arti, cultura e società: Atti del convegno internazionale,* Siena, 28–30 settembre 2003 e 16–18 settembre 2004, ed. Mario Ascheri, Gianni Mazzoni, and Fabrizio Nevola, 107–119. Siena: Accademia senese degli Intronati.

Huppert, Ann C. 2015. *Becoming an architect in Renaissance Italy: Art, science, and the career of Baldassarre Peruzzi.* New Haven/London: Yale University Press.

Kitao, Timothy K. 1974. *Circle and oval in the Square of Saint Peter's: Bernini's art of planning.* New York: New York University Press.

Losito, Maria. 1993. La ricostruzione della voluta ionica vitruviana nei trattati del Rinascimento. *Mélanges de l'École Française de Rome: Italie e Méditerranée* 105(1): 133–175.

Lotz, Wolfgang. 1955. Die Ovalen Kirchenräume des Cinquecento. *Römisches Jahrbuch für Kunstgeschichte* 7: 7–99.

Moscadelli, Stefano. 1991. Maestri d'abaco a Siena tra Medioevo e Rinascimento. In *L'università di Siena: 750 anni di storia*, 207–216. Milan: Silvana.

Ognaretto, Rossella. 1998. I disegni di Baldassarre Peruzzi per Rocca Sinibaldi. *Quaderni dell'Istituto di Storia dell'Architettura* 32 (n.s.): 49–68.

Pagliara, Pier Nicola. 1986. Vitruvio da testo a canone. In *Memoria dell'Antico nell'Arte Italiana*, vol. 3, ed. Salvatore Settis, 5–85. Turin: Einaudi.

Poggi, Paola. 2005. 'Architectonica perspectiva': la prospettiva solida de 'Le Bacchidi' e la voluta ionica di Baldassarre Peruzzi. In *Baldassarre Peruzzi 1481–1536*, ed. Christoph L. Frommel et al., 443–455. Venice: Marsilio.

Rosin, Paul L. 2001. On Serlio's construction of ovals. *The Mathematical Intelligencer* 23(1): 58–69.

Rowland, Ingrid D. 1998. *The culture of the High Renaissance: Ancients and moderns in sixteenth century Rome*. Cambridge/New York: Cambridge University Press.

Scaglia, Gustina. 1994. Drawings of machines, instruments and tools. In *The architectural drawings of Antonio da Sangallo the Younger and his circle*, ed. C.L. Frommel and Nicholas Adams. New York/Cambridge, MA/London: Architectural History Foundation/MIT Press.

Scolari, Massimo. 1994. Giovanni di Gherardo Gherardi, drawing. In *The Renaissance from Brunelleschi to Michelangelo: The representation of architecture*, ed. Henry A. Millon and Vittorio Magnago Lampugnani, 586. Milan: Bompiani.

Serlio, Sebastiano. 1996–2001. *On architecture*. New Haven: Yale University Press.

Swetz, Frank J. 1987. *Capitalism and arithmetic: The new math of the 15th century, including the full text of the treviso arithmetic of 1478*. Trans. D.E. Smith. La Salle: Open Court.

Thoenes, Christof. 1994. St Peter's 1534–46: Projects by Antonio da Sangallo the Younger for Pope Paul III. In *The Renaissance from Brunelleschi to Michelangelo: The representation of architecture*, ed. Henry A. Millon and Vittorio Magnago Lampugnani, 634–648. Milan: Bompiani.

Thoenes, Christof. 1997. Il modello ligneo per San Pietro ed il metodo progettuale di Antonio da Sangallo il giovane. *Annali di architettura* 9: 186–199.

Van Egmond, Warren. 1981. *Practical mathematics in the Italian Renaissance: A catalog of Italian abbacus manuscripts and printed books to 1600*. Florence: Istituto e Museo di Storia della Scienza.

Van Egmond, Warren. 1986. The contributions of the Italian Renaissance to European mathematics. *Symposia Mathematica* 27: 51–67.

Vitruvius. 1999. *Ten books on architecture,* ed. Ingrid D. Rowland and Thomas Noble Howe. Cambridge: Cambridge University Press.

Wallace, William E. 1994. *Michelangelo at San Lorenzo*. Cambridge/New York: Cambridge University Press.

Wurm, Heinrich. 1984. *Baldassarre Peruzzi Architekturzeichnungen: Tafelband*. Tübingen: Ernst Wasmuth.

Wurm, Heinrich. 1992. Baldassarre Peruzzis Entwürfe, 1532. In *Die Kirchen von Siena, vol. 2.1.2: Oratorio della Carità-S. Domenico*, ed. P.A. Riedl and M. Seidel, 512–516. Munich: Bruckmann.

Geometric Survey and Urban Design: A Project for the Rome of Paul IV (1555–1559)

David Friedman

The slow development of survey technology—from the first statement of its geometric principles in the mid-fifteenth century to its application in the administration of property and the design of urban spaces—spans an arc of almost two centuries. One of the landmarks of this progress is a drawing in the Uffizi collection, catalogued under the number 4180A (Fig. 1a, b). It is a large drawing, composed of ten joined sheets, and measuring 117 cm by 133 at its widest points. It is a project for a large building complex on an urban site. The constituent elements identify it as a cloister: the cruciform space of a church, an atrium, and an arcaded court. A "rota" and "parlatoio", located between the two latter spaces allow communication between the cloistered religious and lay visitors. This is not an ideal scheme, and it is the survey that makes it specific. The "Piazza del arco di camillo" to the right side and the "piazza di S. Ma(c)uto" at the bottom left, place the project in Rome, on the site occupied today by the late sixteenth-century structures of the Collegio Romano and the seventeenth-century church of Sant'Ignazio. The drawing represents a project for a convent of Franciscan nuns, or Poor Clares, sponsored by the Marchesa Vittoria della Tolfa and was executed in the period 1555–1559, during the pontificate of the marchioness's uncle, Paul IV Caraffa.[1]

Topographic survey is an invention of the Renaissance. Its underlying geometry was described as early as the mid-fifteenth century in a treatise on mathematics by Leon Battista Alberti. In this text, the *Ludi rerum mathematicarum*, Alberti describes a circular instrument for measuring angles in the plane of the horizon and gives instructions for coordinating observations from two station points to fix the position of a distant landmark: that is, for the system of triangulation that is the foundation of geometric mapping (Fig. 2).[2] The mapping of cities required additional techniques. What the cartographer needed to know about the fabric of cities included not

D. Friedman (✉)
Emeritus, Department of Architecture, School of Architecture,
Massachusetts Institute of Technology, Room 10 – 303, 77 Massachusetts Avenue,
MA 02139, Cambridge, USA
e-mail: dhfriedm@mit.edu

A. Gerbino (ed.), *Geometrical Objects: Architecture and the Mathematical Sciences 1400-1800*, Archimedes 38, DOI 10.1007/978-3-319-05998-3_5,

just the position of things but also their form. In the heavily built up environment of the city the only way to measure form was by the technique that modern survey calls the compass traverse. Raphael described the technique in his report to Leo X, dated 1518–1520, on the project to make a census of the monuments of ancient Rome.[3] It, too, used the circular angle-measuring device known in the sixteenth century as a *bussola*. By combining measurements of orientation and length for every face of a block or section of street it could, in theory, record the internal structure of the city. A sketch plan of the Ponte Sant'Angelo area in Rome from the workshop of Antonio da Sangallo the Younger dated to 1524–5 (UA 1013) records dimensions and

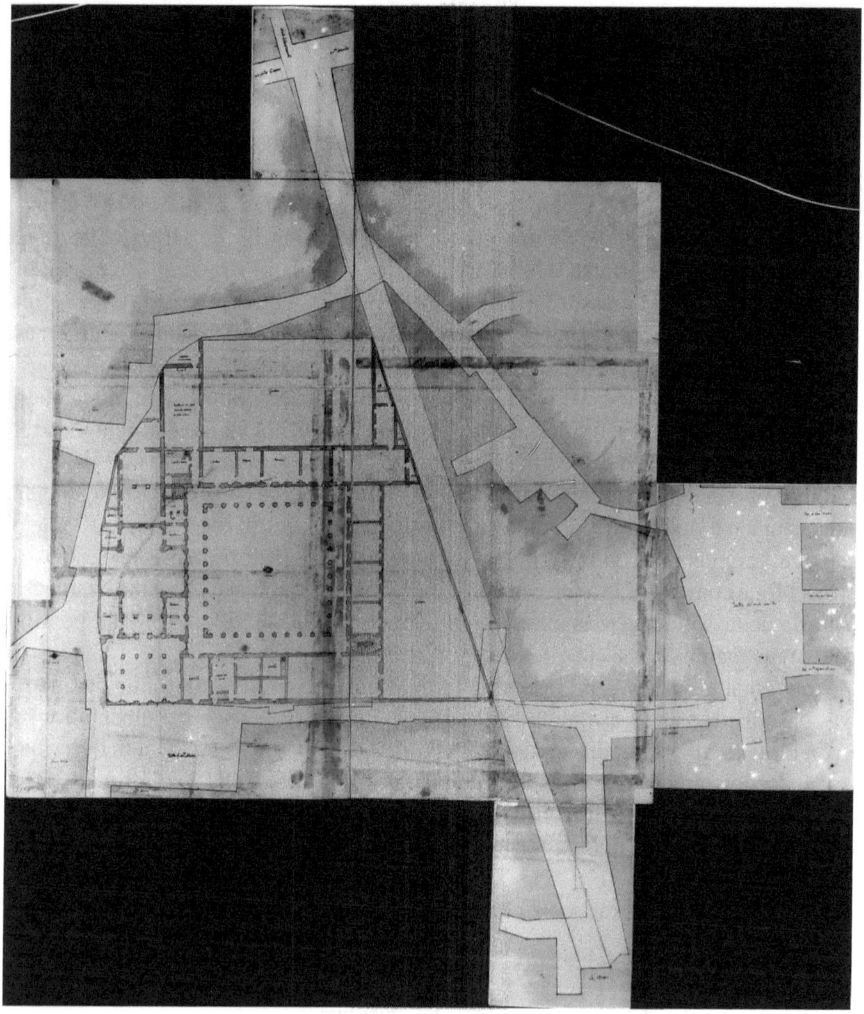

Fig. 1 (**a**) Project for a convent of Poor Clares in Rome, on the site occupied today by the Collegio Romano and Sant' Ignazio 1555–1559, UA 4180 (**b**) Giovanni Battista Nolli, *Nuova Pianta di Roma*, 1748. The area covered by UA 4180

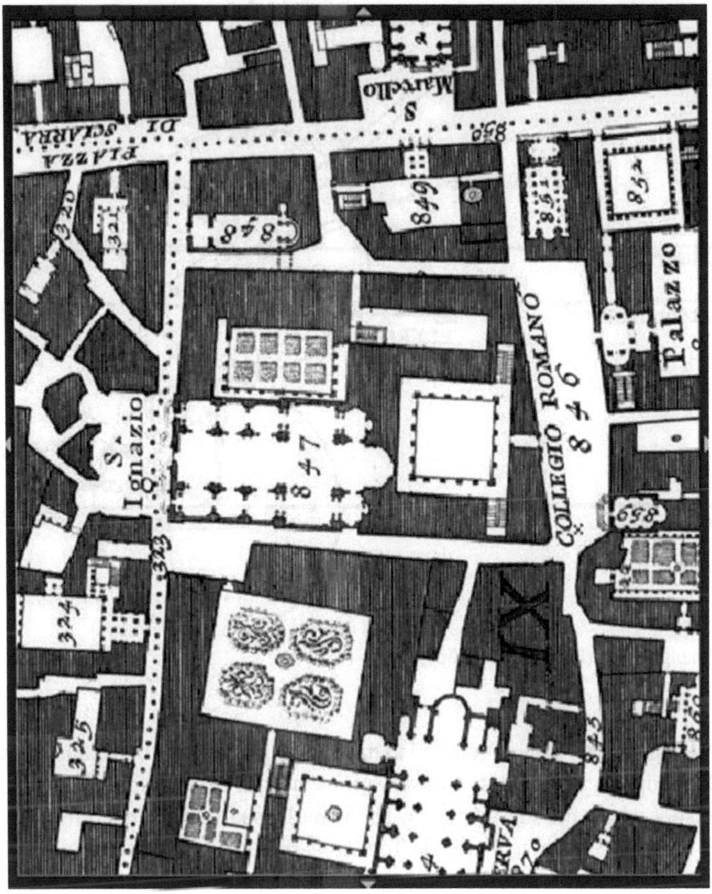

Fig. 1 (continued)

bearings and shows the first stages of this process (Fig. 3a–c).[4] The collection of Sangallo drawings in the Uffizi also preserves a series of working sheets—compass roses—that record observations taken from station points throughout Florence in preparation for the construction by triangulation of a plan of the city (Fig. 4).[5]

All the techniques necessary for mapping the city were in place by the early sixteenth century. Execution, though, lagged far behind theory. In part the problem lay with the instruments, which remained without telescopic sights until the eighteenth century. The compass traverse, in particular, was notoriously inaccurate. The great Ferrarese engineer and cartographer Giovanni Battista Aleotti (1546–1636) testified to the difficulty of this kind of survey:

> Knowledgeable mathematicians (that is map makers) do not hold the bussola to be a reliable instrument. I, honestly, have only rarely, and maybe never, been able to successfully close any plan that I have made using it, and I don't think anyone else does any better. Beyond even the instability of the magnetic compass there are many practical problems (*imperfettioni di mano*) in surveying and drawing the plan of a town or territory.[6]

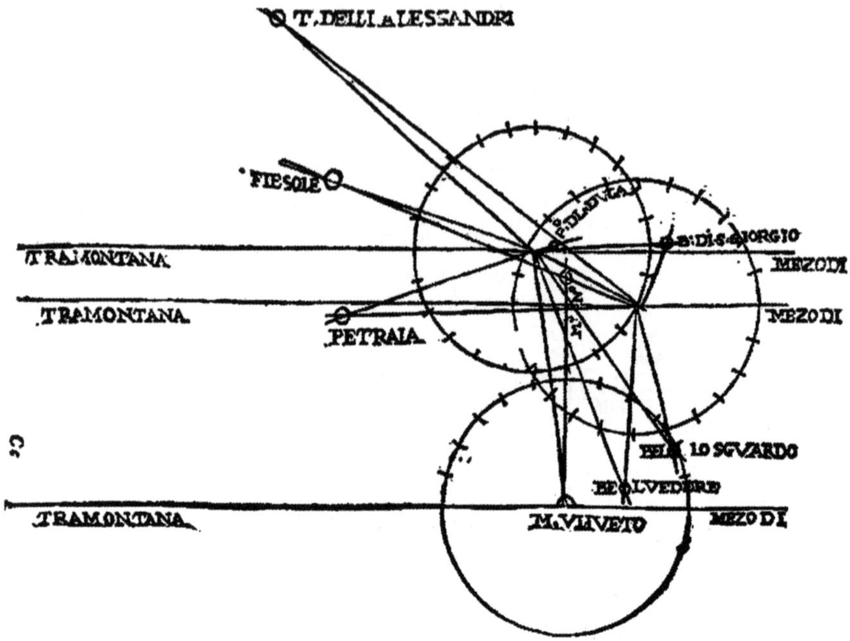

Fig. 2 Triangulation (From Bartoli 1564)

By "closing a plan" Aleotti referred to the successful outcome of a compass traverse, when the endpoint of a series of observations around a closed perimeter coincided with the spot where the survey began. The great Leonardo da Vinci himself was one of those who failed. When Nando de Toni constructed a plan of the walls of Cesena from the survey data recorded in Leonardo's notebooks, he discovered that the artist had made enough errors that the last segment of wall did not meet the first.[7]

The great number of observations required to complete the survey of even a small section of the city made the failure of the compass traverse there almost inevitable. Antonio da Sangallo's survey of four short blocks in the Banchi (Fig. 3) required approximately 80 measurements of distance and bearing to record the course of the streets and the contour of the building fronts that define them. The labor involved in capturing this level of detail goes a long way toward explaining the rarity of drawings of this kind. It is not surprising, then, that Leonardo Bufalini's pioneering plan of Rome of 1551 (Fig. 5a, b), despite seven years of work, does not attempt a precise account of the course of the city's streets. Bufalini gave purchasers of his plan a new, comprehensive view of the city that included the first published survey of the Aurelian wall and, thus, the first realistic representation of the shape of the city. The plan seems to be based on at least two different systems of survey: a triangulation of major monuments and a compass traverse of the wall. The two were not integrated and the forms generated by them on the plan cannot be reconciled. Most streets may not have been surveyed at all but their courses laid down by eye between the points fixed by the two survey systems.[8]

Because of its limited precision and inadequate level of detail and because of its woodcut format and 14 poorly coordinated sheets, Bufalini's *Roma* was not a useful instrument for urban planning. Uffizi 4180A gives a better account of Renaissance survey. It also preserves considerable information about how maps were produced in the mid-sixteenth century. The drawing consists of two discrete parts: the project for the convent and the plan of the building site. The architectural project is a

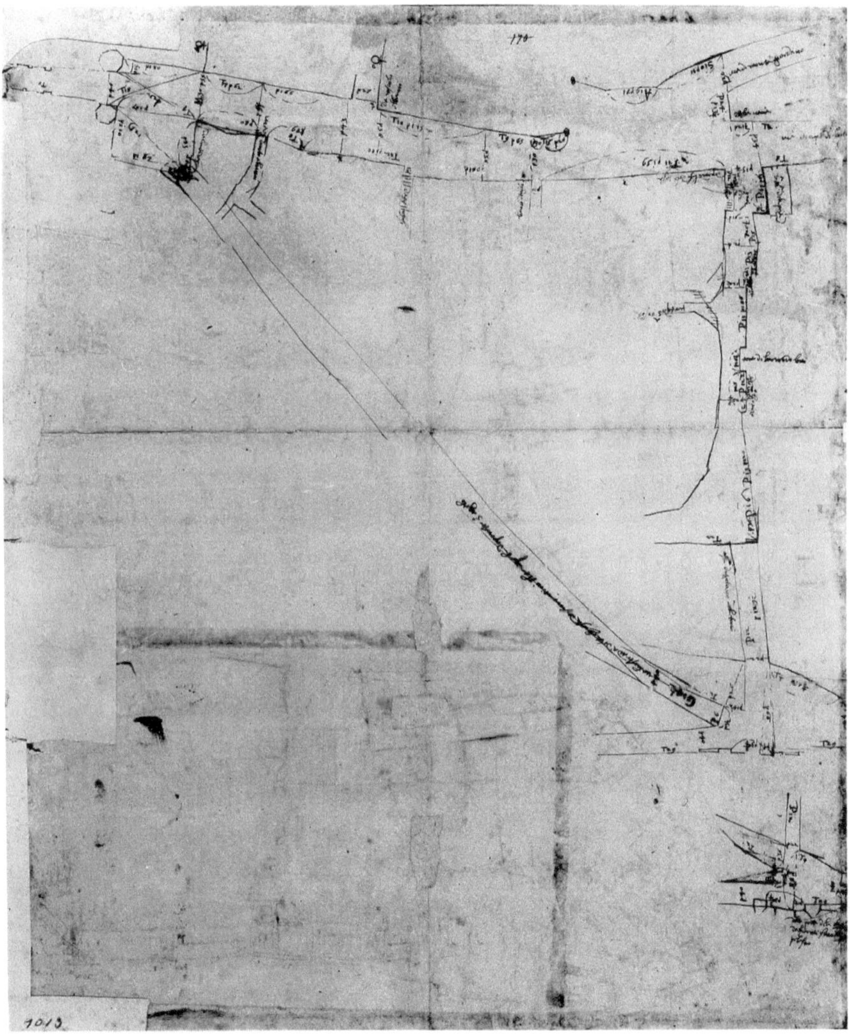

Fig. 3 (**a**) Workshop of Antonio da Sangallo, the younger. Sketch plan of the area of the Banchi, at Ponte Sant'Angelo, Rome, 1524–5 (UA 1013) (**b**) UA 1013, detail (**c**) Area of the Banchi in the view of Rome of Antonio Tempesta, 1593

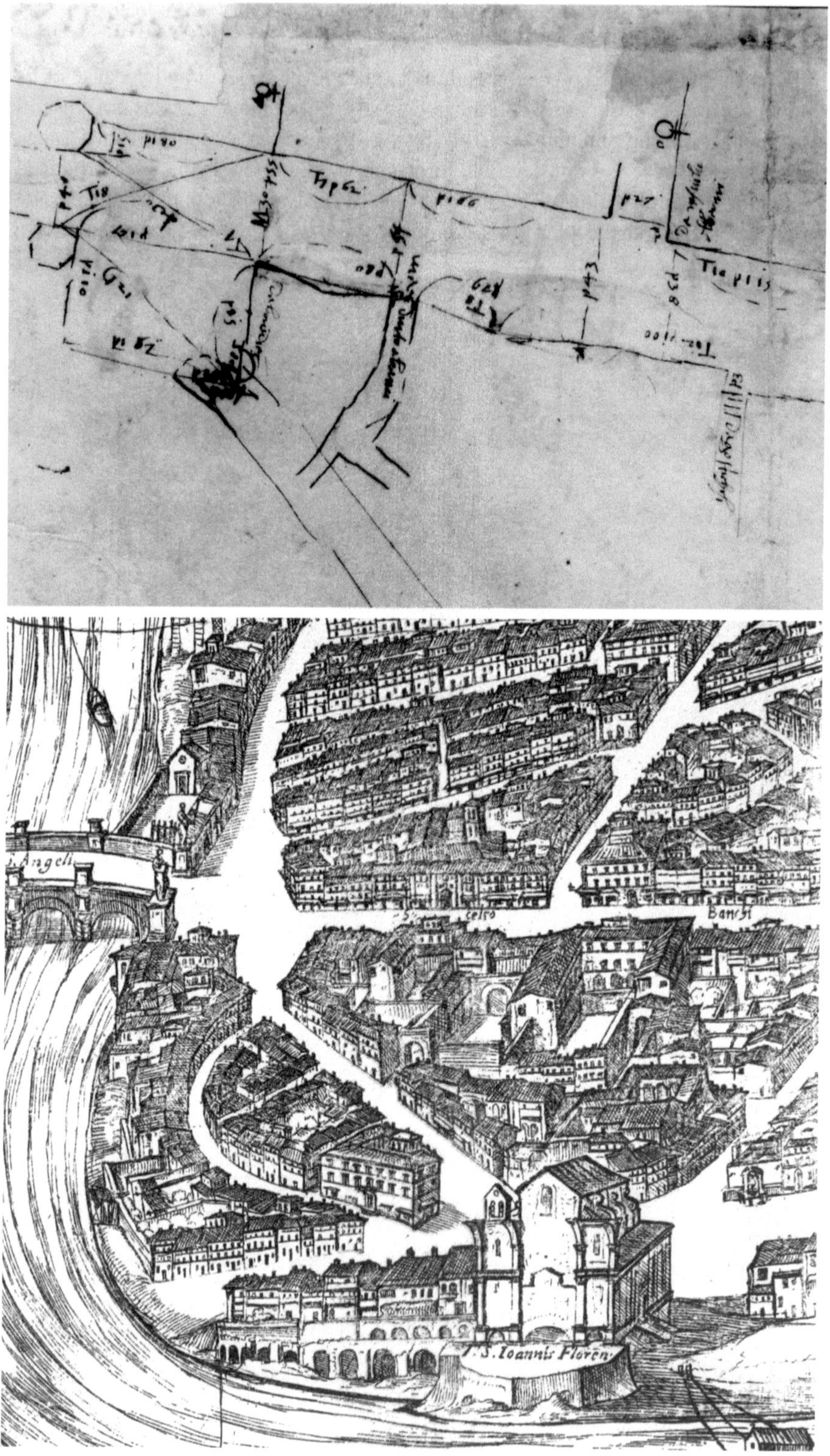

Fig. 3 (continued)

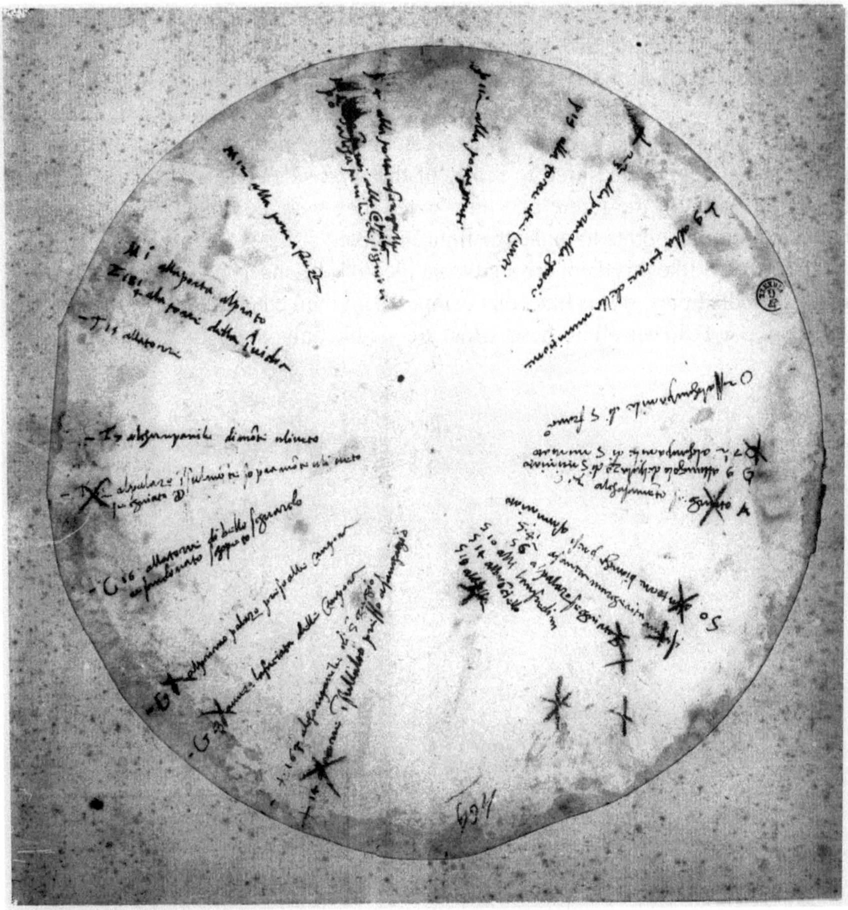

Fig. 4 Workshop of Antonio da Sangallo the Younger, Notes from one station point for a survey of Florence, UA 773

finished work, developed elsewhere and transcribed here line for line. This part of UA 4180 shows no revisions or construction lines and the level of presentation is high. Walls have been reinforced with a light wash and the function of the rooms inscribed. The survey of the site is also handsomely drawn, the wash, applied here with a broad brush, blocking out the built up area beside the streets. This is a drawing meant for presentation outside the workshop, probably to the clients of the project. The survey, however, unlike the project, is not redrawn after an earlier draft. At this stage in the history of urban mapping, the process of transforming survey data into a plan took place on the finished drawing itself. A couple of details speak for this conclusion.

In a publication only a decade old when UA 4180 was drafted, the mathematician Nicolo Tartaglia characterized the process of map-making as one of transferring data from the site to the sheet. Translation was to be as direct as possible, extending to the use of the same instrument for both survey and drafting. North is the constant reference and the magnetic compass fixes it both in the field and on the drafting table. The plan was to be drawn by laying the "*bossolo*" (Tartaglia's spelling of *bussola*) on the drawing surface, aligning it with north using the magnetic compass, and rotating the sighting arm to the bearing recorded in the field for the first side of the figure. A stylus line impressed in the paper recorded the bearing of the sighting arm, and perforations in the paper mark the center of the instrument and the terminal point of the length, measured to scale, of that first side. These steps are repeated around all faces of the figure and only when they were complete was ink applied between terminal points to make the figure visible.[9]

In a process like this there are signs that identify the sheets on which data became image: needle holes, stylus lines that extend beyond inked ones, and, of course, corrections. UA 4180 has all of these. Most are visible only with a magnifying lens but

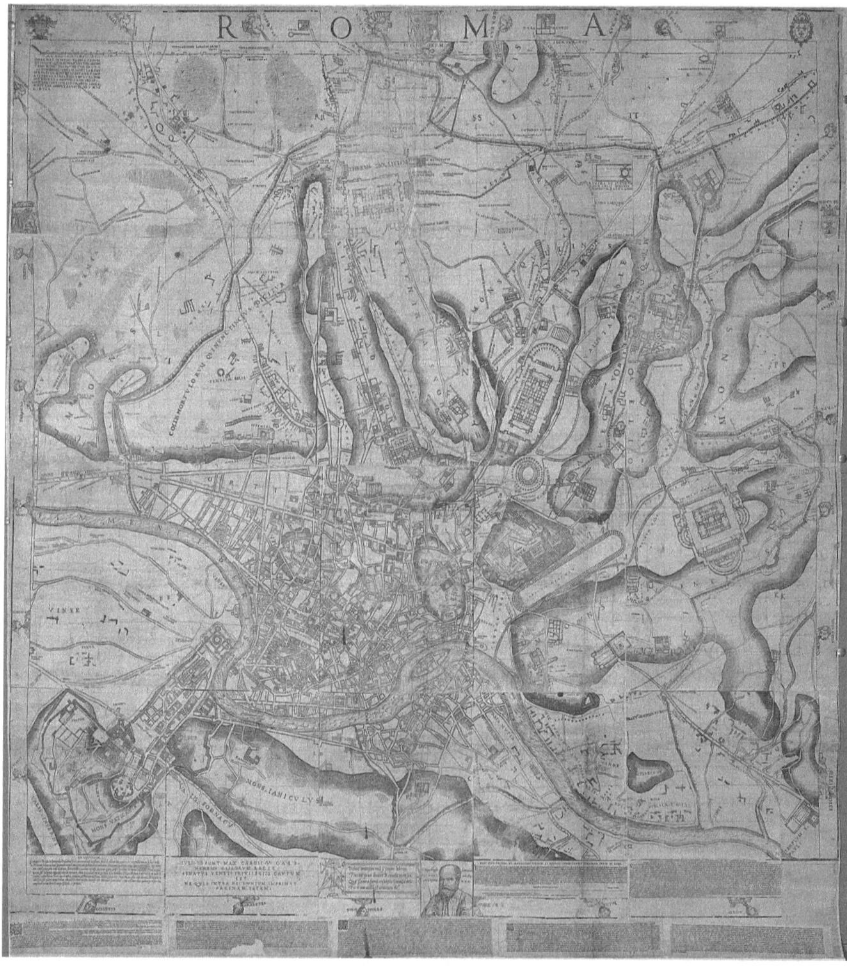

Fig. 5 (**a**) Leonardo Bufalini, Map of Rome, 1560 (original 1551) and (**b**) detail

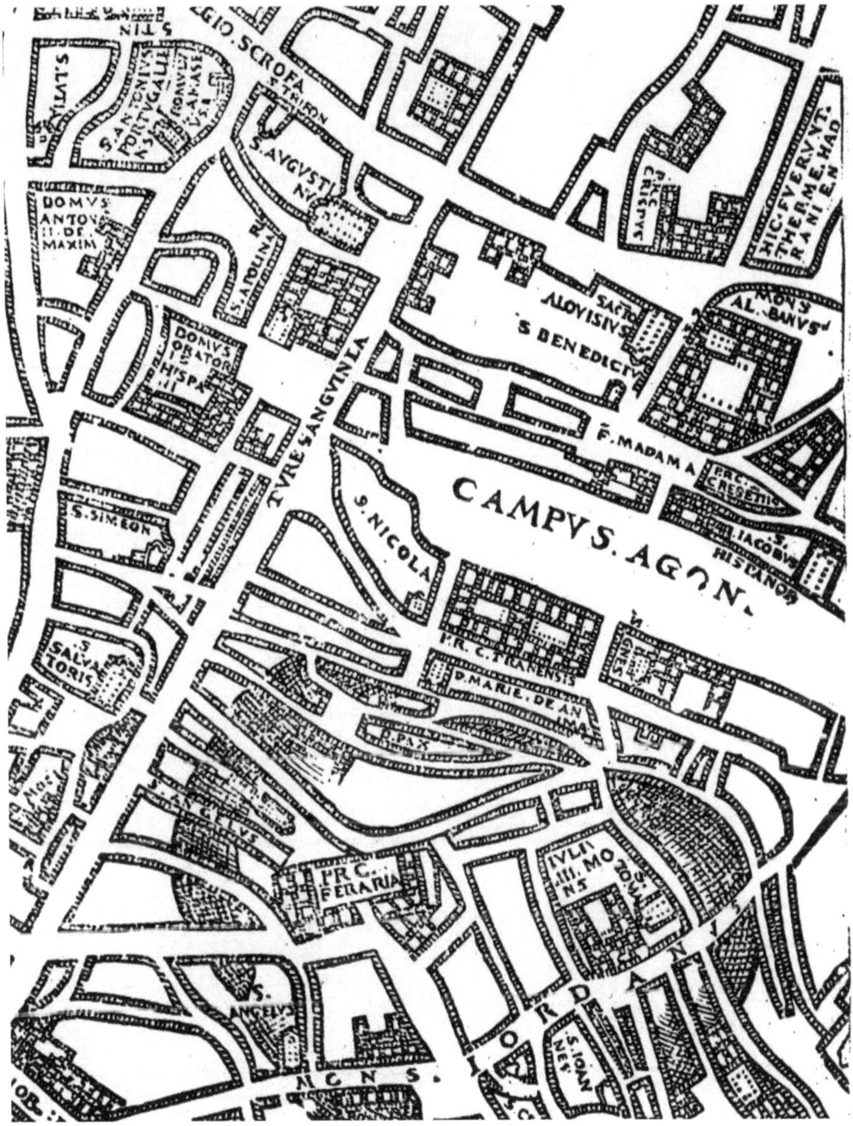

Fig. 5 (continued)

one correction is large enough to see in the photograph of the drawing as a whole (Figs. 6a, b and 7a–c). A long erased line running down the left side and across the bottom of the sheet represents a first attempt to construct the block plan. The executed plan is different from the erased one in a number of ways. In the first place, it is laid on the sheet differently, rotated with respect to the earlier image by about 5°. This adjustment served to coordinate the street facades on the north and west sides of the convent block with the left and bottom edges of the sheet, establishing a

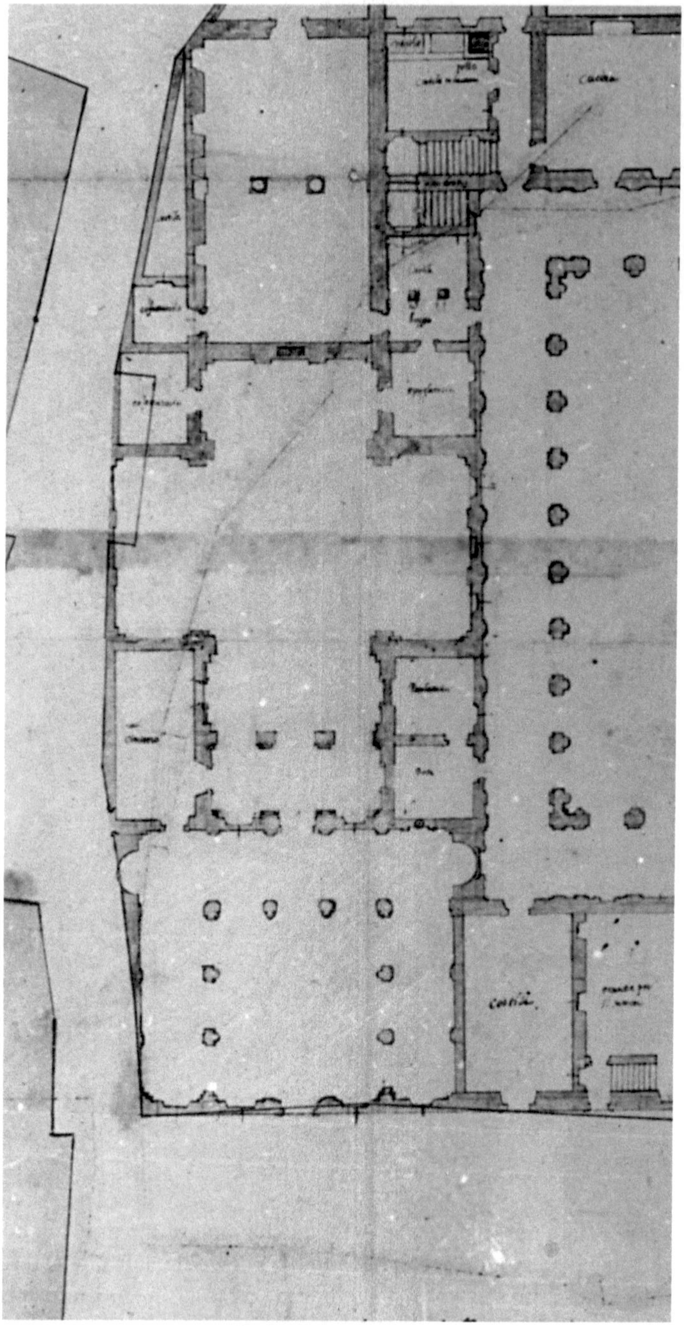

Fig. 6 (**a**, **b**) Details from UA 4180 (1555–1559), showing the failed first effort to construct the image from the survey

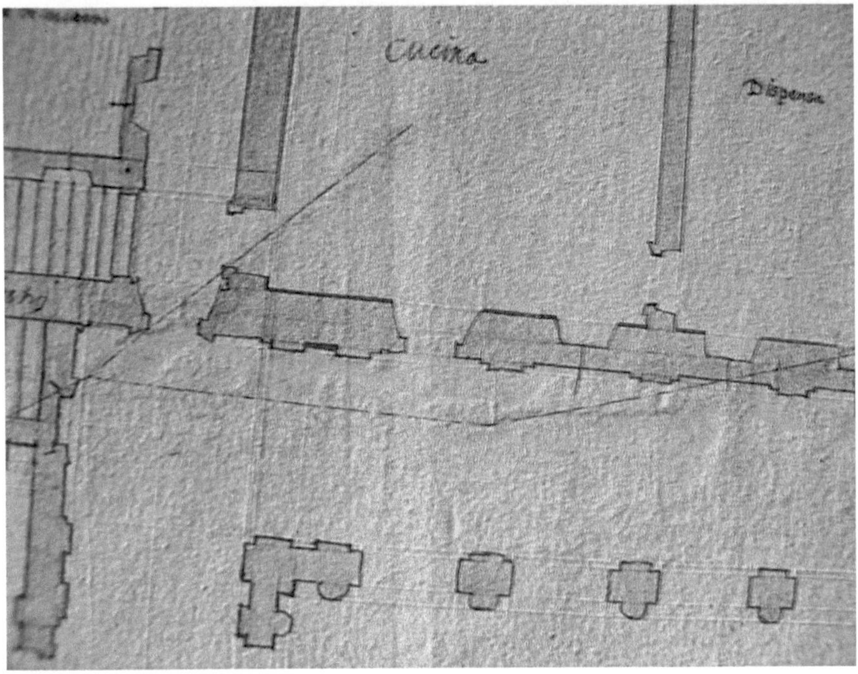

Fig. 6 (continued)

stable compositional base for the mapped image as a whole. The second attempt to lay out the plan also involved moving the building site about 15 cm higher on the page, which further solidified the image by making room for the entire Piazza di San Macuto. But the most fundamental change is the simplest one. The redrawn plan is 50 % larger than the draftsman's first effort.[10] The change in scale meant that the plan had to be constructed anew from the survey data.

The differences between these two versions of the plan illustrate something fundamental about the condition of survey imagery at the middle of the sixteenth century. The difficulties of sizing the image and positioning it on the page show that the draftsman did not know the shape of the blocks before beginning the drawing. There existed no map of the city that he could consult, nor had he transformed his own survey data into a plan before beginning work on UA 4180. For an image intended as a formal presentation drawing, the inability to envision the shape of the project created significant difficulties. The redrafting of the plan shows the author of UA 4180 coming to terms with the surprises inherent in the sixteenth century mapping process.

As both the product of an on-site survey and formal presentation of a design project, UA 4180 has a complex character very different from what we have come to expect from the more rationalized design process of modern practice. The place where the drawing's two natures collide is on the northern and western perimeter of the convent, where the regular forms of church and cloister meet the circumstantial

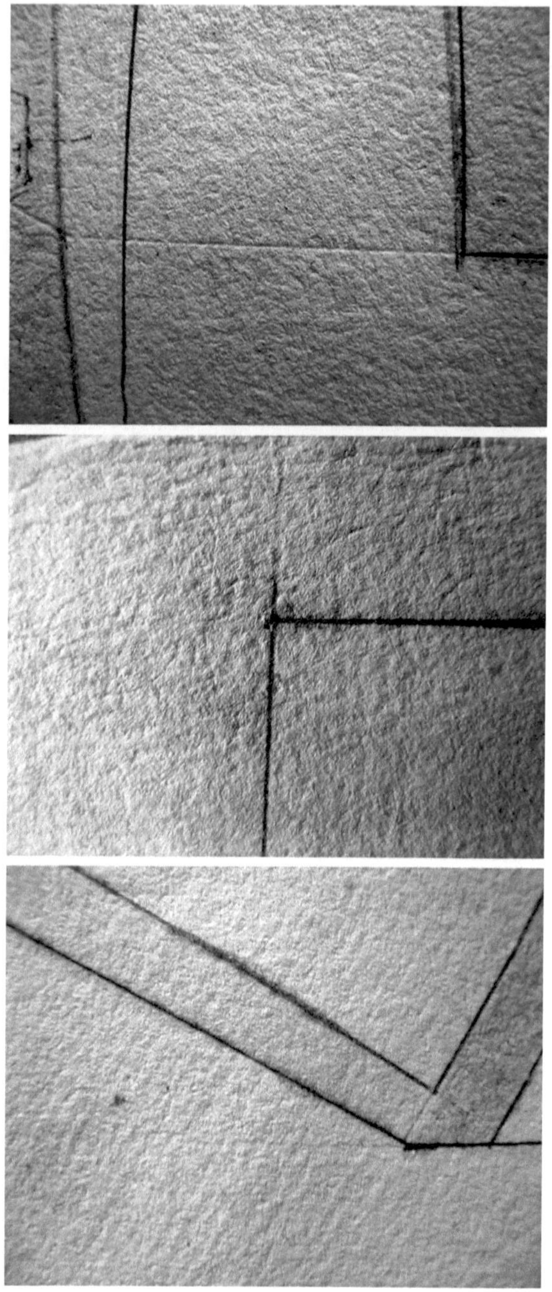

Fig. 7 (**a**–**c**) Details from UA 4180, showing the pinholes and stylus lines used for the construction of the image

shapes of the city. Here, the drawing presents two separate, superimposed plans (Fig. 8a–c). One, with the wall thickness washed in pink, represents the convent project. The other—drafted with a single, un-reinforced line—records the limits of

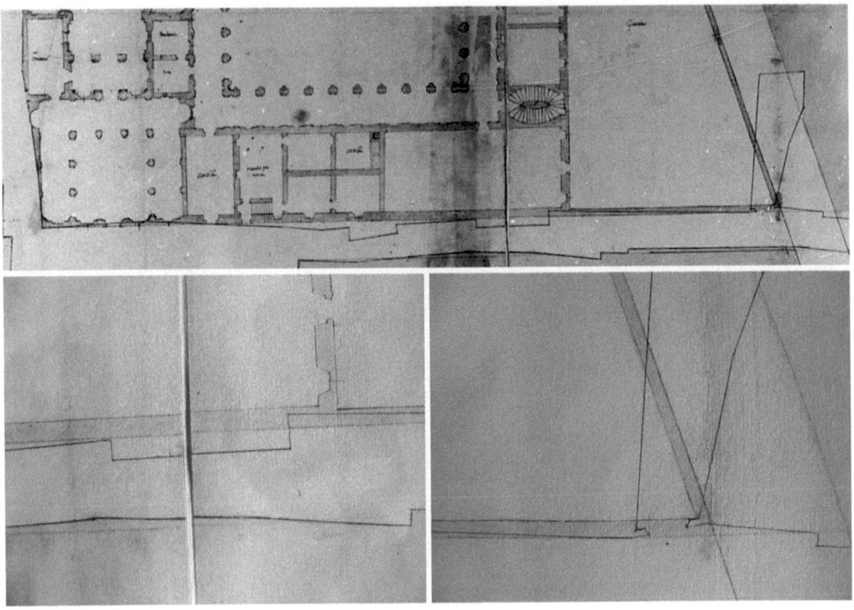

Fig. 8 **(a–c)** Details from UA 4180, showing the existing building lines and the proposed street front of the new project

the existing structures. It is easy to imagine that in an early moment, the plan constructed on UA 4180 consisted only of this line drawing, which was then copied onto a separate sheet where the convent project was worked out. Only when a final design had been established would the internal articulation of the convent have been added to the master sheet. One of the results of this process is the small failure of the convent project to fit the available space. One sees this best at the left of the drawing where the walls of the atrium, church and service rooms have been thinned beyond serviceability to remain within the line defined by the existing property (Fig. 9). In places like this, it is evident that the walls defining the convent were drawn over those outlining the block.

Planning with UA 4180

In modern practice, the plan constructed on UA 4180 would have served to generate two kinds of information important for the design process. In the first place, it would reveal the shape of the property and give a scaled representation of its size. In the second, it would fix the position of the elements of the project in relation to the existing urban fabric. In this latter respect, the most important features of the design were the endpoints of the proposed street that was to define the southern boundary of the convent, the streets and square to the right of the sheet. To accomplish these goals the surveyor had to execute the most difficult of all the operations of his discipline.

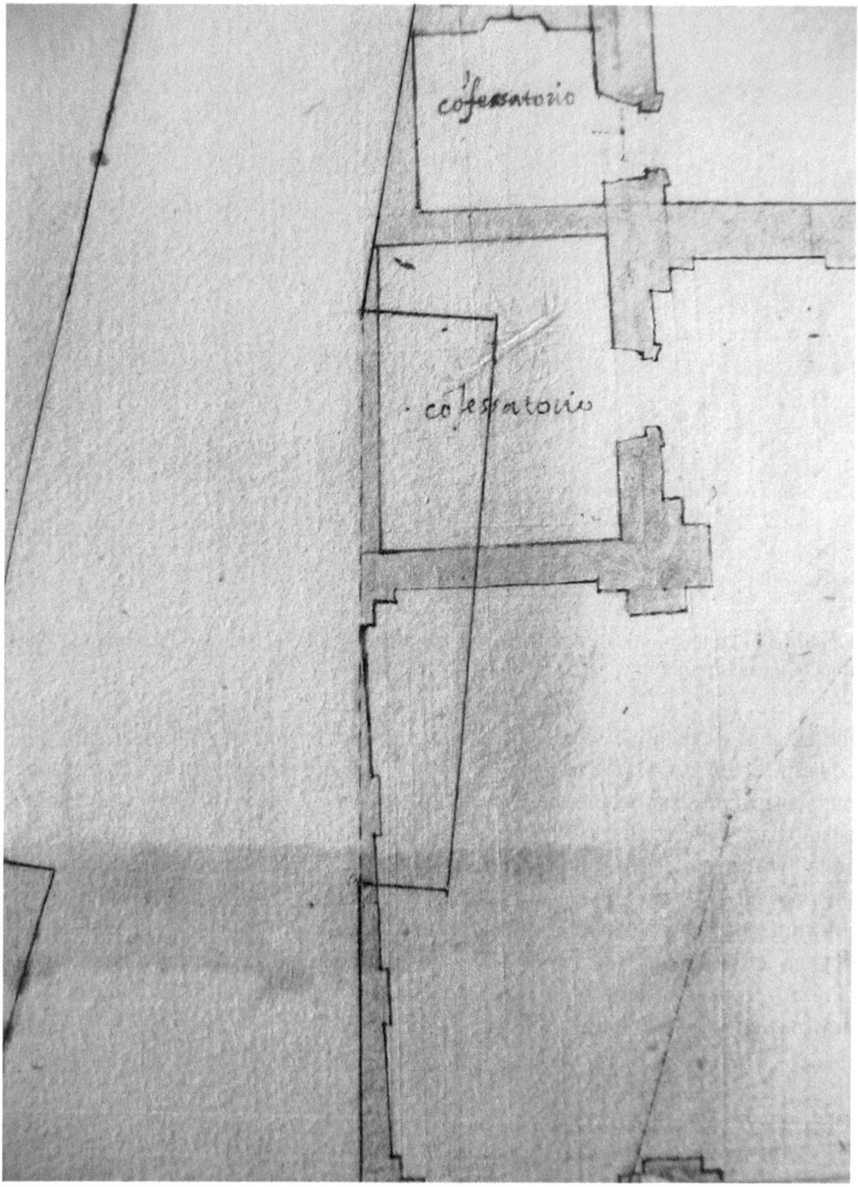

Fig. 9 UA 4180, Detail of the church's left transept and confessionals showing the wall thinned to fit the architectural project within the mapped space of the city block

This is not a plan that could be made by the elegant and relatively accurate methods of triangulation. Only a compass traverse could measure the shape of the streets and blocks at the convent site, and this, as we have seen, was a process notoriously prone to error. The common solution to the open circuit of an imperfect traverse, one

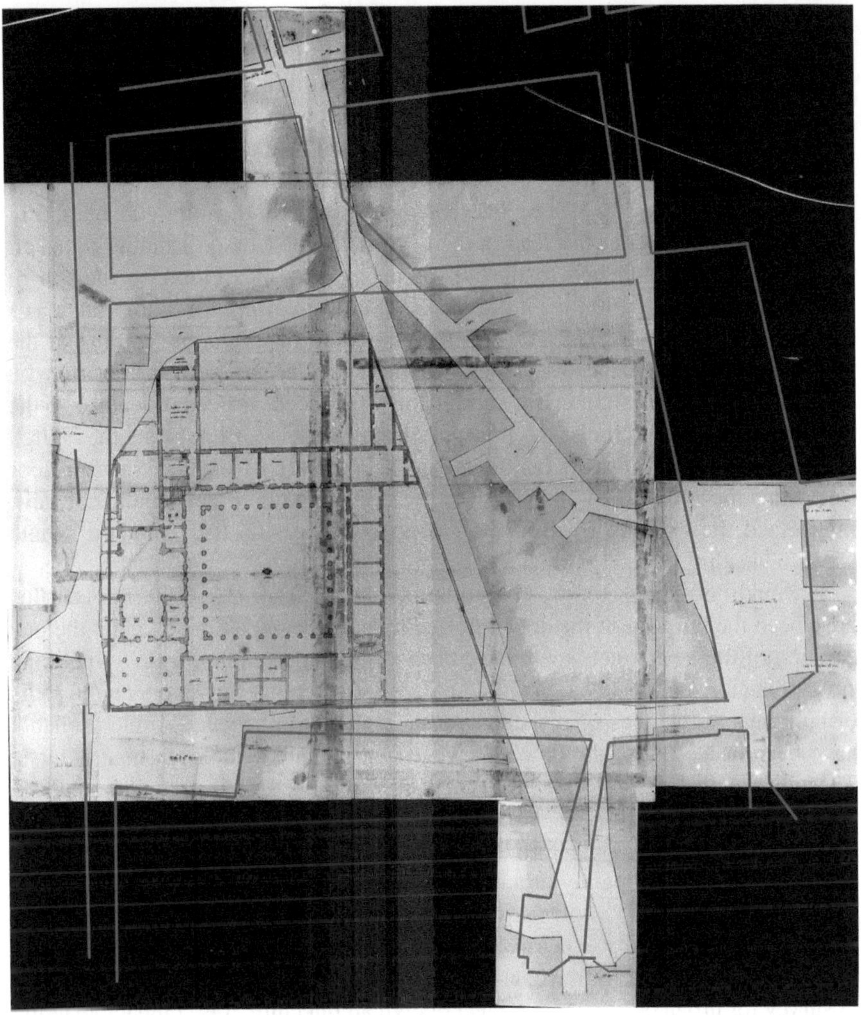

Fig. 10 UA 4180, overlaid with the outline of the Nolli plan, 1748. The scale of the two images has been equalized and the outline positioned on the basis of the Via di Sant'Ignazio frontage (the street in front of the proposed church). North is to the left of the figure

supposes, was to "adjust" angles and segment lengths. The draftsman of UA 4180 did something like this on the right side of his sheet. This is the side least related to either the convent or the new street and therefore the side least likely to have been the starting point for the construction of the plan. It is here that the draftsman would have discovered his problem, and a series of erasures suggest that it was also here that he made the changes that resolved the plan. This adjustment had little effect on the shape of the area to be rebuilt for the convent but at the overall scale, that of the street project, the distortions make an important difference.[11]

By superposing UA 4180 onto the representation of the Collegio Romano area in Giambattista Nolli's 1748 plan of Rome, we get some measure of the accuracy of the sixteenth century survey (Fig. 10). Figure 10 aligns the Via di Sant'Ignazio front of the Collegio Romano represented by Nolli with the *filo*, or building line, proposed in UA 4180 on the west side of the central block. When the scales of the two plans are equalized the comparison reveals the differences. One is the compression of the UA 4180 plan on the east-west axis, from the top to the bottom of the plan. This is most easily read in the position of the Corso. Running right-left on the small flap attached to the top of the plan, it lies closer to the Via di Sant'Ignazio in UA 4180 than it does in Nolli's *Ichnografia*. More important as a measure of the value of UA 4180 for the planning process, is the bearing of the new street. Although the eastern end of the street aligns with its position on the Nolli map, the western end is significantly displaced. Instead of leading to the door in the northern arm of the transept of the Minerva, it intersects the church to the south of the apse. If buildings had been cleared from the transept door following the orientation indicated in the drawing, the street would not have intersected the existing road on the east side of the site, the "Strada a Monte Cavallo" leading to the Quirinal.

It would not have been possible to build directly from UA 4180, nor can this have been the purpose of the drawing. Had it been, a compass rose to plot the orientation of the new street would have been a minimum requirement. To build the street, direct visual observation would have been essential. Surveyors on the roofs of buildings at the ends of the street and at intervals in between would probably have been necessary to direct a somewhat approximate demolition of structures. The only way that the project represented in drawings like UA 4180 could be realized on the ground was through the presence of the architect or a knowledgeable collaborator at the site. The information of this drawing, like contemporary drawings prepared for projects purely architectural in nature, was approximate—essential but not sufficient.[12]

UA 4180 is unusually expansive as an architectural drawing in that it presents a broad physical context for the convent project. It is also a rare document of the use of survey for urban design. As a project of civil architecture, it is comparable to the sketch plan of the Banchi area in Rome (Fig. 3), which seems to have been related to a project to improve the street system at the Ponte San Angelo and the sight lines to the church of San Giovanni dei Fiorentini.[13] A more fully developed design—also by Antonio da Sangallo the Younger's workshop—for the expansion and fortification of the feudal residence of Pratica in the Pontine marshes is more typical of the kind of project for which surveys were produced (Fig. 11).[14] The geometric nature of fortifications based on canon and interdependent bastions made drawings and models essential to military design. Tartaglia speaks of them as if their production were a matter of course.[15] In contrast, the area inside the city walls was less frequently mapped, and design there lagged in its use of survey.[16]

Surveys of urban design sites become more common in the second half of the century, as patrons began to see their utility in the planning process. A painting of

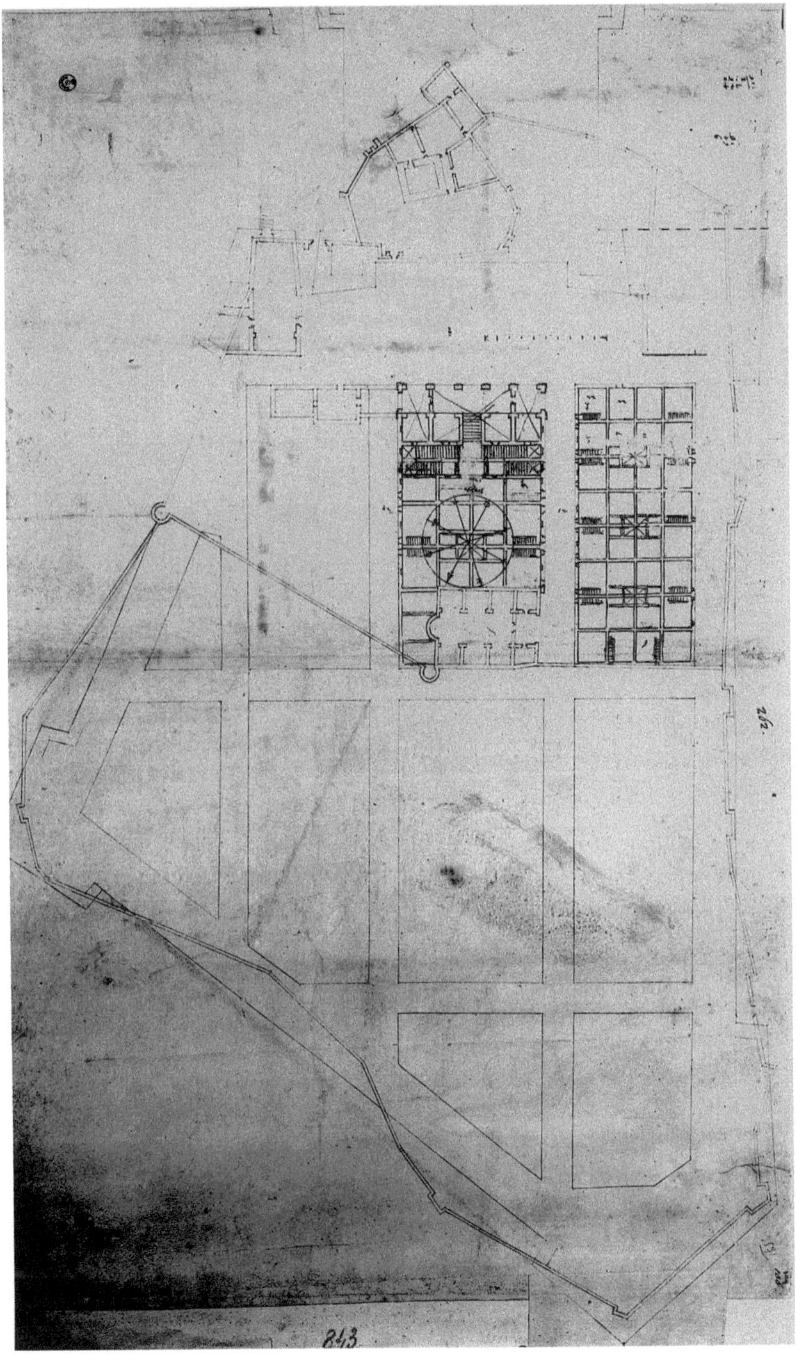

Fig. 11 UA 483, Workshop of Antonio da Sangallo, the Younger. Project for the reconstruction of the town of Pratica (after 1539)

Fig. 12 The Foundation of Cosmopolis, Sala di Cosimo I, Palazzo Vecchio, Florence (1557)

1557 on the ceiling of the Sala di Cosimo I in the Palazzo Vecchio in Florence makes an explicit claim about their use (Fig. 12). The image by the painter and architect Giorgio Vasari celebrates the foundation in 1548 of Cosmopolis (Portoferraio on the island of Elba). Cosimo, surrounded by his advisors, looks at the site from some indeterminate vantage. In his hand, he holds a plan of the project, while before and, symbolically, below him lies the city itself. It, too, is represented in plan.[17] Vasari's painting makes the purpose of such plans explicit: they allow the patron to visualize the city as a whole, as though hovering over it like a demigod. A more literal statement of purpose appears in the dedicatory text of a survey plan of Parma executed between 1589 and 1592 and given a new dedication in 1601. With the plan, "You, Duke Ranuccio, can see the proportions and relationships of the streets to one another and of any street to the body (of the city) as a whole, and if you want to bring the city to its full dignity you will clearly see the places that need to be improved."[18]

With few exceptions, sixteenth-century plans that represent the whole city—like Bufalini's *Roma*—give a schematic and generalized picture of their subject. In that regard, the Cosmopolis plan of Vasari's painting is probably not dissimilar from real working drawings. A plan for the enlargement of the village of Guastalla of 1553 provides an example of this kind of image (Fig. 13). The town, located on the Po

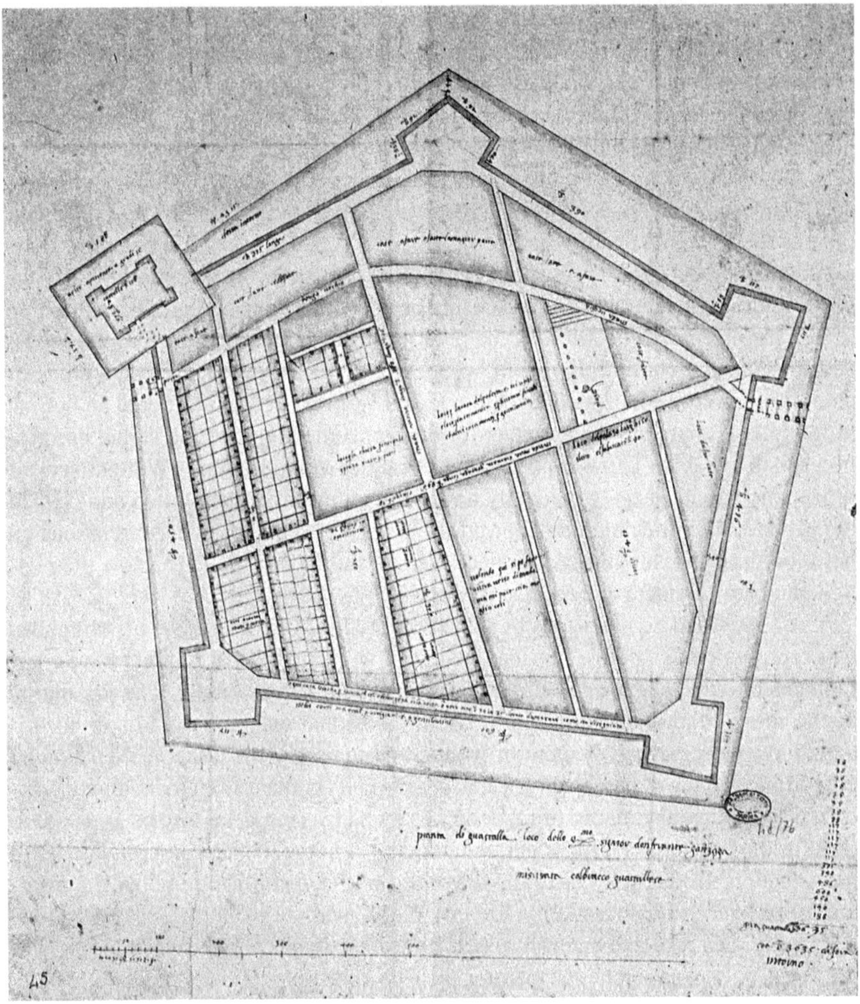

Fig. 13 Domenico Giunti, plan of Guastalla, 1553 (Archivio di Stato di Parma, Raccolta mappa e disegni, volume 48, plan 76)

between Mantua and Parma, was being transformed into a feudal residence by Ferrante Gonzaga, then governor of Milan for Charles V. The drawing is by Domenico Giunti, Ferrante's architect.[19] It is a survey of the defensive perimeter then in the course of construction and also includes a necessarily less concrete and only partially drawn project to develop the area within the new walls. In comparison to UA 4180, the schematic character of the plan is striking. Geometric survey is limited. Inscriptions give bearings for the western curtains at the top of the sheet but for no others. The street plan is defined by dimensions alone. The town as built follows this project in part, but the drawing can only be considered an outline of the scheme. It gives a general picture of the arrangement of blocks, but the orientation of streets and the angles of intersections are significantly different in the town itself. While there is no question that this plan responds to the idiosyncrasies of the Guastalla site, its rough accuracy and limited detail give it an appearance similar to the images that illustrate ideal towns in the literature of military architecture.

UA 4180 is much more precise. It records plan details of individual house fronts and takes as its subject a block and street system of a complexity that no Italian city plan of the sixteenth century had yet addressed. While it was not a blueprint to be executed without further creative intervention, it did establish terms for the consideration of the architectural and urban design project that would not have been possible before the invention of survey. At the practical level, it identified the property to be dedicated to the convent and to the proposed street and marked out the areas of demolition. At the conceptual level, the level of formal design, it offered an outline of the area available to the architect for design, something that the verbal description of sites that we know from medieval practice could not do.[20] Without survey, information about shape would only have been known after demolition had cleared the ground. The drawing also allowed the designer to understand the relationship between the convent and street projects and, at the largest scale, to know the relationship between them and the rest of the city.

UA 4180 was also the means by which the architectural and urban planning idea was represented to a public outside the design workshop. What it might have communicated to that audience is embodied, first of all, in the medium. The translation of the three dimensional city into a two dimensional diagram would still, at this date, have preserved an element of wonder. At the same time, the mathematically based image claimed an objectivity that removed its content from the realm of rhetoric. Spectacular and matter of fact, the survey had very special virtues as an advocate for the project. As it turns out, the drawing is not an entirely honest witness to the situation. In the first place, the drawing presents information about the availability of land. It illustrates an entire city block, washed by a unifying grey-green tint, implying a resource that was not, in fact, in the hands of either the Marchesa or the Pope. A contemporaneous plan for the same site gives what is probably a more accurate picture of the Marchesa's property. Inscribed with the date 1557 and identified as the "Pianta del palazzo di Paolo 4o", it illustrates the area of the houses at San Macuto that the Marchesa proposed to convert into the convent of Poor Clares (Fig. 14).[21] The property mapped there represents less than half of the site covered by the UA 4180 project. It was not until 1581 that the land between the houses of the Marchesa and the Piazza del arco di Camigliano (now the piazza of the Collegio Romano) was acquired from private owners, and then by Pope Gregory XIII for the Collegio Romano, the successor to the convent of Poor Clares as the recipient of the Marchesa's pious charity.[22]

UA 4180 also makes an assertion about the centrality of the site. The project is the central focus of the drawing, of course, but the identification of connections to the Pantheon, to San Marco, to the Minerva, and to Monte Cavallo suggest a similar relation to the city as a whole. Finally, and most importantly, the drawing claims that the convent and street were natural improvements to the site. The blank space

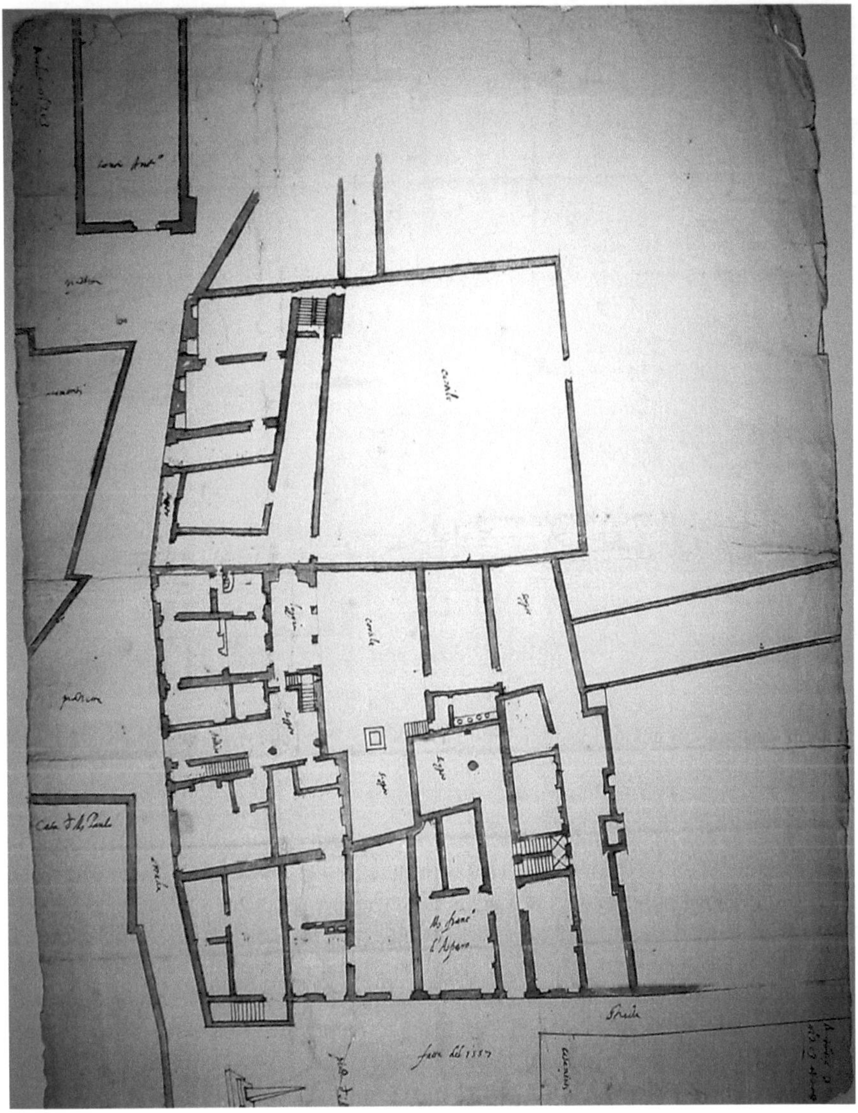

Fig. 14 "Pianta del palazzo di paolo 4o," partially showing the area covered in UA 4180, 1557 (Accademia di San Luca, Fondo Mascarino 2360, plan 1)

of the Piazza di San Macuto gets an architectural focus and the new street (thanks to the slight distortion of the survey) is shown as an organic continuation of the existing road from the Quirinal. For all its technical accomplishment, the plan's rhetorical sophistication also stands out.

The project described on UA 4180 had a mixed fortune. The convent of Poor Clares was very short lived. For the few years that the nuns were in residence at San

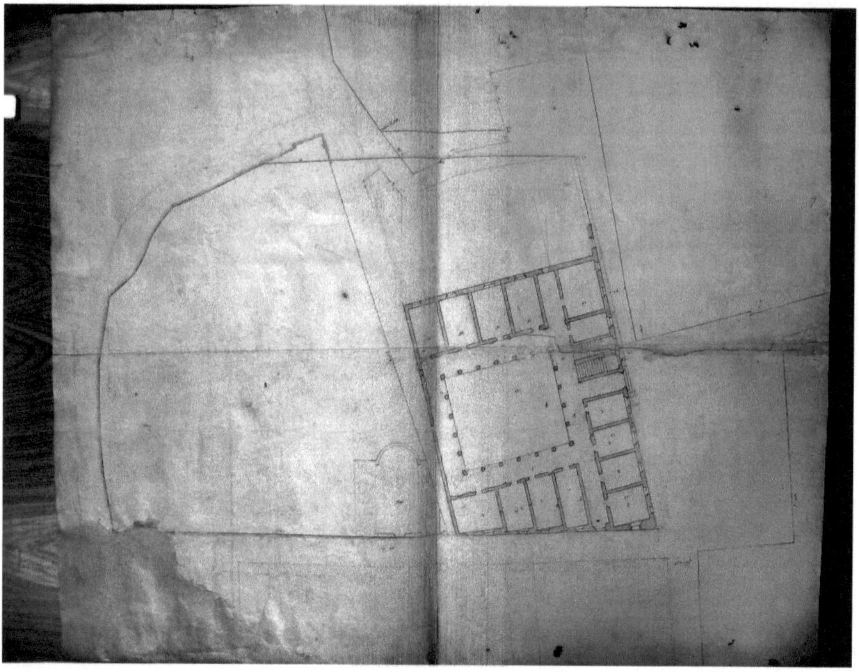

Fig. 15 Early project drawing for the Collegio Romano, Rome (Archivium Romanum Societatis Iesu, Armadio 5). The street at the bottom of the sheet (the west side of the site) is the Via di San Ignazio

Macuto, they occupied quarters adapted from earlier structures. No part of the convent project described in the Uffizi drawings was built. The Marchesa della Tolfa did lay the foundations for a new church—later inherited by the Jesuits—but these occupied a different and less grand position on the convent site than had been imagined in the drawing.[23] The street fared much better. Though not complete in all its parts, the section within the convent block was built as planned. It disappeared under the construction for the Collegio Romano in the late sixteenth century but the earliest drawings for that project include it as one of the boundaries of the site, and plan-views of the city from the 1570s (Mario Cartaro 1576, DuPérac-Lafréry 1577) illustrate it as part if the street system of the area (Fig. 15).[24]

The use of survey did not transform the kind of project that was proposed in UA 4180. The new street connects two terminal points in much the same way that the Via Giulia or the Via Lungara, Julius II's streets through the properties just east and west of the Tiber, had done at the beginning of the century. Plans of the generation of UA 4180 were, nevertheless, pioneers. They taught an audience of designers, building patrons, and civic administrators the value of survey both in design practice and as a tool of representation. Such plans allowed projects to be represented in much more concrete terms than had been possible when verbal description was the only way of communicating information about a design idea. If the imprecision of early survey limited its value for the definition of form, the

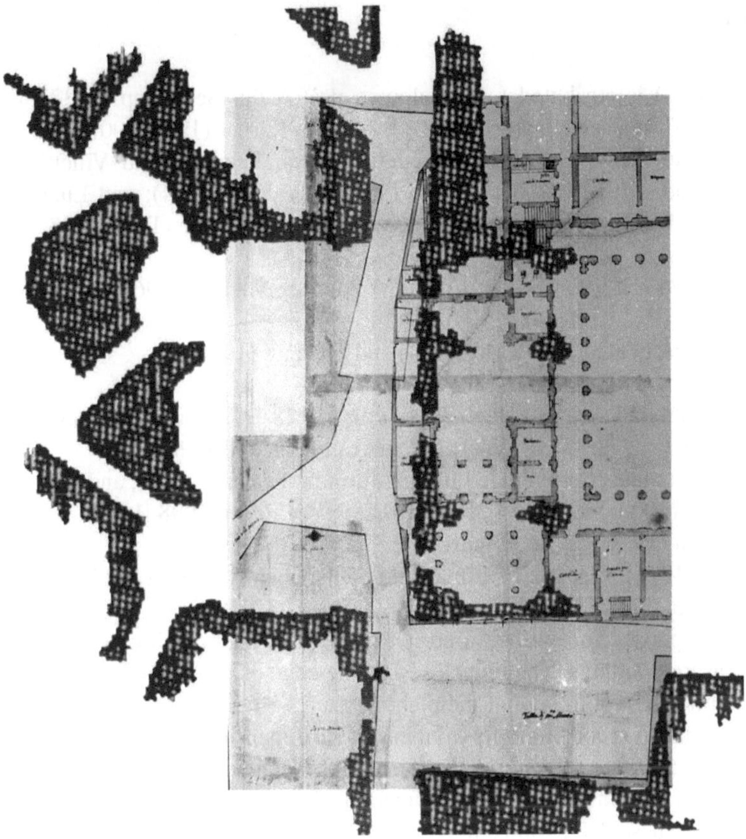

Fig. 16 UA 4180, Detail overlaid with Filippo Raguzzini's Piazza Sant'Ignazio (1727–1736) from Nolli 1748

uniquely abstract vantage on urban space that it offered ultimately opened up unimagined possibilities.

Urban design in the sixteenth century continued to be dominated by orthogonal spaces and straight streets, but a century later, in projects by Bernini, Borromini, and Pietro da Cortona, the topography revealed by survey and displayed in plan would inspire projects that transformed the accidents of the city's street system into coherent formal designs. There is no more spectacular example of this type of Roman urban design than the piazza built on the northwest corner of the Marchesa della Tolfa's site 170 years after the drafting of UA 4180. The circular spaces and triangular buildings of the Piazza Sant'Ignazio that Filippo Raguzzini built for the Jesuits of the Collegio Romano (1727–1736) demonstrate more dramatically than any other Roman square the formal control that drawing and survey introduced to urban design (Fig. 16).[25]

Notes

1. UA 4180 is attributed to Bartolomeo de'Rocchi. See Popp (1937). On the UA4180 site, see Rinaldo (1914); Bösel and Karner (1986–2007, vol. 1, 180–211, especially 180–182); Benedetti (1992); Palmerio and Villetti (1987); Villoslada (1954, 61–67, 93–98, 133–36); Valone (1994); and Lucas (1990, 164–165). On Paul IV more generally, see von Pastor (1901–1953, vol. 14, 56–434).
2. Alberti (1960–1973, vol. 3, 135–173). Also see Vagnetti (1968) and Stroffolino (1999).
3. See Bruschi et al. (1978, 459–484) and Thoenes (1986).
4. Günther (1984, 234–239).
5. Frommel and Adams (1994–2000, vol. 1, 128–30), where the drawings (UA 771r and v, 772r, 773r, 774r) are tentatively dated to 1526.
6. From a manuscript prepared in final form in the early 1630s, and published as Aleotti (2000, 539 (154r)). For the dating, see Rossi (1998, 164).
7. De Toni (1974, 137 and Figure 41).
8. Ehrle (1911). Also see Friedman and Schlapobersky (2005) and Maier (2007).
9. Tartaglia (1606 [1546], 129–131).
10. The upper left, or northeast, corner of the drawing contains two scales. Both measure 100 units. The one closest to the edge of the paper, corresponding to the first attempt to lay out the plan, measures 8.6 cm, the other 12.9 cm. This relationship is consistent for all measurable dimensions of the two plans. There is also a 100 unit scale at the base of the drawing that measures 12.9 cm. Another drawing in the Uffizi collection, UA1900, offers a reduced project for the Marchese della Tolfa site. The convent is smaller and the drawing presents only the property directly affected by the architectural project, with no urban context and no reference to the new street proposed in UA4180. That plan is drawn at the scale of 15.4 cm to 100 palmi, that is, 19 % larger than the plan of UA4180. It presents a simplified version of the site and includes some details not registered in UA 4180 (e.g. a door between the first and second projection on the western face of the property). There are enough discrepancies between that plan and the one registered on UA4180 to conclude that they were constructed separately. Both, however, were made on site.
11. The erasures modify the south face of the convent block, the face that defines one side of the present-day Piazza del Collegio Romano. They are visible at both the top and bottom of that face, a few centimeters to the left of the line that makes the final definition of this side of the block. It is probable that the corner of the block on the western street, the present day via di Sant' Ignazio, had been established early in the projection of the plan. This was an important point in the project and one whose relationship to the area of the convent, and especially to the door of the church, was relatively easy to measure. The faces of the south side of the block, generated in a series that began in the east at the end of a long compass traverse, apparently landed on a point too far to the north and had to be corrected.

12. The drawings for architectural projects of the period had a similar relationship to execution. Those that survive are mostly detail sketches and the plans that we know are often quite different from what was built. Nothing like a full set of working drawings ever existed. See Ackerman (1954). Nor were the architectural projects of Phillip II of Spain, spread across widely scattered sites and centrally controlled by the king himself, recorded in more than a bare minimum of drawings. For the lists of drawings sent to building sites, see Wilkinson-Zerner (1993, 46–62, esp. 58–59). In these cases, as in Renaissance Rome, we must assume that verbal instruction supplemented the drawings.
13. See note 4, above.
14. UA 843 r, dated after 1539. See Frommel and Adams (1994–2000, vol. 1, 151–152). The drawing is inscribed on the verso: "Pratica di me[sser] luca di maximo".
15. Tartaglia (1606 [1546], 69r). For Tartaglia, design, not materials, is the essence of military architecture and design is produced and shared through drawing: "Lo ingegno del huomo, nel fortificare una citta (secondo mio parere) si conosce per la forma, e non per la materia." He tells his interlocutor that he could improve the defenses of Turin in six different ways. To explain them, he states, "a me saria necessario (a volere a sofficientia ben dechiarire, e con ragione dimostrare di cadauno di quelle particularmente sua valuta) a designare varie e diverse piante" (70v). One of the qualities of a good defense is to ensure that the enemy is never closer to the curtain he intends to attack than from a bastion from which he can be attacked. This is a geometric issue and Tartaglia promises to show his interlocutor how it is done with a drawing: "faro figuralmente vedere" (71r). His interlocutor says he will show that he has understood the lesson by making "una pianta designata de mia mano" (71v).
16. The war offices of Italian states collected towns plans of both friend and foe. Concerned exclusively with fortifications, they generally leave the area inside the walls blank. See van den Heuvel (1991, 53–61), citing Biblioteca Nazionale Turin, Ms. q. II. 57 (old signature Serie Atlas C N 5 [Bc. Atl Sala XV]). Also see Lamberini (1988) and Warmoes et al. (2003).
17. See Battaglini (1978, 89–91) and Fara (1997, 3–24). In a letter of 19 June 1549, Cosimo's architect Giovanni Camerini speaks of two plans for the layout of the city's streets and defenses that he has sent to the duke. Fara identifies a drawing of 1553 as reflecting one of those plans (Fara (1997, Figure 33)). It is substantially the same as the plan that Vasari places in Cosimo's hand in the Palazzo Vecchio painting.
18. "…potrà vedre le proportioni e le corrispondetie che hanno tutte le strade et borghi fra loro et ciascuna a tutto il corpo di quella et volendola ridurre al suo vero decoro, chiaramente conoscerà i luoghi che rimuovere si dovrebbono per ridurrla a perfetione…" Adorni (1980, 34). Also see, Uluhogian (1983).
19. The plan measures 372 mm by 428. It is preserved in Raccolta mappe e disegni, volume 48, plan 76 (previous catalogued as volume 70), Archivio di Stato di Parma. Also see Soldini (1992–3) and Storchi (1999).

20. This represented a considerable improvement over the uncertainty about boundaries that reigned in the era of purely verbal description. When land was cleared for the Piazza Maggiore in Bologna at the beginning of the thirteenth century, not all of the properties within the perimeter of the project were identified before demolition began. A second set of purchases was necessary to gain control of the site and to regularize its perimeter. See Bocchi (1995–1998, vol. 2, 11–16, esp. 16). For late medieval urban design systems see Friedman (2009).
21. Fondo Mascarino 2363, plan 1, Accademia nazionale di San Luca, Archivio storico. The drawing measures 58 by 44 cm. It carries the inscription, presumably of a later date: "Casa del Colegio de Gesuiti alla Guglia di S. Mauto". See Marconi et al. ([1974], 16).
22. The land was expropriated by Gregory with a *Motu Proprio* of 13 July 1581. Cerchiai (2003, 63).
23. Bösel and Karner (1986–2007, vol. 1, 181).
24. The Collegio Romano drawings are held in Armadio 5, Archivium Romanum Societatis Iesu, Rome. They have been published in Lucas (1990, 164–65, cat. no. 96). For the Rome plans, see Frutaz (1962), Plate 244 for the relevant detail of the Cartaro plan, Plate 250 for the detail of the DuPérac-Lafréry plan.
25. The surviving eighteenth century plan of the piazza was produced in 1731 for the *Maestri di Strade*, the agency in charge of the physical city, as one of a series of plans of public spaces. Disegni e piante, c. 80, no. 240, Archivio di Stato di Roma It is illustrated in Habel (1981, Figure 16). Also see Connors (1989, 279–294).

Photographic Credits

Uffizi, Gabinetto dei Disegni e Stampe (by permission of the Ministero dei beni e delle attività culturali e del turismo): Figs. 1, 3–4, 6–11, 16
British Library: Fig. 5
Musei Civici Fiorentini (Ministero dei beni e delle attività culturali e del turismo): Fig.12
Archivio di Stato di Parma: Fig. 13
Accademia di San Luca: Fig. 14
Archivium Romanum Societatis Iesu, Rome: Fig. 15

References

Ackerman, James. 1954. Architectural practice in the Italian Renaissance. *Journal of the Society of Architectural Historians* 13(3): 3–41.

Adorni, Bruno. 1980. Iconographia della città di Parma (1589–1592). In *'Io Smeralda Smeraldi ingegnero et perito della congragatione dei cavi del Parmigianino.' Territorio, città, offizio nel Ducato di Parma 1582–1634*, ed. Carlo Quintelli and Luca Monica. Parma: Grafiche STEP.

Alberti, Leon Battista. 1960–1973. *Opere Volgari*, ed. Cecil Grayson. Bari: Laterza.

Aleotti, Giovanni Battista. 2000. *Della scienza et dell'arte del ben regolare le acque*. Modena: Panini.

Bartoli, Cosimo. 1564. *Del modo di misurare le distantie, le superficie, i corpi, le piante, le prouincie, le prospettiue, & tutte le altre cose terrene, che possono occorrere a gli huomini, secondo le vere regole d'Euclide, & de gli altri più lodati scrittori*. Venice: F. Franceschi.

Battaglini, G.M. 1978. *Cosmopolis: Portoferraio medicea, storia urbana 1548–1737*. Rome: Multigrafica.

Benedetti, Sandro. 1992. La prima architettura gesuita a Roma: Note sulle chiese dell'Annunziata e sul Collegio Romano. In *L'architettura della Compagnia di Gesù in Italia, XVI-XVIII secolo: atti del convegno, Milano, 24–27 ottobre 1990*, ed. L. Patetta and S. Della Torre, 57–68. Genova: Marietti.

Bocchi, Francesca (ed.). 1995–1998. *Bologna*. Bologna: Grafis.

Bösel, Richard, and Herbert Karner. 1986–2007. *Jesuitenarchitektur in Italien 1540–1773*. Vienna: Österreichische Akademie der Wissenschaft.

Bruschi, Arnaldo, et al. (eds.). 1978. *Scritti rinascimentali di architettura*. Milan: Il Polifilo.

Cerchiai, Claudia (ed.). 2003. *Il Collegio Romano dalle origini al Ministero per i beni e le attività culturali*. Rome: Istituto Poligrafico e Zecca dello Stato, Libreria dello Stato.

Connors, Joseph. 1989. Alliance and enmity in Roman Baroque urbanism. *Römisches Jahrbuch der Bibliotheca Hertziana* 25: 207–294.

De Toni, Nando. 1974. I rilievi cartografici di Leonardo per Cesena ed Urbino contenuti nel manoscritto "L" dell'Istituto di Francia. In *Leonardo da Vinci: Letture vinciane I–XII (1960–1972)*, ed. Paolo Galluzzi, 133–148. Florence: Giunti-Barbèra.

Ehrle, Francesco. 1911. *Roma al tempo di Giulio III: La pianta di Roma di Leonardo Bufalini del 1551*. Rome: Danesi.

Fara, Amelio. 1997. *Portoferraio: architettura e urbanistica 1548–1877*. Turin: Edizione della Fondazione Giovanni Agnelli.

Friedman, David. 2009. Urban design without drawing. In *Arnolfo's Moment: Acts of an international conference*, Florence, Villa I Tatti, May 26–27, 2005, ed. David Friedman, Julian Gardner, and Margaret Haines, 125–146. Florence: Leo S. Olschki.

Friedman, David, and Paul Schlapobersky. 2005. Leonardo Bufallini's orthogonal *Roma* (1551). *Thresholds* 28: 10–16.

Frommel, Christoph L., and Nicholas Adams (eds.). 1994–2000. *The architectural drawings of Antonio da Sangallo the Younger and his circle*. New York/Cambridge, MA/London: Architectural History Foundation/MIT Press.

Frutaz, Amato Pietro (ed.). 1962. *Le piante di Roma*. Rome: Istituto di studi romani.

Günther, Hubertus. 1984. Das Trivium vor Ponte S. Angelo: Ein Beitrag zur Urbanistik der Renaissance in Rom. *Römisches Jahrbuch für Kunstgeschichte* 21: 164–251.

Habel, Dorothy. 1981. Piazza S. Ignazio, Rome in the 17th and 18th centuries. *Architectura* 11(1): 31–65.

Lamberini, Daniela. 1988. Funzione di disegni e rilievi delle fortificazione nel Cinquecento. In *L'architettura militare veneta del Cinquecento*, ed. Sergio Polano, 48–61. Milan: Electa.

Lucas, Thomas (ed.). 1990. *Saint, site, and sacred strategy: Ignatius, Rome and Jesuit urbanism: Catalogue of the exhibition: Biblioteca Apostolica Vaticana*. Città del Vaticano: Biblioteca apostolica vaticana.

Maier, Jessica. 2007. Mapping past and present: Leonardo Bufalini's plan of Rome (1551). *Imago Mundi* 59(1): 1–23.

Marconi, Paolo, Angela Cipriani, and Enrico Valeriani (eds.). 1974. *I disegni di architettura dell'Archivio storico dell'Accademia di San Luca*. Roma: De Luca.

Palmerio, Giancarlo, and Gabriella Villetti. 1987. L'isolato del convento di S. Maria sopra Minerva in Roma alla fine del Cinquecento. *Architettura storia e documenti* 1–2: 5–33.

Popp, A.E. 1937. Unbeachtete projekte Michelangelos. *Münchener Jahrbuch der Bildenden Kunst*: 12: 439–443.

Rinaldo, Ernesto S.I. 1914. La signora Vittoria della Tolfa marchesa della Valle e la donazione fatta al Collegio Romano. In *La fondazione del Collegio Romano. Memorie storiche*, 61–67. Arezzo: Cooperativa Tipografica.

Rossi, Massimo. 1998. La cartografia aleottiana. In *Giambattista Aleotti e gli ingegneri del rinascimento*, ed. Alessandra Fiocca, 161–187. Florence: Leo S. Olschki.
Soldini, N. 1992–3. La costruzione di Guastalla. *Annali di architettura* 4–5: 57–87.
Storchi, Stefano. 1999. *Guastalla: La costruzione di una città*. Reggio Emilia: Diabasis.
Stroffolino, Daniela. 1999. *La città misurata. tecniche e strumenti di rilevamento nei trattati a stampa del Cinquecento*. Rome: Salerno editrice.
Tartaglia, Nicolò. 1606 [1546]. *Quesiti et inventioni diverse*. Venice: segno del Lione.
Thoenes, Christof. 1986. La 'lettera' a Leone X. In *Raffaello a Roma. Il convegno del 1983*, 373–381. Rome: Edizione del Elefante.
Uluhogian, F. Miani. 1983. *Le immagini di una città: Parma (secoli XV-XIX). Dalla figurazione simbolica alla rappresentazione topografico*. Parma: La Nazionale.
Vagnetti, Luigi. 1968. La *Descriptio urbis romae* di Leon Battista Alberti. *Quaderno dell'istituto di elementi di relievo dei monumenti di Genova* 1: 25–78.
Valone, Caroline. 1994. Piety and patronage: Women and the early Jesuits. In *Creative women in medieval and early modern Italy*, ed. Ann Matter and John Coakley, 157–184. Philadelphia: University of Pennsylvania Press.
van den Heuvel, Charles. 1991. *'Papiere Bolwercken': De introductie van de Itliaanse stede- en vestingbouw in de Nederlanden (1540–1609) en het gebruik van tekeningen*. Alphen aan den Rijn: Canaletto.
Villoslada, R. Garcia. 1954. *Storia del Collegio Romano dal suo inizio (1551) alla soppressione della Compagnia di Gesù (1773)*. Rome: Apud aedes Universitatis Gregorianae.
von Pastor, Ludwig. 1901–1953. *The history of the Popes: From the close of the middle ages*. London: Kegan Paul, Trench, Trübner.
Warmoes, Isabelle, Emilie d'Orgeix, and Charles van den Heuvel (eds.). 2003. *Atlas militaires manuscrits européens XVIe-XVIIIe siècles: Forme, contenu, contexte de réalisation et vocations: Actes des 4es journées d'étude du Musée des plans-reliefs, Paris, Hôtel de Croisilles, 18–19 avril 2002*. Paris: Musée des Plans-Reliefs.
Wilkinson-Zerner, Catherine. 1993. *Juan de Herrera: Architect to Phillip II of Spain*. New Haven: Yale University Press.

Part III
The Baroque Institutional Context

Our next two essays look at the new religious orders of the Catholic Reformation and the institutional settings they established for architectural teaching and practice. As Susan Klaiber shows, orders such as the Jesuits, Theatines, and Barnabites founded a durable, twofold rationale for linking architecture and mathematics. This bond was, in the first place, pedagogical. Beginning with the Jesuits' influential *Ratio Studiorum*, the orders routinely offered special instruction in geometry within the Aristotelian framework of mixed mathematics. Architecture—like surveying, navigation, hydraulic engineering, and fortification—was suited to the active, "apostolic" spirituality of Catholic reform, which valued utility and worldly experience in the training of novices and missionaries. The orders also promoted this link on an administrative level, by recruiting their own members—often mathematicians—to official posts as building supervisors, responsible for the construction of new churches and mission houses throughout Europe. This twin program of mathematical education and new building provided a firm disciplinary foundation for the art, while also giving rise to a new class of building practitioners. Unlike their secular counterparts, the orders' architects were equally trained in both the classroom and at the construction site. As Klaiber demonstrates, this intellectual and institutional entwining helps to explain the novel approaches to structural design, iconography, and architectural theory that we see in the work of Guarino Guarini and many analogous figures.

This Baroque intellectual culture had a truly global reach, spreading as far and as widely as the orders themselves. Kirsti Andersen's article takes an in-depth look at one Jesuit scholar-practitioner. Andrea Pozzo was not an architect, but his chosen craft was inextricably tied to building. *Quadratura*—the art of illusionistic ceiling painting—drew upon real spaces to produce fanciful and dizzying visual allegories. A prolific artist, Pozzo was also one of the most influential in his field. Andersen considers his renowned *Perspectiva pictorum et architectorum* (1693–1700), successor to a long line of similar *trattati*. This hybrid book reflects Pozzo's own background, shaped both by Jesuit mathematical pedagogy and extensive practical experience. Although conceived as a basic instructional text and as a showcase for the author's own paintings, the book is nonetheless written in a forbidding Latin and its content expressed in complicated geometrical constructions. It is clear that Pozzo saw his own artistic virtuosity as inherently geometrical. In this sense, he followed in the footsteps of many earlier writers. While the *Perspectiva* recalls similar instructional texts by painters and architects such as Alberti, Piero della Francesca, and Vignola, it also revels in a formal and visual complexity that would have appealed to mathematicians like Guidobaldo del Monte and Girard Desargues.

Architecture and Mathematics in Early Modern Religious Orders

Susan Klaiber

For most of us, a familiar image from Raphael's *School of Athens* serves to illustrate our intuitive notions about the links between early modern architecture and mathematics. The artist's portrait of the great Renaissance architect Bramante as the geometer Euclid recalls the medieval traditions of Gothic architects and master masons using geometry (Fig. 1). Moreover, the inclusion of Zoroaster and Ptolemy—identified by celestial and terrestrial globes—in the group huddled around Euclid/Bramante further seems to associate geometry and architecture with astronomy, vaguely echoing the medieval quadrivium of arithmetic, music, geometry, and astronomy. In short, the architect as mathematician (or mathematician as architect) operating within a larger group of quantifiable crafts and sciences seems obvious, and not particular to the early modern world. Yet a closer look at a well-defined culture which produced such individuals illuminates much about the period's understanding of both architecture and mathematics.

The religious orders traditionally associated with the Counter Reformation, such as the Jesuits, Theatines, and Barnabites, provide rich material for investigating the relationship between architecture and mathematics, and they nurtured a specific type of priest-architect. In 1595, the architect Vincenzo Scamozzi remarked about the Venetian Theatines: "these fathers... are both good mathematicians and they understand architecture...," implicitly linking the priests' ability in these two fields.[1] This paper considers why this relationship flourished in the Italian Seicento and how the intellectual culture of these orders promoted architectural activity.

S. Klaiber (✉)
Independent Scholar, Rychenbergstrasse 271, 8404 Winterthur, Switzerland
e-mail: sklaiber@bluewin.ch

A. Gerbino (ed.), *Geometrical Objects: Architecture and the Mathematical Sciences 1400-1800*, Archimedes 38, DOI 10.1007/978-3-319-05998-3_6,

Fig. 1 Raphael, *School of Athens*, Vatican, Stanza della Segnatura, c. 1510–1511, detail with Euclid/Bramante

The Orders

Two major factors account for the centrality of this phenomenon in the seventeenth century. First, the Seicento witnessed a dramatic expansion of the counter-reformational orders all across Europe, but particularly in Italy. This expansion created a pressing need for new churches and for designers or construction supervisors.[2] Second, as the century progressed, the gradual organization (and reorganization) of knowledge in the wake of the new sciences influenced these orders as they consolidated their educational programs for their members. In these curricula, architecture became systematized as a discipline related to applied mathematics. The orders maintained ambivalent—not entirely hostile—relationships to the new sciences, which to a great extent had their center in Italy with Galileo.[3] While the relationship between mathematics and architecture also developed in the secular world, in other religious orders, and outside of Italy—figures like Christopher Wren, Ignazio Danti, or the Belgian Jesuit François d'Aiguillon spring to mind—the new orders' institutional qualities as significant centers of mathematical learning and architectural patronage made them crucibles for the development of a mathematical approach to architecture, especially in Italy.[4]

How were the Jesuits, Theatines, and Barnabites trained in mathematics and the sciences? The Jesuits' *Ratio studiorum* (first edition 1586) outlined the course of study pursued by the order's aspiring priests.[5] The Theatines and Barnabites at first summarized their curricula in their constitutions; the Barnabites followed with a document similar to the *Ratio studiorum* in 1666.[6] In all three orders, the basic seven-year curriculum covered three years of philosophy and four (or sometimes five) of theology.[7] For the Jesuits and Theatines, the first year of philosophy dealt with logic; the second year with natural philosophy (via Aristotle's *Physics* and *De caelo*) as well as Euclid's *Elements*; the third year with metaphysics. The four years of theology were chiefly based on Aquinas.

Beyond the *Elements*, which comprised the basic mathematical education for most future priests, aspects of astronomy, perspective, optics, music, and mechanics were also treated in what was essentially an updated version of the medieval quadrivium.[8] For talented students, however, the Jesuits offered supplemental tutoring, exploring these topics in depth as well as other "mixed" or applied mathematics topics, such as surveying, navigation, instrument making, hydraulic engineering, and both civil and military architecture.[9] The term refers to the "mix" of abstract mathematical concepts with quantifiable, empirical properties of the real world. Christoph Clavius, professor of mathematics at the Collegio Romano, promoted this concept from the very beginning of his tenure as an integral part of the Jesuit curriculum.[10]

The second edition of the *Ratio studiorum* (1591) prescribed a kind of occasional mathematical forum to be held at the Collegio Romano that would present material related to the private lessons of individual students. Michael John Gorman has recently studied the *problemata* presented in this forum under the guidance (and disguised authorship) of Christoph Grienberger, professor of mathematics at the Collegio Romano off and on between 1595 and 1633.[11] Among the thirteen problems published and analyzed by Gorman, two treat broadly architectural topics: one on geometry and architectural design with specific reference to the Jesuits' Collegium nobilium in Bologna (1588–1601), the other speculating on the dimensions of Egyptian pyramids (based on descriptions of the Seven Wonders of the Ancient World) in the manner of Juan Bautista Villalpando. In the first, Grienberger also comments on a structural curiosity like the leaning Torre degli Asinelli in Bologna: "Without doubt that Bolognese structure had an outstanding mathematician as its architect by whose vigilance Geometry has come to inhabit that tower."[12] These *problemata* seem to confirm the place of architecture in the advanced mathematical tutoring offered at the Collegio, an innovation in architectural education that produced a new type of architect, emerging not from a background in crafts or the building trades, but rather from a scholarly approach supplemented by practical experience at the order's construction sites.

The Jesuit Antonio Possevino's *Bibliotheca Selecta* (first edition Rome, 1593) marks the first written record of architecture's place within the order's mathematical world. Possevino's monumental study, closely related to the discussions surrounding the early versions of the *Ratio studiorum*, offers summaries of virtually all fields of knowledge, with suggestions for further reading.[13] Book XV deals with mathematics, and within this mathematical section Possevino devoted three chapters to architecture, declaring at the outset that architecture is discussed immediately after the chief mathematical disciplines, since it depends on these, and in fact is perfected through their leadership.[14] In his encyclopedic ambition to present the knowledge necessary for Jesuits, Possevino concentrates on critical discussions of the architectural texts of antiquity, including both Vitruvius and the Bible, with its description of the Temple of Solomon. Possevino supplements these sources with references to Alberti, Palladio, and Barbaro's commentary on Vitruvius. The relevant booklist also cites the perspective works of Barbaro, Dürer, and Ignazio Danti. These treatises form the basic literature for early Jesuit architects.

Early inventories of Theatine and Barnabite libraries dating to around 1600—just a few years after Possevino—indicate that architectural and perspective treatises also played a role in the intellectual life of other counter-reformational orders. The larger Theatine libraries, such as Sant'Andrea della Valle and San Silvestro al Quirinale, both in Rome, San Paolo Maggiore in Naples, Santa Maria della Ghiara in Verona, or San Siro in Genoa, each possessed several of the standard architectural books in various editions, including those of Palladio, Serlio, Vitruvius (including Barbaro's edition), Alberti, and Labacco. The same works are listed in the Barnabites' libraries at San Paolo at Piazza Colonna in Rome, Santa Maria della Corona in Pavia, and San Paolo in Casale Monferrato. Other books, such as Alberti's *Della pittura*, Martino Bassi's *Dispareri in materia di architettura et perspettiva* (Brescia, 1572), Dürer's *Unterweysung der Messung* (Nuremberg, 1525), and Vignola's *Regola delli cinque ordini d'architettura* (Rome, 1562) also appear in single libraries of both orders.[15]

The Jesuits ultimately institutionalized the link between mathematics and architecture. From the early Seicento, a centralized architectural policy generally cast the professor of mathematics at the Collegio Romano as the order's *consiliarus aedificiorum*. The *consiliarus* reviewed designs for all Jesuit churches and houses across Europe before presenting them to the order's general for final approval. Beginning in 1613, the general required submission of plans in duplicate, so that one copy could be kept in the order's central archive in Rome; this is the source of the Jesuit plan collection now at the Bibliothèque Nationale, Paris. The *consiliarus* usually edited plans for structural flaws and budgetary extravagance rather than aesthetic concerns. As Vallery-Radot points out, editing the designs for style was uncommon since the plans sent to Rome for approval were rarely accompanied by sections and elevations. Typical reasons given for the rejection of a design were "inconvenience", "crowding", and "errors". Suggestions for revisions included adding windows or replacing a vault with a lighter, less expensive wooden ceiling.[16] The Barnabites, though not the Theatines, also had a similar office, the *prefetto delle fabriche*.[17]

As Steven Harris has shown, the Jesuits' involvement with the sciences must be seen in a specific theological, vocational, and ideological context. Rejecting the apparent contradiction between these pursuits, Harris links Jesuit scientific work with the order's ideology of "apostolic spirituality," an active engagement in "worldly labor, performed in service to their fellow-men and for the honor and greater glory of God."[18] Harris classifies six major categories of Jesuit scientific publications: (1) Aristotle's natural philosophical books; (2) Euclidean geometry and "mixed" mathematics; (3) astronomy; (4) experimental and natural philosophy; (5) natural history; and (6) medical and pharmaceutical topics.[19] Noting a combination of classical Aristotelianism and a new empiricism in these works, Harris suggests the empirical aspects stemmed from the order's active experience in the world. Although such empiricism characterized much seventeenth-century science, Harris argues that the order enthusiastically adopted only those forms of scientific endeavor that also proved useful in the Jesuits' three major "apostolates" or spheres of activity: education, European courts, and foreign missions. He concludes that for the Jesuits a "...supraconfessional doctrine of the sanctity of mundane labor, in conjunction with a high esteem for learning and reason, provided fertile ground for the acceptance and development of active-empirical forms of early modern science."[20] The consequence of this was that "...those forms of scientific activity that Catholic princes found either useful (e.g. navigation, surveying, hydraulics, military architecture) or entertaining (the 'virtuoso' sciences, curiosity cabinets, and telescopic astronomy) became part of the Jesuit scientific repertoire."[21] Of course, civil architecture also counted among the activities useful for Catholic princes, so it is no surprise to find it within the purview of Jesuit science.

The Architects

Who were these early modern priest-mathematician-architects? We now survey six careers, exemplary for the range of activities in which these men operated. The phenomenon, however, includes dozens of other figures, a few of whom are briefly mentioned in notes below.[22] Their work covers the spectrum of higher, "mixed", and practical mathematics in which architecture was embedded. Some priests excelled more in one area than others—perhaps geometry, astronomy, perspective, or indeed architecture—while others worked across the entire spectrum. Well-known figures such as Orazio Grassi or Guarino Guarini have entered history as notable astronomers or architects, with only cursory mention of their other work in the broader world of seventeenth-century mathematics. Contemporaries, however, seem to have perceived all their efforts as various parts of a single discipline.[23]

The first generation of Jesuit architects—Giuseppe Valeriano (1542–1596), Giovanni Tristano (c. 1505–1575), and Giovanni De Rosis (1538–1610)—received architectural training and experience before entering the order.[24] Although Valeriano assisted Possevino in composing the architectural chapters of the *Bibliotheca Selecta,* there is little further evidence to suggest that these men viewed architecture as a primarily intellectual or mathematical undertaking. The first accomplished Theatine architect, Francesco Grimaldi (1543–1613), also came from a background in the crafts.[25] It seems the immediate needs of the orders for churches in these early years outweighed any impulses to enhance the status of architecture. Yet the groundwork laid by the first generation of Jesuit architects was instrumental. Actively running a number of construction sites, they provided the hands-on training required to turn academically-trained priests into practicing architects. Although the tradition of priest- or lay brother-architects from a crafts background continued through the seventeenth and on into the eighteenth century, it gradually waned as the numbers of scholar-architects increased. The systematization of architecture as a mathematical discipline became consolidated throughout the course of the Seicento.

Orazio Grassi (1583–1654) illustrates virtually all aspects of a mathematical-architectural career within the Jesuit order.[26] He was an advanced student who devoted an entire supplemental year to mathematics during his studies at the Collegio Romano (1605–1606). In 1612, he led an unsuccessful attempt to found a Jesuit architectural academy there, perhaps even offering courses under its auspices for a short time. He was professor of mathematics at the Collegio Romano from 1616 to 1624 and 1626 to 1628, serving thus for 10 years as *consiliarus aedificiorum*. During the latter period he designed the Jesuits' Sant'Ignazio (Fig. 2) in Rome and an adjacent wing of the Collegio Romano. His other major church for the order was San Vigilio in Siena. His manuscript Vitruvius commentary survives in Milan, and an album in Rome of his (mostly autograph) drawings mixes architectural and perspectival studies with further scientific material in cartography, instruments, and natural history. Grassi worked in other areas of mathematics as well, notably astronomy. Writing under the pseudonym Lotario Sarsi, he gained particular notoriety as Galileo's adversary in their dispute over the nature of comets, culminating in the latter's publication of *Il Saggiatore* (Rome, 1623).[27]

Another such figure, of a generation or two later, was Francesco Eschinardi (1623–1703), a professor of mathematics at the Collegio Romano and thus the *consiliarus aedificiorum* during his tenure. Historians of science today emphasize the importance of Eschinardi's work in optics or in the development of the thermometer. Yet he published two architectural treatises: *Architettura civile* (Terni, 1675) and *Architettura militare* (Rome, 1684), both under the pseudonym Costanzo Amichevoli (Fig. 3). He also produced, under his own name, a learned commentary on Giovanni Battista Cingolani's map of the Roman countryside, *Topografia geometrica dell'agro romano* (Rome, 1692), which included a concise guide to the city.[28]

In contrast to Grassi, Eschinardi seems to have had little impact on actual building within the Jesuit order—as *consiliarus aedificiorum*, he is known only for his work overseeing the early planning stages of the Jesuit church in Vercelli—but his involvement in the art took place in the same context of contemporary mathematical science. Eschinardi's architectural treatises were apparently written in connection with the Roman Accademia Fisico-Matematica, sponsored by Monsignor Giovanni Giustino Ciampini in his palace near Piazza Navona; Costanzo Amichevoli was Eschinardi's academic pseudonym.[29] The scope of the academy's activities was defined as philosophical, medical, mathematical, and mechanical, all subsumed under the heading "Fisico-Matematica". Here the category "mechanical" included the "mixed mathematical" disciplines such as optics, horology, civil and military architecture, and the use of perspective in painting, sculpture, and theater. Eschinardi's treatises are unremarkable in content—his treatment of the orders relies chiefly on Vignola—but the context of their origins sheds further light on architecture's systematization within the intellectual culture of the Jesuit order.

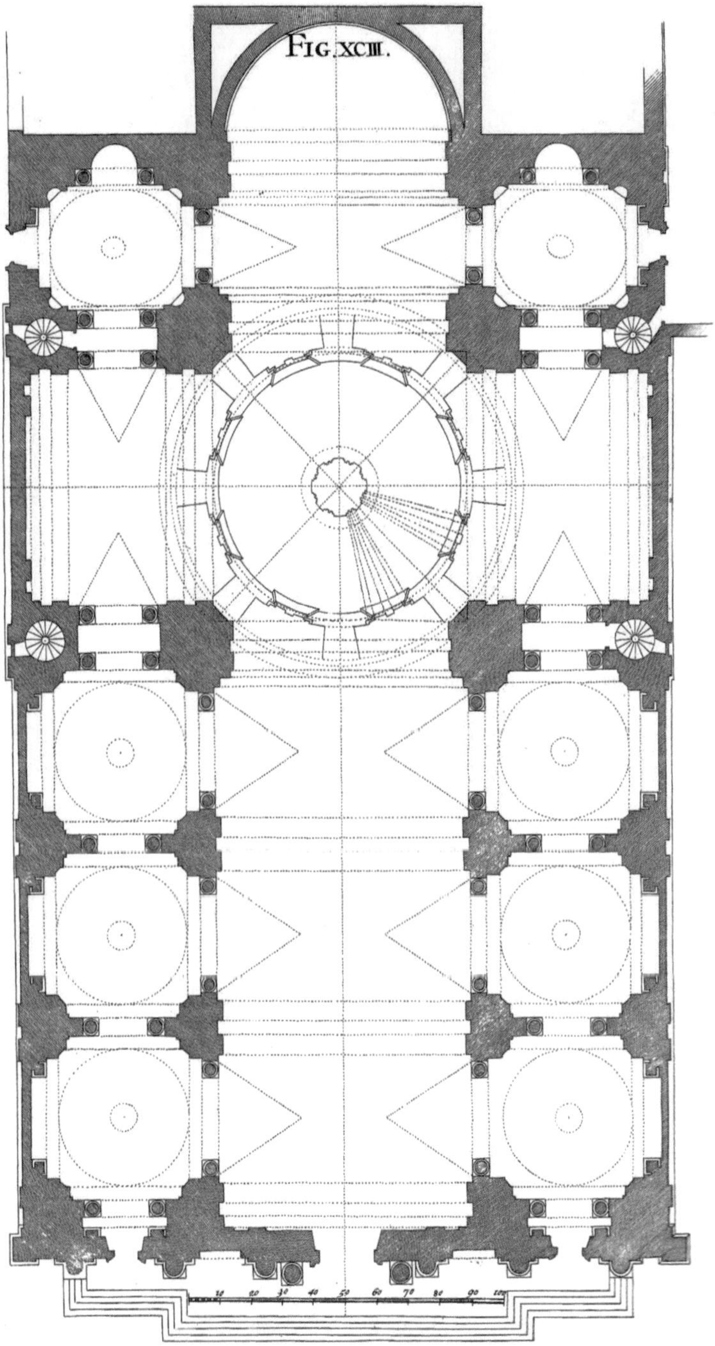

Fig. 2 Orazio Grassi, Sant'Ignazio, Rome, 1626–1628, plan (From Pozzo (1707, repr. 1989, 200))

Fig. 3 Title page, Eschinardi [Costanzo Amichevoli] (1675)

Perhaps the most famous Jesuit architect of the Seicento was Andrea Pozzo (1642–1709), who came to architecture via another mathematical discipline, perspective painting.[30] His built works consist mainly of altars and other church furnishings, though he also designed or remodelled several provincial churches for the

Jesuits such as the Gesù in Montepulciano or Sant'Ignazio in Dubrovnik. Having risen from a background as a craftsman and as a lay brother, Pozzo lacked a priest's scholarly education. Yet his theoretical aspirations demonstrate how much the academic culture of mathematics had permeated the Jesuit order by the late seventeenth century. His two-volume treatise *Perspectiva pictorum et architectorum*, (Rome, 1693–1700) was published in a parallel Latin-Italian text, presumably aimed at a dual audience of erudite scholars and vernacular practitioners. The treatise chiefly addresses the problems of painting fictive architectural settings in perspective. Although not a work of architectural theory *per se*, Pozzo repeatedly emphasizes the similarities and connections between perspectival painting and architectural design. Both, for instance, utilize the same drawing skills: "The Geometrical Plan… is no less necessary for painting a Design in Perspective, than it is for raising a Structure with Solid Materials."[31] The accompanying plates demonstrate how such plans are transformed geometrically into perspectival constructions and how Pozzo relied on geometry for both perspective painting and architectural designs (Fig. 4).

Among the Barnabites, the two most prominent priest-architects were Lorenzo Binago (1554–1629) and Giovanni Ambrogio Mazenta (1565–1635). Binago served for many years as *prefetto delle fabriche*, but Mazenta more clearly reflects the intellectual culture of the counter-reformational orders. The patrician Mazenta joined the Barnabites at the relatively late age of 25, ultimately rising to become the order's father general from 1612 to 1618. He and his two brothers Guido and Alessandro all demonstrated an early interest in architecture, and Giovanni Ambrogio was apparently encouraged by his early patron Federico Borromeo and the latter's Accademia degli Accurati, devoted to the "exact sciences and architecture." This academic experience thus recalls the case of Eschinardi, but unlike the Jesuit, Mazenta was a prolific designer and builder. His works stand chiefly in Bologna. The innovative designs for the Barnabites' San Paolo, San Salvatore (Fig. 5), and the nave of the cathedral of San Pietro confirm his standing as one of the leading priest-architects of the century.

The erudite and wide-ranging Mazenta studied Leonardo manuscripts, corresponded with Cassiano dal Pozzo, and directed Barnabite colleges in Pisa and Bologna. He also wrote *pareri* on the restoration of the Pantheon and the Lateran basilica. Mazenta cannot be considered a full-blooded mathematician as Grassi or Guarini. None of his scholarly works deal specifically with mathematics and architecture, nor does he seem to have taught mathematics within the order. Yet when we look carefully at his entire career, we see telltale signs of an architect trained in mixed mathematics. He designed and supervised construction of bridges, barracks, fortifications, as well as the harbor mole at Livorno for Grandduke Ferdinando I Medici in 1600–1602. He was an expert consultant during a dispute between Bologna and Ferrara regarding a tributary of the Po. Hydraulics also played a role in his unexecuted designs for enormous columns crowned with crosses to be placed at major intersections in Milan. These crossroad monuments would have continued Carlo Borromeo's post-Tridentine project of punctuating Milan's urban fabric with crosses erected throughout the city, while also serving practical functions as reservoirs and clepsydrae.[32]

The Theatine Guarino Guarini (1624–1683) emerged as the greatest architectural talent in this group of seicento mathematical priests, and he is best known today for

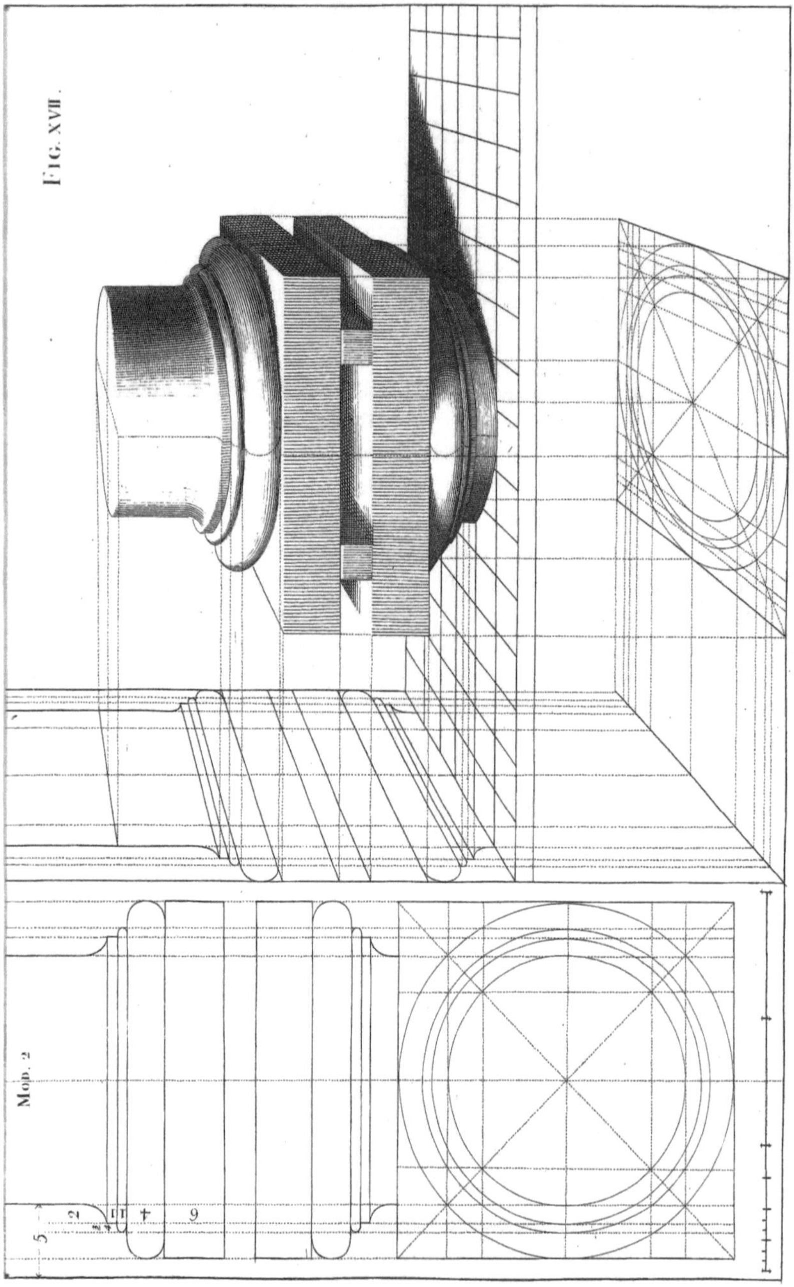

Fig. 4 Perspective study of Doric base (From Pozzo 1707, Plate 17)

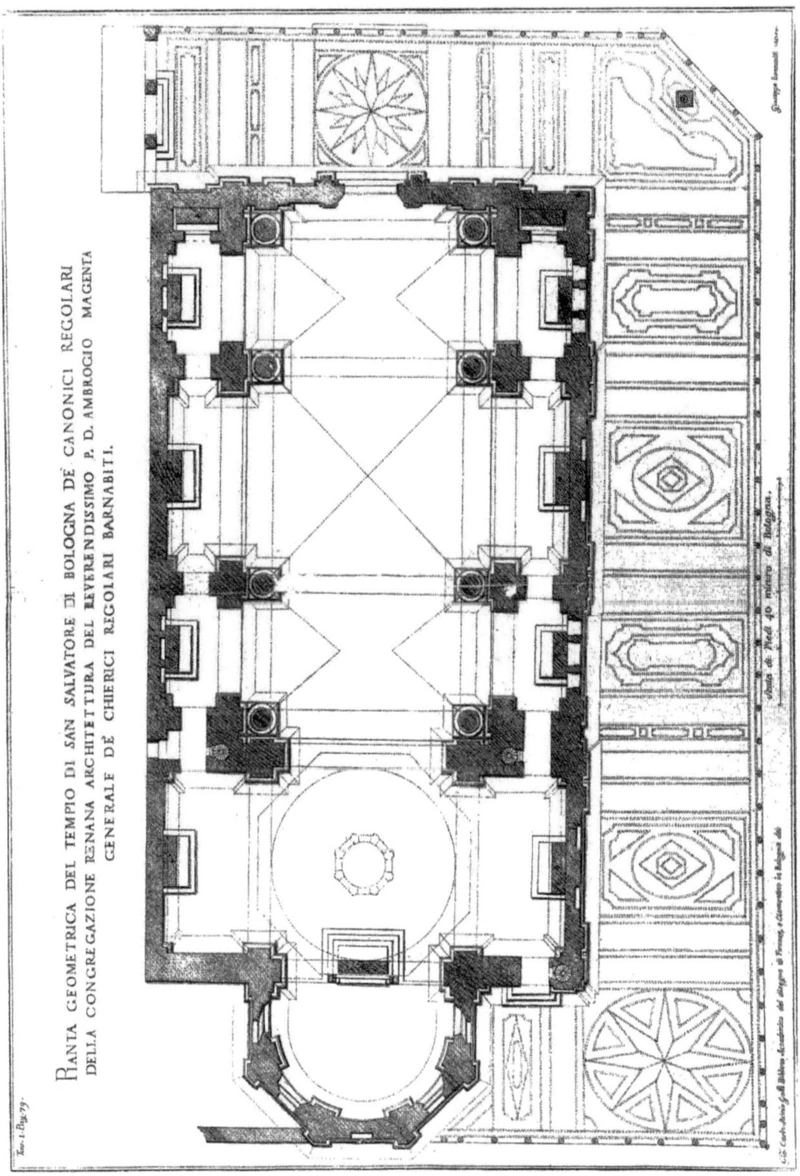

Fig. 5 Giovanni Ambrogio Mazenta, San Salvatore, Bologna, 1605–1623, plan (From Trombelli 1752, Plate 1)

Fig. 6 Engraving of Guarino Guarini and Gaetano Fontana (From Bianchi 1768, 108)

his works in Turin: the church of San Lorenzo, the chapel of the Holy Shroud, and the Palazzo Carignano.[33] Although he was never as prominent a scholar as Grassi, the early modern age nonetheless considered him primarily as a mathematician. One eighteenth-century biography described Guarini and fellow Theatine Gaetano Fontana in just such terms: "they were formed by nature to adorn sanctity, the sciences, and particularly the mathematical disciplines…."[34] The accompanying engraving reinforces the point, showing his colleague taking measurements from a celestial globe, while Guarini applies a set square to the design of a building (Fig. 6).

Here, astronomy and architecture illustrate different aspects of "the mathematical disciplines" within the Theatine intellectual world.

Like other priest-architects, Guarini received practical training at his order's various building sites, but extensive travel within Italy (including Sicily) and France greatly augmented this experience. Paris, in particular, offered Guarini direct contact with the architecture of Louis Le Vau and François Mansart, as well as the great French mathematical and constructive tradition of stereotomy. During his entire career, Guarini was producing architecture for the order and court patrons, as well as learned treatises on philosophy and mathematics.

Most of Guarini's writings can be understood as textbooks for the various subjects within the Theatine curriculum. In 1665, he published the *Placita Philosophica*, a compendium of Aristotelian logic, physics, astronomy, and metaphysics. In the broad field of mathematics, both pure and "mixed", we find three astronomical publications, as well as the *Euclides adauctus* (Turin, 1671), an encyclopedic treatment of the *Elements* studied in the second year of the Theatine curriculum, supplemented with further material.

Guarini's three architectural publications, however, belong to a different category. The *Modo di misurare le fabriche* (Turin, 1674), the *Trattato di fortificatione* (Turin, 1676), and the posthumous *Architettura civile* (Turin, 1737) are also works of "mixed" mathematics, yet they clearly sought a larger audience than his scholarly books, since they were written in the vernacular. Like Pozzo's perspective treatise, they straddle academic and practical genres. The *Modo di misurare* (a handbook for *stimatori* or construction surveyors) includes a mini-odometer for measuring irregular surfaces such as those of vaults (Fig. 7), demonstrating Guarini's interest in mathematical instruments and their applications. Although the principle can be traced back to Vitruvius, and larger odometers were commonly used for surveying land, Guarini apparently thought this application deserved wider dissemination.

Of the five *trattati* comprising Guarini's *Architettura civile*, only the third conforms more or less to the traditional structure and content of architectural treatises on the orders. The other four sections are overwhelmingly mathematical in character, presenting geometric axioms, surveying principles, mensuration, instructions for constructing geometric figures with straightedge and compass, and the basics of stereotomy. Indeed, if Guarini's mathematical understanding of architecture is intuitively visible in his domes, and explicitly stated in his treatise, it is almost palpable in his drawings. A plan for the Palazzo Carignano in Turin—pitted with scoring, compass pricks, and construction lines—captures him in the process of design, "thinking" with straightedge and compass (Fig. 8).

Guarini's Theatine colleague Girolamo Vitale (1624–1698) is the last of our six examples. A friend of Guarini's since 1640—they were novices together at San Silvestro al Quirinale in Rome—Vitale was not a practitioner and played little or no role in the construction of any buildings for the order. Nonetheless, Vitale arrived at architectural theory through mathematics, conceiving the second edition of his *Lexicon mathematicum* (Rome, 1690) as a dictionary of pure and applied mathematics.[35] Approximately a third of the entries treat the vocabulary of civil and military architecture, and several of these are illustrated with plates (Fig. 9). The *Lexicon*

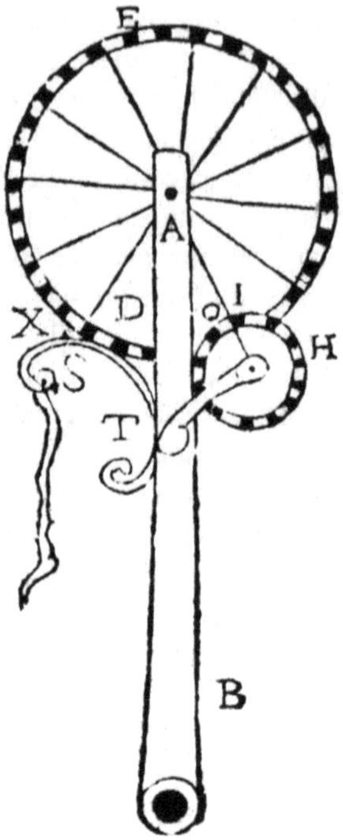

Fig. 7 Handheld odometer (From Guarini 1674, 48)

presents an updated treatment of common architectural topics for readers, citing standard architectural writers like Vitruvius, Alberti, Barbaro, Palladio, Serlio, and Vignola as well as more recent authors such as Bernardino Baldi and Gioseffe Viola-Zanini. In addition to examples from antiquity, Vitale uses contemporary buildings to illustrate his points. St. Peter's and Bernini's colonnade in Rome make appearances, as do many churches from the counter-reformational orders: Sant'Andrea della Valle, Santa Maria in Vallicella, the Gesù, Sant' Ignazio and Sant'Andrea al Quirinale, all in Rome, and Santi Apostoli and San Paolo Maggiore in Naples.[36] The prominent position of architecture in Vitale's *Lexicon* demonstrates the broad definition of mathematics supported by the counter-reformational orders, a definition that included both architectural theory and practice.

Architecture, Science, and Vocation

As Steven Harris has argued, for these priest-architects, the intellectual and the spiritual were inseparable. Architecture, too, was part of this mix. An example involving both Guarini and Vitale illuminates this interplay. Vitale's *Viaggio al cielo di S. Gaetano Thiene* (Rome, 1671) documents a novena marking the 1671 canonization of San Gaetano, cofounder of the Theatine order. Each day of the novena—a nine-day cycle of prayers and devotions—was named after a heavenly body, and traced events in Gaetano's saintly life as stages on his way to heaven and canonization.

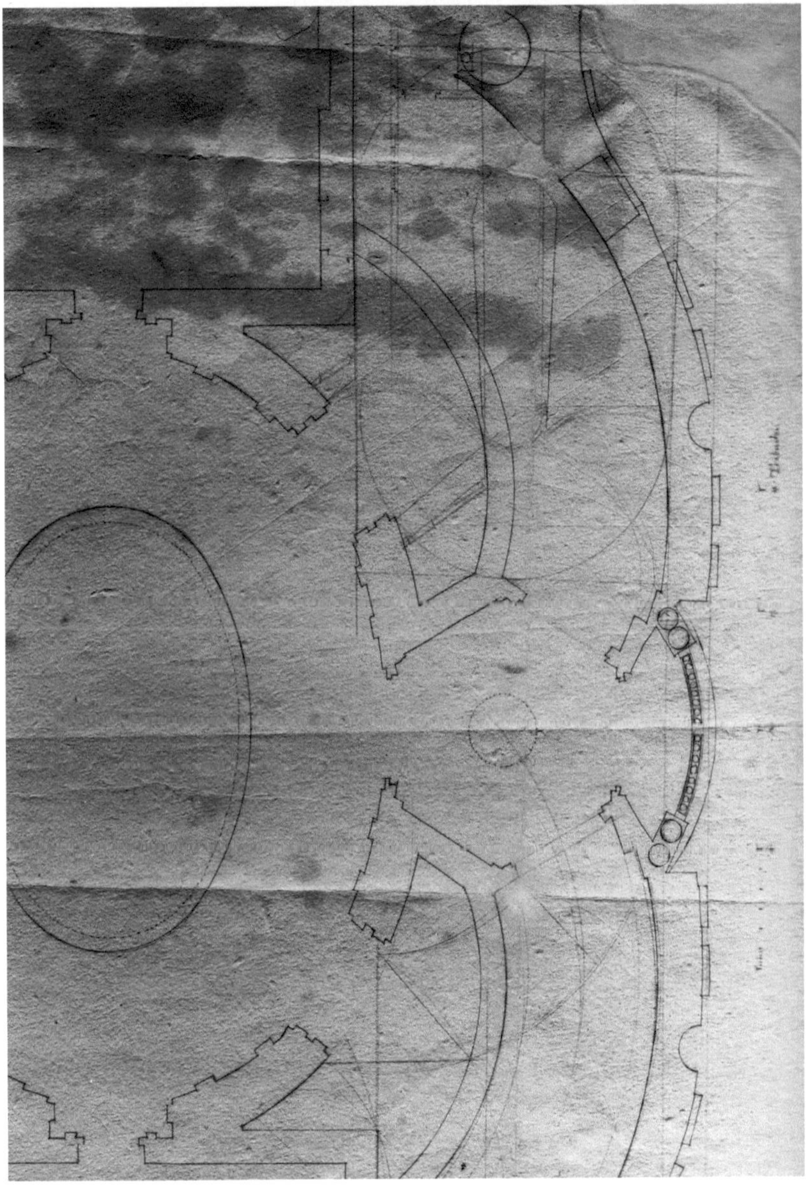

Fig. 8 Guarino Guarini, study for Palazzo Carignano, Archivio di Stato di Torino, Azienda Savoia-Carignano, cat. 53, mazzo 1, fasc. 9, n. 30 r, detail in raking light

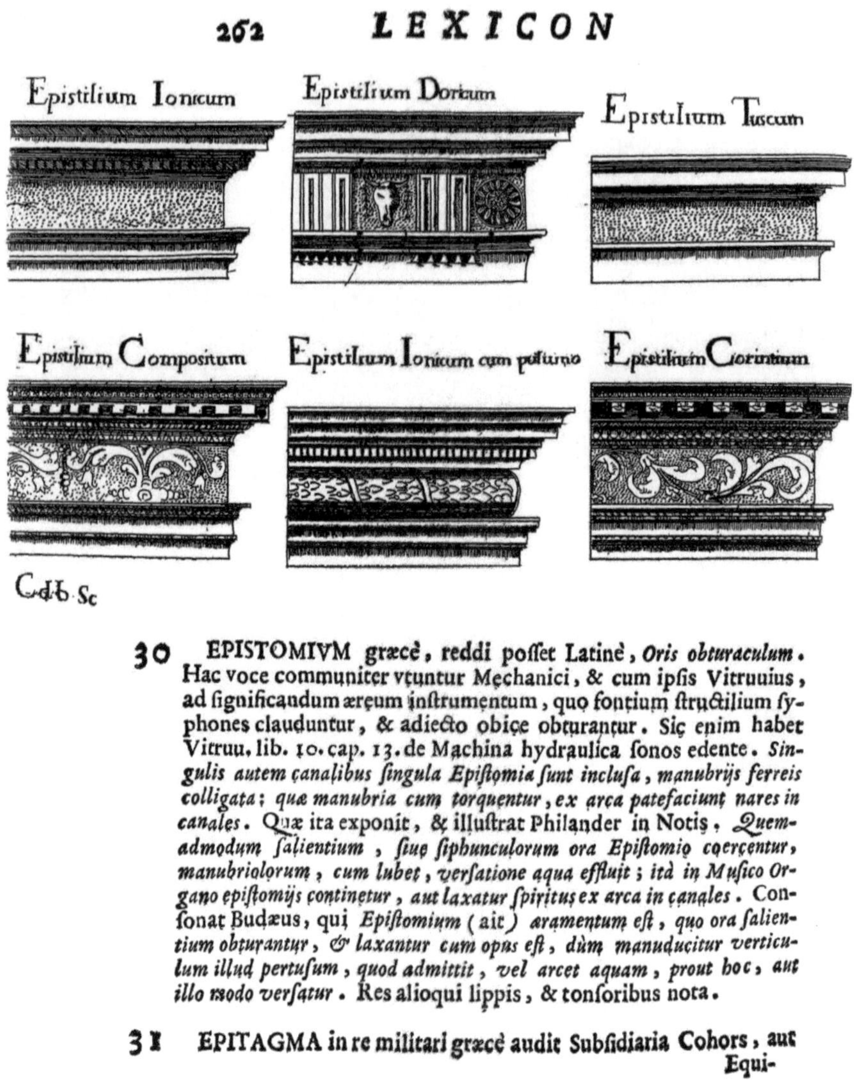

262 LEXICON

Epistilium Ionicum
Epistilium Doricum
Epistilium Tuscum
Epistilium Compositum
Epistilium Ionicum cum pulvino
Epistilium Corintium
Cdb Sc

30 EPISTOMIVM græcè, reddi poſſet Latinè, *Oris obturaculum*. Hac voce communiter vtuntur Mechanici, & cum ipſis Vitruuius, ad ſignificandum æreum inſtrumentum, quo fontium ſtructilium ſyphones clauduntur, & adiecto obice obturantur. Sic enim habet Vitruu. lib. 10. cap. 13. de Machina hydraulica ſonos edente. *Singulis autem canalibus ſingula Epiſtomia ſunt incluſa, manubrijs ferreis colligata; quæ manubria cum torquentur, ex arca patefaciunt nares in canales.* Quæ ita exponit, & illuſtrat Philander in Notis, *Quemadmodum ſalientium, ſiue ſiphunculorum ora Epiſtomio coercentur, manubriolorum, cum lubet, verſatione aqua effluit; ità in Muſico Organo epiſtomijs continetur, aut laxatur ſpiritus ex arca in canales.* Conſonat Budæus, qui *Epiſtomium* (ait) *æramentum eſt, quo ora ſalientium obturantur, & laxantur cum opus eſt, dùm manuducitur verticulum illud pertuſum, quod admittit, vel arcet aquam, prout hoc, aut illo modo verſatur.* Res alioqui lippis, & tonſoribus nota.

31 EPITAGMA in re militari græcè audit Subſidiaria Cohors, aut Equi-

Fig. 9 Plate illustrating entry "Epistilium," (From Vitale 1690, 262)

The frontispiece visualizes this process: San Gaetano rises toward heaven, indicated by a Ptolemaic diagram of planetary orbits above (Fig. 10). The novena and this image seem to have inspired Guarini's design for the unexecuted church of San Gaetano in Vicenza (1675), in which the angel-borne saint rises from the altarpiece to a frescoed representation of heaven in the dome (Fig. 11). Vitale may have

Fig. 10 Teresa del Po, engraved frontispiece of Vitale (1671)

intended the connection between the devotion and this iconography. He claimed that his novena presented a view of San Gaetano that was "foreshortened and in perspective," much in the same way that the saint would appear in the altarpiece and the frescoed dome of Guarini's church, and probably to be surrounded by the same astronomical references. Here, devotion is inextricably intertwined with astronomy, perspective and architecture, central aspects of the priests' work in mathematics and mathematical natural philosophy.[37]

Such references were by no means unusual. Vitale and Guarini were building on a deeply ingrained tradition of visual and spiritual allegory coupled with mathematical practice. A half century earlier, Mazenta's Milanese water column designs united architecture, mechanics, and hydraulics with the iconography of the Holy Cross and Moses' staff. Orazio Grassi had used a similar metaphor to link the use of mathematical instruments with religious belief, when he compared the salvation of the soul entrusted to the Virgin Mary (here in the roll of *Stella Maris*) with the salvation of the navigator who relies on his measurement of the heavens.[38] Although this particular example does not involve architecture, the metaphor reveals the mentality of the priest-mathematicians who sought to reconcile the methods and results of their scientific pursuits with their religious vocations. Pozzo, too, employed such rhetoric in the preface of his perspective treatise. "My Advice," he addressed his readers, "is that you chearfully begin your Work with a Resolution to draw all the Points thereof to that true Point, the Glory of GOD…."[39]

Fig. 11 Section and half-plan of San Gaetano, Vicenza (From Guarini 1686, Plate 26)

The work of both Grassi and Pozzo spanning the Seicento in Rome focuses our attention on the incubator of the priest-mathematician-architect phenomenon: the Collegio Romano, with its associated church of Sant'Ignazio, stood at the center of this development throughout the century. They—along with the activities pursued inside the complex—form a "unified cultural field," as Joseph Connors has termed it (adapting Bourdieu), asking: "Will we be able one day to look at Pozzo's frescoes, Kircher's museum, the distillery, and the mathematics classes and see their interaction? Will Grassi's architecture and his astronomy ever be studied together?"[40] The understanding of architecture as a part of mathematics arose from this cultural field. Embodied in the *consiliarus aedificiorum* and combined with the recognition of the utility of architecture in realizing the order's aims, it permeated not only the entire Jesuit order, but also most other counter-reformational orders during the course of the seventeenth century.

Most priests who benefited from the orders' architectural-mathematical education had undistinguished careers and thus remain largely unknown. Serving as building superintendents, local consultants, or occasional designers, their individual contributions may have been modest, but their cumulative effect was nonetheless important. Consider, for example, the Theatine Antonio Spinelli (1630–1706), provost of the order's Munich house, who discovered a serious error during the construction of the Munich Theatinerkirche. Spinelli replaced the secular architect Agostino Barelli as construction supervisor and went on to design the Theatine library in Munich.[41] On the other hand, the mathematical culture of the orders also produced authors like Eschinardi or Vitale, who apparently considered themselves qualified to publish works of architectural theory despite little or no practical experience as architects. Such scholars remain relatively unknown in the history of architecture, as does the impact of their writings on the wider reception of the art, defined no longer as a trade but as an object of formal study and quasi-scholarly discipline. The Savoyard Jesuit Claude-François Milliet Dechales (1621–1678), for example, was a prolific and widely translated author on mathematical topics. His encyclopedic *Cursus seu mundus mathematicus* (Lyon, 1674) aimed to provide a complete mathematical course of instruction, eliminating, in the process, the confusion caused by reliance on many different authors.[42] The 31 chapters of the work cover all aspects of mathematics, from Euclid and trigonometry to pyrotechnics and astronomy. Four are devoted to architecture, pointing again to the subject's consolidation within the Jesuit mathematical curriculum by the second half of the century.

But what separated a major priest-architect like Guarini from a competent priest serving as a local building superintendent, apart from the elusive quality "talent"? In his treatise, Guarini affirms that architecture, "as a discipline which utilizes measurements in all of its operations, depends on geometry." But the statement is quickly qualified: "architecture, even if it depends on mathematics, nonetheless is a flattering art, which does not offend the senses in order to satisfy reason." For Guarini, architecture's reliance on mathematics is interdependent with the pleasure it gives the senses. He thus qualifies it as both an art as well as a branch of mathematical science.[43] Here we reach the limits of what "mixed mathematics" could offer seicento architects.

It was perhaps inevitable that the tradition, as it continued into the next century, would be transformed. The orders continued to produce priest-practitioners in the mold of a Grassi or Guarini. The Somaschan priest Francesco Vecelli (1695–1759) provides a case in point. As the librarian at Santa Maria della Salute in Venice and later *preposito generale* of his order, he designed two Somaschan churches: Sant'Agostino, Treviso and Santa Croce, Padova, as well as the libraries of the Salute and the Camaldolesi on Mattia near Murano.[44] But over the course of the eighteenth century, the relationship between architecture and mathematics within the orders gradually merged with a modern understanding of engineering. The Jesuit scientist and polymath Roger Boscovich (1711–1787), professor of mathematics at the Collegio Romano was brought in to consult on structural problems at the dome of St. Peter's. The Barnabites Francesco De Regi (1720–1794) and Paolo Frisi (1728–1784) similarly advised on engineering questions at Milan Cathedral and the sanctuary at Rho. They published on mathematics and engineering education, statics, and Gothic architecture.[45]

A distinct architectural culture emerges from this survey. It was, in the first place, international, the European and global expansion of the orders acting to spread this culture far and wide. It was also comprehensive, addressing both theory and practice within the orders' programs of mathematical education. On both levels, the historical consequences of this culture were profound. Although the orders were not "schools of architecture" in any strict sense, the training they offered their members can nonetheless be viewed as an early kind of professional architectural education. Oriented to lecture hall rather than the building site, this form of instruction predated the foundation of various architectural academies in the late Seicento. On a physical level, too, the orders' architectural impact was considerable. Priest-architects provided designs for churches and houses for their orders all across Europe, particularly in smaller provincial towns. Like the Jesuit complex of San Vigilio in Siena—built by a succession of Jesuit architects including Grassi and Pozzo—such buildings shaped substantial portions of the fabric of early modern cities. Priest architects attracted noble patronage as well. Working in the "court apostolate," gifted architects like Mazenta or Guarini built churches outside of their orders as well as secular buildings and even fortifications. The orders' mathematical-architectural culture thus nurtured a rich variety of architects ranging from pure theoreticians through competent builders in the early modern vernacular idiom all the way to—in the exceptional case of Guarini—a master of the Baroque.

Notes

An early version of this paper appeared as part of Chapter Two in my dissertation, "Guarino Guarini's Theatine Architecture," Columbia University, 1993. Further research on the topic was supported by a J. Paul Getty Postdoctoral Fellowship in the History of Art and the Humanities in 1994–95.

1. "essi Padri... sono anche buoni Mathematici et intendono anche l'Architettura...," Quoted in Gallo (1958–59, 120).
2. For a survey of architecture in the counter-reformational orders in Italy, see Bösel (2003).
3. The literature on seventeenth-century Jesuit science and education is extensive. See for example the following recent contributions for further bibliography: Harris (1996); O'Malley et al. (1999–2006); and Feingold (2003a). The literature on Theatine and Barnabite science is paltry by comparison, but see Masetti-Zannini (1967) and Bianchi (1993). On early modern science in Rome, see now Romano (2008).
4. This phenomenon was by no means an invention of the seventeenth century. Renaissance priest-architects with mathematical or engineering backgrounds include the Franciscan Fra Giovanni Giocondo (1433–1515) and the Dominicans Danti (1536–1586), and Giuseppe Donzelli. The latter, known as "Fra Nuvolo", was active in Naples from c.1600 to 1630. The Augustinian

Giovanni Maria da Bitonto (1586-?) excelled at perspective constructions, such as the high altar tabernacle of the Barnabites' San Paolo in Bologna or the perspective colonnade at Palazzo Spada in Rome, where he worked with Borromini. For further details and bibliography on these and other figures, see Klaiber (1993, 43–46).

5. The 1586 edition was solely for internal use in the order; the first public edition dates from 1599. All editions also outlined the order's educational programs for secular students. On the *Ratio studiorum*, see Pachtler (1887–1894) and Lukács (1965–1992, esp. vol. 5, *Ratio atque institutio studiorum Societatis Iesu*).
6. See the Theatines' *Constitutiones congregationis clericorum regularium* (1604) with many subsequent editions; the Barnabites' *Constitutiones clericorum regularium S. Pauli decollati* (1617), and Gorini (1666).
7. To some extent the comments here on the Theatines and Barnabites proceed by analogy with the Jesuits, given the relative lack of research on their educational systems and scientific enterprises in comparison with the larger order. When appropriate material for comparison is available, however, the similarities between the three orders are confirmed (perhaps also due to the Theatines and Barnabites modelling themselves on the successful Jesuits).
8. Pachtler (1887–1894, vol. 2, 256, 348). For an exhaustive account of the early evolution of the Jesuit mathematical curriculum, see Romano (1999). On the practice of Jesuit mathematical education, see de Dainville (1954); Cosentino (1970, 1971); and Baroncini (1981). Compare also Dear (1995, ch. 2).
9. On the Jesuits and military architecture, see now De Lucca (2012).
10. Harris (1989, 42, note 23).
11. Gorman (2003). Grienberger was the first professor of mathematics at the Collegio Romano to serve as *consiliarus aedificiorum*, and he himself designed a few buildings for the order: the college at Aurillac and Santo Spirito in Sora, see ibid., p.70. For the attribution of the latter church, see Bösel and Karner (1986–2007, vol. 1, 294–295).
12. Cited in Gorman (2003, 23, 71, and 109, n. 96).
13. Possevino (1593). On Possevino and mathematics, see Romano (1999, 146–153). On Possevino and architecture, see Tessari (1983); Balestreri (1990); McQuillan (1992); Kiene (1996); Oechslin (1999, 213–214); and Carpo (2001, 113–118).
14. "De Architectura, post principes Mathematicas disciplinas dicen dum est, quandoquidem illę ab his pendent, earumq. ductu perficiuntur," Possevino (1593, vol. 2, 207). Possevino's architectural chapters are titled: "Architecturae origo. Cap. XVI", "An Aedificandi ratio peti debeat ex uno Vitruvio. Num item ex Salomonici Templi, quae olim extabat structura. Cap. XVII", and "Architecturae partes, atque divisio: Quaenam spectanda priusquàm aedificia inchoëntur, praesertim ea quae ad viros religiosos attinent. Cap. XVIII"
15. The inventories are found in Vat. Lat. 11267 (Theatines) and 11300 (Barnabites), Biblioteca Apostolica Vaticana, Rome. They belong to a group of such inventories catalogued (but not transcribed) in Lebreton and Fiorani (1985). Janis Bell kindly informed me of the existence of these MSS. Jesuit libraries were not

included in the group of inventories, and other orders (for instance the Padri Somaschi and the Cappuccini, Vat. Lat. 11275 and 11326 respectively) had no notable holdings of architectural books at the time. For the books listed here, see Vat. Lat. 11267, fols. 25v, 134r, 167r, 183r, 412r, 489r and Vat. Lat. 11300, fols. 89v, 95v, 109v, 113r, 120v, 133v, 134r, 136v, 157v, 165r, 166r.

16. On the *consiliarus aedificiorum* and for the Jesuit plan collection, see Vallery-Radot (1960, 8*–11*). For a list of the mathematics professors at the Collegio Romano see Villoslada (1954, 335). According to my own count, identifiable architectural activity—architectural designs, significant work as a consultant or building superintendent, or authorship of architectural publications—can be attributed to approximately 8 out of these 34 professors of mathematics at the Collegio Romano between 1553 and 1773, beyond their usual responsibilities as *consiliarus aedificiorum*. Finally, the work of Richard Bösel is fundamental for any consideration of Jesuit architecture: Bösel and Karner (1986–2007).
17. On the Barnabite office of the *prefetto delle fabriche* and the practice of architecture within the Barnabite order, see Repishti (1991), Gauk-Roger ([1991]); and Gatti Perer and Mezzanotte (2002).
18. Harris (1989, 48–49).
19. Harris (1989, 42 and n. 23).
20. Harris (1989, 60).
21. Harris (1989, 56).
22. For the Cinquecento, see note 4, above. Important seventeenth-century figures not mentioned in this study include the Jesuits Étienne Martellange (1569?–1641), a lay brother, the priest François Derand (1588–1644), and the Belgian François d'Aiguillon (1567–1617). For further information and bibliography on these men, see Klaiber (1993, 40–42, 67, notes 72–75). Among the Italian Theatines, Bernardo Castagnini (*ca.* 1603–1658) was one of Guarini's architectural mentors; the two men worked together at San Vincenzo, Modena in the early 1650s. Castagnini presented a design for the *casa* at San Vincenzo in 1646; at least one corridor and the library seem to have been completed to his design. He had earlier worked on a remodelling at San Bartolomeo, Bologna, begun in 1632. Documents describing this campaign refer to the priest and another Theatine involved with the project as "assai intendente delle matematiche," again stressing the connection between mathematical knowledge and architectural or engineering skill. See Sandonnini (1890). On Castagnini's work in Bologna, see Ravaglia (1909). The early Barnabite Pier Paolo D'Alessandro (1514–1591) built the cupola and upper loggia at Santa Maria di Canepanova, Pavia, and left manuscripts on painting, architecture, "poesia," and mathematics. See Boffito (1933–1937, vol. 1, 19–20), and Scotti (1985).

    Spanish priest-mathematician-architects include the Cistercians Juan Caramuel de Lobkowitz (1606–1682) and his teacher, Angel Manrique (1577–1649), Bishop of Badajoz. On Caramuel, see Filippo Camerota in the present volume (A Scientific Concept of Beauty in Architecture: Vitruvius Meets Descartes, Galileo, and Newton). A Spanish Augustinian, Fray Lorenzo de San Nicolás (1595–1679), wrote *Arte y uso de architectura*, 2 vols.,

(Madrid, 1639–1664), a treatise which emphasized the role of arithmetic and geometry in architecture and, significantly, contained Spanish translations of Euclid's fifth and seventh books. A second edition of Fray Lorenzo's first volume, published in 1667, added a translation of Euclid's first book, which then appeared in all subsequent editions. Fray Lorenzo also enjoyed a prolific career as an architect, building several churches throughout Spain. See Thieme and Becker (1929, vol. 23, 393); and de Llaguno y Amirola (1829, vol. 4, 20–26).

23. See, for example, the contemporary descriptions of Guarini's career in Klaiber (1993, 1–3).
24. On Tristano, see Pirri (1955); Bösel and Karner (1986–2007, vol. 1, 129–133, 164–165, 181–182, 212–213). On De Rosis: Pirri and Di Rosa (1975); Bösel (1991). On Valeriano, see Pirri (1970); and Bösel (1996).
25. Grimaldi is best known for the order's Sant'Andrea della Valle, Rome (begun 1591) where he worked with Giacomo della Porta under the patronage of Cardinal Alfonso Gesualdo. Most of Grimaldi's works survive in Naples however, including three Theatine churches and the Cappella del Tesoro di San Gennaro in the cathedral. Unfortunately, the details of Grimaldi's architectural training remain obscure, and no publications by him are known. On Grimaldi, see Savarese (1986) and Hibbard (1961).
26. On Grassi, see Bösel (2004), with complete further bibliography.
27. For an illustrated catalogue of Grassi's Roman album, see Bösel (2004, 59–310). On Grassi and Galileo, see for instance P. Redondi (1987) and the recent reassessment in Feingold (2003b).
28. This work appeared under the title *Espositione della carte topographica Cingolana dell'agro romano* (Rome, 1696). Eschinardi also taught at the Collegio Inglese and was therefore well-known to English travelers in Rome, including early members of the Royal Society. On Eschinardi, see Middleton (1966, 54–55); Muccillo (1993); Payne (1999, 157, 167); Cook (1999, esp. 180); and Cook (2004). On Eschinardi's *Cursus physicomathematicus* see Feldhay and Heyd (1989). For Eschinardi and the Jesuit church at Vercelli, see Bösel and Karner (1986–2007, vol. 2, 411–424); for a further example of his work as a consultant (at Pozzo's Cappella di Sant'Ignazio in the Gesù, Rome) see Levy (2004, 235 n.21).
29. On this academy, see Middleton (1975).
30. On Pozzo, see the following recent works, including older bibliography: De Feo and Martinelli (1996); Battisti (1996). A convenient summary of Pozzo's architectural work is provided by Kelly (1982). Also see Andersen's contribution in this volume (The Master of Painted Architecture: Andrea Pozzo, S. J. and His Treatise on Perspective).
31. Here quoted from the first English edition, Pozzo (1707; reprint 1989, 206).
32. On the Accademia degli Accurati, see Grammatica (1919, 14). Further on Mazenta, see. Mezzanotte (1961); Gatti Perer and Mezzanotte (2002) both with earlier bibliography, and now Stabenow (2011). On Mazenta's writings, see Boffito (1933–1937, vol. 2, 451–463). On the monumental columns for Milan, see Schofield (2004, 79–120, esp. 92–94).
33. For the most recent overview of Guarini's career, see Dardanello et al. (2006), with a complete bibliography of earlier studies.

34. Bianchi (1768, 108).
35. On Vitale and his *Lexicon* see Vezzosi (1780, vol. 2, 481–484); Masetti-Zannini (1967); and now Rabassini (2012). In the preface to the second edition Vitale himself criticized the much shorter first edition of his work (Paris: Ludovic. Billaine, 1668).
36. Vitale also mentions the Roman churches Sant'Agnese, Santa Maria della Pace, Santa Maria degli Angeli, and San Pietro in Montorio; villas at Frascati and Caprarola; and a prominent non-Italian example, Hagia Sophia.
37. See Klaiber (2006).
38. Bösel (2004, 24–25).
39. [*sic*]: the Latin gives "lineas" for "point". See Pozzo (1707; reprint 1989, 12).
40. Connors (1999–2006).
41. On Spinelli, see Dischinger (1988, vol. 1, 141–42, 145–46); and Klaiber (1993, 36, 64 n. 60).
42. Claretta (1878, vol. 2, 585), referring to a 1674 letter from Dechales to the duke. On Dechales, see Dainville (1947–1948); De Backer et al. (1890–1932, vol. 2, cols. 1040–44).
43. Guarini (1737, I.ii, I.iii.intro, 3).
44. See Pilo (1964). Vecelli's manuscript "Problemi di geometria pratica," dealing with geometry and fortifications, survives in a posthumous copy (1767) at the University of Pennsylvania.
45. For Boscovich, see the article by Pascal Dubourg Glatigny in this volume (Epistemological Obstacles to the Analysis of Structures: Giovanni Bottari's Aversion to a Mathematical Assessment of Saint-Peter's Dome (1743)). For De Regi and Frisi, see Boffito (1933–1937, on De Regi: vol. 1, 640–644; on Frisi, vol. 2, 72–98). Also see Baldini (1998) and A. Bianchi (1993, 143–164).

Photographic Credits

Wikimedia Commons / public domain: Fig. 1
Author: Figs. 2, 4, 7, 11
Museo Galileo / Istituto e Museo di Storia della Scienza, Florence: Fig. 3
Stabenow (2011, 194): Fig. 5
Archivio Generale dei Teatini, Rome: Figs. 6, 10
Archivio di Stato di Torino, Photograph: Paolo Robino: Fig. 8
Hathi Trust / University of Michigan / public domain: Fig. 9

References

Baldini, Ugo. 1998. Paolo Frisi. Dizionario Biografico degli Italiani. http://www.treccani.it/biografie.
Balestreri, Isabella. 1990. L'architettura negli scritti della Compagnia di Gesù. In *L'architettura della Compagnia di Gesù in Italia XVI-XVIII sec*, ed. L. Patetta, I. Balestreri, C. Coscarella, and D. Zocchi, 19–26. Brescia: Grafo.

Baroncini, G. 1981. L'insegnamento della filosofia naturale nei collegi italiani dei Gesuiti (1610–1670): un esempio di nuovo aristotelismo. In *La "Ratio studiorum": Modelli culturali e pratiche educative dei Gesuiti in Italia tra Cinque e Seicento*, ed. G.P. Brizzi, 192–202. Rome: Bulzoni.

Battisti, Alberta (ed.). 1996. *Andrea Pozzo*. Milan-Trent: Luni.

Bianchi, Ignazio Lodovico. 1768. *Le immagini di alcuni uomini e alcune donne per pietà illustri della Congregazione de'cherici regolari*. Venice.

Bianchi, A. 1993. *L'instruzione secondaria tra barocco ed età dei lumi: Il Collegio di San Giovanni alle Vigne di Lodi e l'esperienza pedagogica dei Barnabiti*. Milan: Vita e Pensiero.

Boffito, G. 1933–1937. *Scrittori Barnabiti*. Florence: Olschki.

Bösel, Richard. 1991. Giovanni De Rosis. Dizionario Biografico degli Italiani. http://www.treccani.it/biografie.

Bösel, Richard. 1996. Giuseppe Valeriano. Oxford Art Online. www.oxfordartonline.com.

Bösel, Richard. 2003. L'architettura dei nuovi ordini religiosi. In *Storia dell'architettura italiana: Il Seicento*, vol. 1, ed. A. Scotti Tosini, 48–69. Milan: Electa.

Bösel, Richard. 2004. *Orazio Grassi architetto e matematico gesuita*. Rome: Argos.

Bösel, Richard, and Herbert Karner. 1986–2007. *Jesuitenarchitektur in Italien 1540–1773*. Vienna: Österreichische Akademie der Wissenschaft.

Carpo, Mario. 2001. *Architecture in the age of printing*. Cambridge, MA: MIT Press.

Claretta, G. 1878. *Storia del regno e dei tempi di Carlo Emanuele II duca di Savoia*. Genoa: R. Istituto de'sordo-muti.

Connors, Joseph. 1999–2006. Reflections. In *The Jesuits: Cultures, sciences, and the arts 1540–1773*, 2 vols., ed. J.W. O'Malley, G.A. Bailey, S.J. Harris, and T.F. Kennedy, 1, 705–706. Toronto: University of Toronto Press.

Constitutiones clericorum regularium S. Pauli decollati. 1617. Milan: Apud hær. Pacifici Pontij, & Ioan. Baptistam Piccaleum.

Constitutiones congregationis clericorum regularium. 1604. Rome: Stefano Paulini.

Cook, Alan. 1999. Our foreign correspondence. *Notes and Records of the Royal Society of London* 53(2): 179–182.

Cook, Alan. 2004. Rome and the Royal Society 1660–1740. *Notes and Records of the Royal Society of London* 58(1): 3–19.

Cosentino, G. 1970. Le matematiche nella 'Ratio studiorum' della compagnia di Gesù. *Miscellanea storica ligure* n.s. 2(2): 169–213.

Cosentino, G. 1971. L'insegnamento delle mathematiche nei collegi gesuitici nell'Italia settentrionale. *Physis* 13: 205–217.

Dardanello, Giuseppe, Susan Klaiber, and Henry A. Millon (eds.). 2006. *Guarino Guarini*. Turin: Allemandi.

De Backer, Augustin, Aloys De Backer, and Carlos Sommervogel. 1890–1932. *Bibliothèque de la compagnie de Jésus*, 10 vols. Brussels/Paris: Schepens/Picard.

de Dainville, François. 1947–1948. Foyers de culture scientifique dans la France méditerranéenne du XVIe au XVIIIe siècle. *Revue d'histoire des sciences et de leurs applications* 1: 289–300.

de Dainville, François. 1954. L'enseignement des mathématiques dans les Collèges Jésuites de France du XVIe au XVIIIe siècle. *Revue d'histoire des sciences* 7, 8: 6–21, 109–123.

De Feo, Vittorio, and Valentino Martinelli (eds.). 1996. *Andrea Pozzo*. Milan: Electa.

de Llaguno y Amirola, Eugenio. 1829. *Noticias de los arquitectos y arquitectura de España desde su restauración*. Madrid: En la Imprenta Real.

De Lucca, Denis. 2012. *Jesuits and fortifications: The contribution of the Jesuits to military architecture in the Baroque Age*. Leiden: Brill.

de San Nicolás, Lorenzo. 1639–1964. *Arte y uso de architectura*. Madrid.

Dear, Peter. 1995. *Discipline and experience: The mathematical way in the scientific revolution*. Chicago: University of Chicago Press.

Dischinger, Gabriele. 1988. *Zeichnungen zu kirchlichen Bauten bis 1803 im Bayerischen Hauptstaatsarchiv*. Wiesbaden: Reichert.

Eschinardi, Francesco [pseud. Costanzo Amichevoli]. 1675. *Architettura civile, ridotta a metodo facile e breve*. Terni: B. Arnazzini.

Eschinardi, Francesco [pseud. Costanzo Amichevoli]. 1684. *Architettura militare, ridotta a metodo facile e breve col modo distinto di formare ciascuna parte*. Rome: A. Bernabò.
Eschinardi, Francesco [pseud. Costanzo Amichevoli]. 1696. *Espositione della carte topographica Cingolana dell'agro romano*. Rome.
Feingold, Mordechai (ed.). 2003a. *The new science and Jesuit science: Seventeenth century perspectives*. Dordrecht: Kluwer.
Feingold, Mordechai. 2003b. The grounds for conflict: Grienberger, Grassi, Galileo, and Posterity. In *The new science and Jesuit science: Seventeenth century perspectives*, ed. Mordechai Feingold, 121–157. Dordrecht: Kluwer.
Feldhay, R., and M. Heyd. 1989. The discourse of pious science. *Science in Context* 3(1): 109–142.
Gallo, R. 1958–59. Vincenzo Scamozzi e la chiesa di S. Nicolò da Tolentino di Venezia. *Atti dell'Istituto veneto di scienze, lettere ed arti: Classe di scienze morali e lettere* 117: 103–122.
Gatti Perer, M.L., and G. Mezzanotte. 2002. Lorenzo Binago e la cultura architettonica dei Barnabiti. *Arte Lombarda* [special issue] 134.
Gauk-Roger, N. The architecture of the Barnabite Order 1545–1659, with special reference to Lorenzo Binago and Giovanni Ambrogio Mazenta. Ph.D. dissertation, Cambridge [1991].
Gorini, M. 1666. *Exterarum scholarum disciplina apud clericos regulares S. Pauli....* Milan: Francesco Vigoni.
Gorman, Michael John. 2003. Mathematics and modesty in the Society of Jesus: The problems of Christoph Grienberger. In *The new science and Jesuit science: Seventeenth century perspectives*, ed. Mordechai Feingold, 1–120. Dordrecht: Kluwer.
Grammatica, Luigi. 1919. *Le memorie su Leonardo da Vinci di Don Ambrogio Mazenta*. Milan: Alfieri e Lacroix.
Guarini, Guarino. 1674. *Modo di misurare le fabriche*. Turin: Heredi Gianelli.
Guarini, Guarino. 1686. *Dissegni d'architettura civile, et ecclesiatica*. Turin: Paulino.
Guarini, Guarino. 1737. *Architettura civile*. Turin: Gianfrancesco Mairesse.
Harris, S.J. 1989. Transposing the Merton thesis: Apostolic spirituality and the establishment of the Jesuit scientific tradition. *Science in Context* 3(1): 29–65.
Harris, S.J. 1996. Jesuits and the knowledge of nature. *Early Science and Medicine* [special issue] 1(3).
Hibbard, Howard. 1961. The early history of Sant'Andrea della Valle. *Art Bulletin* 43: 289–311.
Kelly, C. 1982. Pozzo, Andrea. In *Macmillan encyclopedia of architects*, 4 vols., ed. Adolf K. Placzek, 3, 469–472. New York: Macmillan.
Kiene, M. 1996. Bartolomeo Ammannati et l'architecture des jésuites au XVIe siècle. In *Les jésuites à l'âge baroque (1540–1640)*, ed. L. Giard and L. de Vaucelles, 183–206. Grenoble: J. Millon.
Klaiber, Susan. 1993. Guarino Guarini's Theatine architecture. Ph.D. dissertation, Columbia University.
Klaiber, Susan. 2006. L'iconografia di San Gaetano nel progetto per Vicenza. In *Guarino Guarini*, ed. Giuseppe Dardanello, Susan Klaiber, and Henry A. Millon, 404–411. Turin: Allemandi.
Lebreton, M., and L. Fiorani. 1985. *Codices Vaticani Latini: Codices 11266–11326, Inventari di biblioteche religiose italiane alla fine del cinquecento*. Vatican City: Bibliotheca Vaticana.
Levy, Evonne. 2004. Das 'Jesuitische' der jesuitischen Architektur. In *Die Jesuiten in Wien: Zur Kunst- und Kulturgeschichte der österreichischen Ordensprovinz der "Gesellschaft Jesu" im 17. und 18. Jahrhundert*, ed. H. Karner and W. Telesko, 231–241. Vienna: Österreichische Akademie der Wissenschaften.
Lukács, L. (ed.). 1965–1992. *Monumenta paedagogica Societatis Iesu*, 7 vols. Rome: apud "Monumenta Historica Soc. Iesu".
Masetti-Zannini, G.L. 1967. I Teatini, la nuova scienza e la nuova filosofia in Italia. *Regnum Dei* 23(89–90): 3–150.
McQuillan, James. 1992. Antonio Possevino's 'Architectura' and the Baroque. *Scroope: Cambridge Architecture Journal* 4: 45–50.
Mezzanotte, G. 1961. Gli architetti Lorenzo Binago e Giovanni Ambrogio Mazenta. *L'arte* 26: 231–271.

Middleton, W.E. Knowles. 1966. *A history of the thermometer*. Baltimore: Johns Hopkins University Press.
Middleton, W.E. Knowles. 1975. Science in Rome, 1675–1700, and the Accademia Fisicomatematica of Giovanni Giustino Ciampini. *British Journal for the History of Science* 8: 138–154.
Muccillo, Maria. 1993. Francesco Eschinardi. Dizionario Biografico degli Italiani. http://www.treccani.it/biografie.
O'Malley, J.W., G.A. Bailey, S.J. Harris, and T.F. Kennedy (eds.). 1999–2006. *The Jesuits: Cultures, sciences, and the arts 1540–1773*, 2 vols. Toronto: University of Toronto Press.
Oechslin, Werner. 1999. 'Architectura est scientia aedificandi': Reflections on the scope of architectural and architectural-theoretical literature. In *The triumph of the Baroque*, ed. H.A. Millon, 206–217. Milan: Bompiani.
Pachtler, G. 1887–1894. *Ratio Studiorum et Institutiones Scholasticae Societatis Jesu*, 4 vols. Berlin: A. Hofmann.
Payne, Alina A. 1999. Architectural criticism, science, and visual eloquence: Teofilo Gallaccini in seventeenth-century Siena. *Journal of the Society of Architectural Historians* 58(2): 146–169.
Pilo, G.M. 1964. Notizie sull'architetto somasco Francesco Vecelli. *Bollettino del Centro Internazionale di Studi d'Architettura Andrea Palladio* 6(1): 311–314.
Pirri, Pietro. 1955. *Giovanni Tristano e i primordi dell'architettura gesuitica*. Rome: Institutum Historicum S.J.
Pirri, Pietro. 1970. *Giuseppe Valeriano S.I., architetto e pittore 1542–1596*. Rome: Institutum Historicum S.J.
Pirri, Pietro, and Pietro Di Rosa. 1975. Il P. Giovanni de Rosis (1538–1610) e lo sviluppo dell'edilizia gesuitica. *Archivum Historicum Societatis Iesu* 44: 3–104.
Possevino, Antonio. 1593. *Bibliotheca selecta: qua agitur de ratione studiorum in historia, in disciplinis, in salute omnium procuranda. Cum diplomate Clementis VIII. Pont. Max.* Rome: Ex Typographia Apostolica Vaticana.
Pozzo, Andrea. 1693–1700, 2nd ed. 1702–1723. *Perspectiva pictorum et architectorum/Prospettiva de'pittori e architetti*. Rome: Typis Joannis Jacobi Komarek Bohemi apud S. Angelum Custodem.
Pozzo, Andrea. 1707; reprint New York: Dover, 1989. *Rules and examples of perspective proper for painters and architects, etc.* London: J. Senes, etc.
Rabassini, Andrea. 2012. Il lessicografo dell'astrologia. Note biografiche su Girolamo Vitale (1623/24-1698). *Bruniana & Campanelliana* 18(2): 675–681.
Ravaglia, E. 1909. Il portico e la chiesa di S. Bartolomeo in Bologna. *Bollettino d'arte* 3: 110–111.
Redondi, Pietro. 1987. *Galileo Heretic*. Trans. Raymond Rosenthal. Princeton: Princeton University Press.
Repishti, F. 1991. *Formula del offitio del Prefetto delle fabriche apresso delli Chierici Regolari della Congregazione di S. Paolo* secondo Lorenzo Binago. *Arte Lombarda* 96–97: 137–140.
Romano, Antonella. 1999. *La contre-réforme mathématique: constitution et diffusion d'une culture mathématique jésuite à la Renaissance (1540–1640)*. Rome: École Française de Rome.
Romano, Antonella (ed.). 2008. *Rome et la science moderne: entre Renaissance et Lumières*. Rome: École française de Rome.
Sandonnini, T. 1890. Il Padre Guarino Guarini modenese. *Atti e Memorie delle RR. Deputazioni di storia patria per le provincie modenesi e parmensi* ser. 3: 5, part 2: 491–492.
Savarese, Silvana. 1986. *Francesco Grimaldi e l'architettura della Controriforma a Napoli*. Rome: Officina Edizioni.
Schofield, Richard. 2004. Architecture and the assertion of the Cult of Relics in Milan's Public Spaces. *Annali di architettura* 16: 79–120.
Scotti, A. 1985. Tommaso [p. Pietro Paolo] D'Alessandro. *Dizionario biografico degli italiani*. http://www.treccani.it/biografie.
Stabenow, Jörg. 2011. *Die Architektur der Barnabiten: Raumkonzept und Identität in den Kirchenbauten eines Ordens der Gegenreformation 1600–1630*. Munich: Deutscher Kunstverlag.
Tessari, A.S. 1983. Antonio Possevino e l'architettura. *Archivum Historicum Societatis Iesu* 52: 247–261.

Thieme, Ulrich, and Felix Becker (eds.). 1907–50. *Allgemeines Lexikon der bildenden Künstler von der Antike bis zur Gegenwart*. Leipzig: Engelmann.

Trombelli, Giovanni Crisostomo. 1752. *Memorie istoriche concernenti le due canoniche di S. Maria di Reno, e di S. Salvatore insieme unite*. Bologna: per Girolamo Corciolani ed eredi Colli s. Tommaso d'Aquino.

Vallery-Radot, J. 1960. *Le recueil de plans d'édifices de la Compagnie de Jésus conservé à la Bibliothèque nationale de Paris*. Rome: Institutum Historicum S.I.

Vezzosi, A.F. 1780. *I scrittori de' Cherici Regolari detti Teatini*. Rome: Propaganda Fide.

Villoslada, R. Garcia. 1954. *Storia del Collegio Romano dal suo inizio (1551) alla soppressione della Compagnia di Gesù (1773)*. Rome: Apud aedes Universitatis Gregorianae.

Vitale, Girolamo. 1671. *Viaggio al cielo di S. Gaetano Thiene*. Rome: Ignatio de' Lazari.

Vitale, Girolamo. 1690. *Lexicon mathematicum*, 2nd ed. Rome: Vannacci.

The Master of Painted Architecture: Andrea Pozzo, S. J. and His Treatise on Perspective

Kirsti Andersen

The Art of Perspective does,
with wonderful Pleasure, deceive the Eye,
the most subtle of our outward Senses.[1]

Andrea Pozzo's illusionistic work is well-known among historians of architecture and art, as is his *Perspectiva pictorum et architectorum*, published in the last decade of the seventeenth century.[2] Historians have presented the general outline of the treatise and have discussed its valuable descriptions of the author's own designs.[3] The book's place, however, in the history of the literature on perspective has received far less attention. This article examines the style and content of the *Perspectiva* in relation to the broader tradition of perspective writings in Italy. Being a Jesuit played an essential role for Pozzo's self-understanding; hence it is also natural to ask how common it was for men in holy orders to write on the subject.

Life and Work

Pozzo's background is notable in that it was both intellectual and practical. At the age of 17 he was apprenticed to a painter in his native town Trento, but unlike most young painters of the time he had studied before, having attended the Jesuit School in the same town, where he became proficient in Latin. He continued his training as a painter in Como and Milan, where, after a brief period as a novice in the Discalced Carmelites (1661–1662), he joined the Society of Jesus in December 1665.

K. Andersen (✉)
Emerita, Centre for Science Studies, Aarhus University, van Reigersbergstraat 232, NL1052 WS Aarhus, Denmark
e-mail: kirsti.andersen@ivs.au.dk

A. Gerbino (ed.), *Geometrical Objects: Architecture and the Mathematical Sciences 1400-1800*, Archimedes 38, DOI 10.1007/978-3-319-05998-3_7,

Thereafter, Pozzo's life and work would be directed to the glory of the Jesuits, especially that of its founder, Ignatius of Loyola. He would remain a lay brother of the order to his death.[4]

Pozzo showed an early aptitude for the illusionistic decoration of architectural interiors. A commission of 1676 to decorate the interior of San Francesco Saverio in Mondovi featured a *trompe l'oeil* cupola, painted on a canvas stretched across the vault of the crossing. His abilities in this particular genre, known as *quadratura*, were quickly recognized. In 1681, the Jesuit Father-General called him to Rome, where he would be engaged in the decoration of the order's principal churches: Il Gesù and Sant'Ignazio. It was in this early Roman period that Pozzo was commissioned to provide a self-portrait for Cosimo III de'Medici's ducal collection, another indication of the painter's rising status.[5] Pozzo was also employed in creating ephemeral decorations to be used at ecclesiastical festivals in Il Gesù, one of which he later included in the *Perspectiva* (Fig. 1). Another early Roman project involved the restoration of the corridor linking the Gesù to the rooms in which Ignatius had lived at the end of his life. Pozzo supplemented the architecture of the gallery—a small staircase and some diminutive windows—with a rich painted decoration of architectural ornaments, paintings, and textiles. Among the many playful elements in the gallery are painted legs hanging from feigned architecture. Richard Bösel has described Pozzo's composition as "a *tour de force* of illusionism in which the deception and its disclosure are complementary parts of the artistic concept."[6]

From 1684, Pozzo's primary responsibility was the decoration of Sant'Ignazio, adjacent to the Collegio Romano. Although begun in 1626 and opened for worship in 1650, the church never received its dome. Mounting financial concerns, changes of the original design, and worries about the potentially dominating exterior profile of the planned dome all contributed to interrupt the project. Pozzo solved the problem of the missing element in a manner reminiscent of his early expedient at Mondovi, that is, with a "virtual" cupola surmounted by a lantern, painted on an 18-m wide canvas stretched over the crossing of the nave (Fig. 2). This success was awarded with further prestigious commissions, particularly for the nave ceiling in the same church (Fig. 3) and, in 1695, for the main altarpiece of il Gesù. Pozzo's involvement with Sant'Ignazio also extended to a number of minor projects for altar designs and frescos.

During this period, Pozzo found the time to lecture on painting and architecture at the Collegio Romano. Some of the material from his course is likely included in the *Perspectiva*.[7] His workshop also appears to have been responsible for the imaginative frescos in the refectory of the then-Minim convent on the Trinità dei Monti in Rome, dating to the 1690s.[8]

From the end of the seventeenth century, Pozzo was increasingly engaged in Jesuit churches outside Rome. He redecorated old buildings and provided proposals for new ones, like San Francesco Saverio in Trento, which was rebuilt according to his designs. By 1702, he had acquired a name throughout the Catholic world. In that year, he was invited to Vienna by Emperor Leopold I, to whom he had dedicated the first volume of his *Perspectiva*.

Fig. 1 A *teatra sacra* designed by Pozzo for Il Gesù in 1685 showing the wedding at Cana, "for the Solemnity of exposing the Holy Sacrament". It consisted of several stage flats (see Fig. 8) and was lit by candles (From Pozzo 1707, Figure 71)

In Vienna, he was engaged in several projects, the most important of which was the renovation of the *Jesuitenkirche* also called the *Universitätskirche*. Completely transforming the interior, he provided it with what was by now his characteristic device: a painted dome. Although he planned to return to Italy to take part in a reconstruction of the Jesuit church in Venice, Pozzo died in Vienna, where he was buried in the newly redecorated *Jesuitenkirche*.[9]

Fig. 2 Andrea Pozzo, the painted dome of Sant'Ignazio, Rome, 1685. The canvas has a diameter of about 18 m

Pozzo's art constitutes a symbiosis of painting and architecture, characterized by the continuation and elaboration of real spaces into fictive ones. Helen Hills has aptly characterized his style.

Despite its geographical dispersion and variety, certain themes unite his work. Illusionism, perspective, drama, manipulating light and creating impressions of forbidding majesty were his perennial concerns, mustered in support of Catholic faith and of the Jesuits in particular. He was indebted to the theatre: much of his painting exploits theatrical devices and stage requisites, such as the proscenium arch, the curtain, the *quadratura* backdrop, and painted "actors" stepping out from the painted wings.[10]

In Pozzo's brand of *quadratura*, the physical building serves to lend structural and decorative features to the virtual one, which, in turn, elevates, expands, and allegorizes the prosaic, material form of the physical structure. This reciprocal relationship is guaranteed by the elaborate geometry of the perspective construction.[11]

Fig. 3 Andrea Pozzo, *The Apotheosis of St. Ignatius*, c. 1690–1694. Sant'Ignazio, Rome

Authors and Approaches in the Literature on Perspective

The technique of geometrical perspective emerged among a circle of Florentine artists in the Quattrocento, to become a much-applied method of representation in painting, graphic art, and decoration during the fifteenth and sixteenth centuries. All sorts of practitioners—from architects and painters to engineers, furniture and stage designers, sculptors, goldsmiths, stonecutters, woodcarvers, etchers, and engravers—used some form of perspective. Apprentices in these disciplines would have learned the rudiments of the technique as a matter of course, presumably from their masters.

Besides learning by doing, it was possible to become acquainted with perspective through textbooks. These books were produced by a significantly narrower range of practitioners, dominated by architects and painters. The group of authors also included mathematicians, at first very few, but with their number increasing markedly from about 1600. From the very beginning, the literary styles and intended audiences of the theoreticians and the practitioners differed considerably. This difference is well illustrated by two earliest important quattrocento texts on the subject, by Leon Battista Alberti and Piero della Francesca. Both authors were clearly inspired by the painters and architects with whom they were in close contact, but their approaches had little in common.

In *De pictura* (*On Painting*, 1435, first printed in 1540), Alberti described for the first time geometrical perspective—without using the word perspective—and also instructed his readers on how to construct the perspectival image of a grid of squares. His idea was that the grid could serve as a sort of coordinate system in the picture plane to help painters organize their compositions. His main concern, however, was to show that painting was an academic activity on a par with the *artes liberales*, and despite claiming to speak "as a painter… to painters," he presumably wrote the book more for the sake of potential patrons.[12]

The major part of *De pictura* discusses painting in antiquity and advocates classical themes as appropriate subjects, leaving only a few pages for the technique of painting. To these, Alberti devoted a couple of applications of the grid, but not enough to fully train the reader to represent various forms in perspective. Nor did Alberti provide any mathematical proofs or explanation of why his proposed construction led to the claimed result.

Piero, who was both a painter and a mathematician, aimed to expose both the technique and the geometrical secrets of perspective to his readers. In his *De prospectiva pingendi*—written in Italian despite its Latin title—he carefully guided his readers through all the steps of his constructions and provided many different examples of perspectival shapes. His attempt to explain not only the *why* of perspective but also the *how* was less successful, but interesting because he initiated a development which eventually made the theory of perspective a subdiscipline of geometry.[13]

Most of the other non-academic authors had intentions closer to Piero's than to Alberti's, though without the same level of detail and using fewer examples. Nor did they typically attempt to provide geometrical proofs or justifications for the constructions they presented. These authors usually followed a scheme that included some, and sometimes all, of the following items: one or more methods of perspective construction, a number of examples of shapes and forms in perspective, perspective instruments, examples of scenography (constructions for perspectival stage sets), and anamorphoses. The favourite objects for exercises were polygons, circles, polyhedra, crosses, columns, arches, vaults, and simple rooms with a few windows.

Although most authors took the time to explain their perspective constructions, one wonders how well they were understood by the uninitiated. It seems unlikely that many early modern practitioners would have picked up the technique solely by following the text. It is more probable that they would have mastered it partly by consulting the explanatory diagrams, and partly by adapting the book's contents to their own workshop practice. In other words, most of the books were only useful for those already familiar with perspective constructions.

As an ecclesiastic writing on perspective, Pozzo was by no means alone. We know of at least twelve other authors among his fellow Jesuits, a reflection, no doubt, of the order's great interest in integrating practical mathematics into its curriculum.[14] One writer, Jean Dubreuil, was particularly successful. His most popular book—the first of three he would publish on this and related subjects—was known simply as the *Jesuit's Perspective*. Written in French and reissued often, it was translated into English twice and once into German.[15] Pozzo would achieve even more success with his *Perspectiva*, but Dubreuil's example must have shown him the potential audience for such books. Of the other religious orders, the Theatines[16] and the Minims[17] each could count two authors on perspective, and the Dominicans, the Augustinians, and the Camillians each one.[18]

Among the authors of perspectival manuals, architects comprised a surprisingly large group. Before Pozzo's *Perspectiva*, eight books entirely devoted to perspective had been published in Italy, and it is remarkable that five of those were written by architects.[19] That is not to mention the architects who touched on perspective in their architectural treatises. Pozzo could have known at least five examples of printed works in this genre.[20] It is also surprising that these authors did not substantially distinguish themselves from other practitioners. Although they may have included more architectural examples—Sebastiano Serlio for one—architects as a whole took no identifiable approach to explaining how to proceed with perspective constructions. That is not to say that professional differences did not sometimes arise. One fairly well-known painter, Pierre-Henri Valenciennes, writing around 1800, declared himself unsatisfied with the architects' textbooks, lumping them with those of mathematicians. This group, he claimed, did not understand the needs of young painters.[21] Judging from the contents of his own rather wordy manual on perspective, he apparently believed that students of painting were less familiar with geometrical forms than students of architecture and mathematics, but in other respects, his techniques were not unfamiliar.

Mathematicians, on the other hand, did treat perspective differently. Although many of these authors claimed to write for painters and other practitioners, they typically presented the subject in forbidding style of formal geometry—that is, with theorems and proofs.

Among the mathematicians, Guidobaldo, Marchese del Monte stands out as the only important contributor to the geometrical theory of perspective to address his fellow scholars. His work, *Perspectivae libri sex*, published in 1600, was inspired by those perspective authors, like Piero, who had attempted to penetrate the geometry behind the technique. Guidobaldo is mostly known as the patron, collaborator, and friend of Galileo Galilei, but his own work on perspective was revolutionary. Not only did *Perspectivae libri sex* establish the precedent for treating the subject in purely formal terms, in it Guidobaldo coined the concept of a general vanishing point, which became the basis for all further geometrical developments in the field.[22]

Guidobaldo's achievement represented a high point in the Italian literature on perspective. In other parts of Europe, interest in and understanding of the subject increased throughout the seventeenth century, reaching its highpoint in the eighteenth. However, in Italy fascination waned. The Italian tradition of perspective treatises was mainly kept alive by stage designers and painters, who specialized in illusionistic paintings. In addition, a few architects, such as Giuseppe Galli-Bibiena and Giovanni Battista Piranesi, published collections of impressive perspective architectural engravings, accompanied by very little or no text at all.[23] Their didactic function is fulfilled primarily by the images themselves, which are distinguished by their complexity and virtuosity.

Pozzo's *Perspectiva*

Perspectiva pictorum et architectorum was published in two volumes of which the first appeared in 1693 and the second in 1700. The books were printed with parallel Latin and Italian texts, presumably to add a quasi-scholarly appeal. The first volume immediately attracted widespread attention. Translated into English as early as 1707, it appeared over the years in more than thirty editions and in at least nine languages.[24] The book was even translated into Chinese in the period when the Jesuits brought western science and art to China. The Chinese did not show much interest in the *Perspectiva*, but for architects and other practitioners elsewhere it became a standard work.

The style of the *Perspectiva* is distinctive. Pozzo had all his explanations printed as figure captions, which contain virtually no mathematics, nor much, if any, guidance for constructing the figures.[25] His diagrams, on the other hand, are more numerous, better drawn, more illustrative, and in general far more impressive than usual.

The lack of direct instruction makes the *Perspectiva* a difficult text, an aspect that Pozzo's English translator, John James, commented on: "… the Brevity or Silence of our Author," he explained, "writing in a Country where the Principles of this Art are more generally known than with Us, had no need to insist so long on some things, as may be thought necessary to *Beginners*."[26] Not only were most of Pozzo's

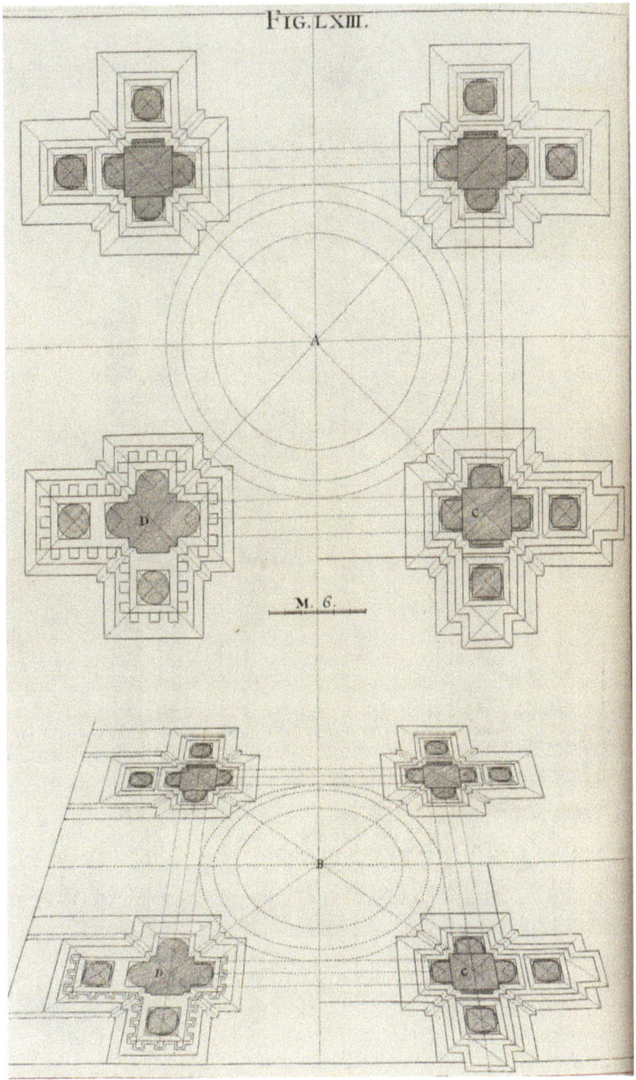

Fig. 4 "Plan of a Square Design" and its perspective image (From Pozzo 1707, Figure 63)

examples extremely elaborate, he typically left out the construction lines. Nor did he always provide all the information required to replicate his constructions. Two of the book's illustrations, concerning the perspective construction of an elaborate domed tabernacle, are characteristic of this approach. The intervening steps between the plan (Fig. 4) and the perspective elevation (Fig. 5) are only alluded to. Although Pozzo may have been overly optimistic about what some of his readers

Fig. 5 "Square Design in Perspective," (From Pozzo 1707, Figure 64)

could achieve, his aim was to challenge them to replicate the constructions for themselves: "If you long to profit from this art quickly, do not waste time on mere speculations … but seize the compass and the ruler and work, and in this way you will feel spurred to proceed further and further, not only to draw the figures in this book, but to invent some even better."[27]

Apart from basic geometrical forms, Pozzo's examples of perspectival drawings are all architectural, proceeding from columns, column bases, capitals, entablatures, and cornices to larger architectural ensembles for the five orders. Although these

Fig. 6 A "machine" consisting of two "ranges of frames," (From Pozzo 1707, Figure 61)

forms belonged to the standard repertoire of the *quadratura* painter, they also served to satisfy the book's second principal aim: to showcase Pozzo's own work. A tabernacle, for example, designed in the form of a theatrical set piece—what Pozzo called a "machine"—recalls an altar he designed for San Francesco Saverio in Mondovi (Fig. 6). A pronounced interest in sacred theatre and stage sets in general recurs throughout the book.[28] His two major commissions for Sant'Ignazio—the *trompe l'œil* dome and nave ceiling—also receive in-depth treatment, as does an octagonal cupola (Fig. 7), a reminder perhaps of his early work for San Francesco Saverio in Mondovi.

The second volume of the *Perspectiva* continues along the same lines, with perspectival constructions relating to the author's more recent activities. A long section on altars, for example, includes specimens that Pozzo had recently completed or was still working on, including a commission for an altar in Sant'Ignazio dedicated to the titular saint. This section is supplemented by further proposals of increasing variety. Among other designs, Pozzo also included proposals for a decorative urn and for new church facades—in particular for San Giovanni Laterano in Rome. He even provided drawings for an entire church. As for secular architecture, the design of a triumphal arch and a number of fortifications stand out.

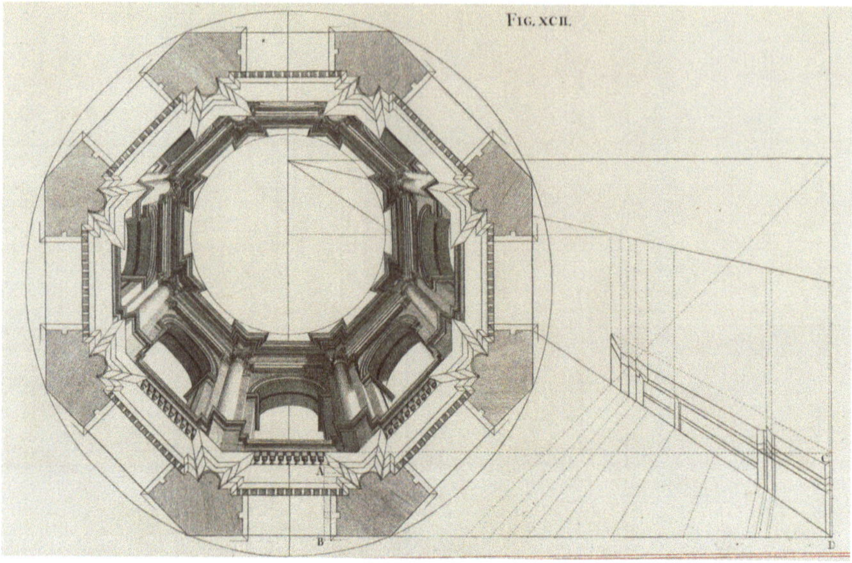

Fig. 7 An octagonal cupola in perspective (From Pozzo 1707, Figure 92)

The title page of the *Perspectiva* is addressed to painters and architects, in that order. The combination is in one sense curious, since the role of perspective in the two crafts was very different. It was in the art of *quadratura*, however, that they were combined. The book was not primarily intended for beginners in the craft, but for a specialist audience dedicated to architectural painting. Pozzo's foremost aim was to stress the skill and care required for this genre. To properly draw architecture in perspective, he held, painters had to have knowledge of architecture itself: "Whence you may perceive, that for designing things of this kind, the Painter ought to have no less Skill in Architecture, than is requir'd for the Execution of solid Works."[29]

The Two Methods

In principle, there is no limit to the number of possible methods for constructing the perspective image of a point, and throughout history an impressive number have been proposed. Only two, however, became really popular: the so-called distance point construction (Fig. 8), and one based on a plan and elevation of the object to be cast into perspective, incorporating the picture plane and the eye point for the composition.[30]

These methods had long been known and their comparative benefits debated. The contrast proved to be important, for it stimulated questions about the content and character of perspective as a mathematical subject. Piero had included both techniques in his *De prospectiva pingendi*. Although he found that the easiest way

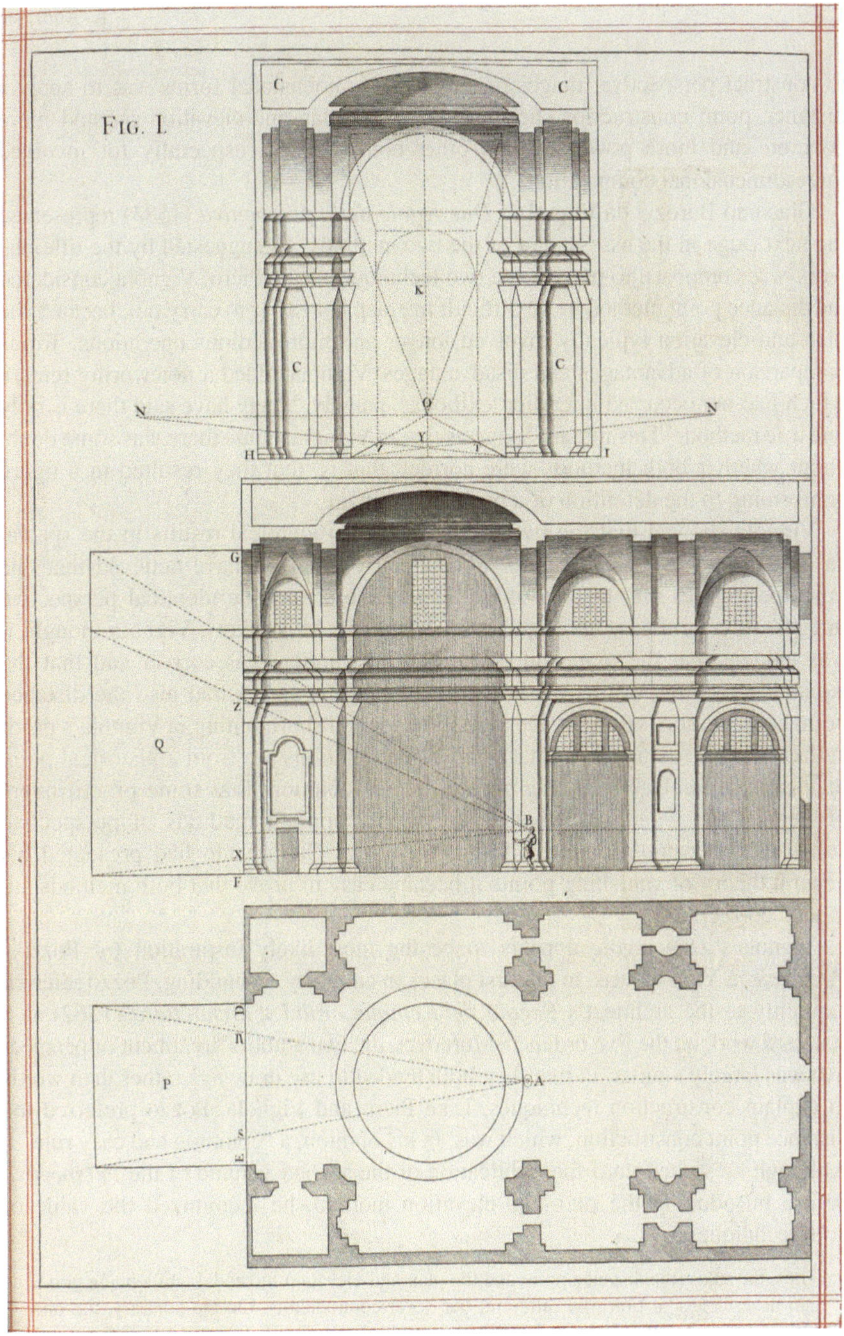

Fig. 8 A distance-point construction on the end wall of a church. The longitudinal section in the middle shows the viewer's eye point at *B* and the picture plane *ZF*. The plan of the church at bottom shows the eye point *A* and the picture plane as *ED*. The lateral section at the top shows the elevation of the picture plane, with the principal elements of the distance point construction. The point O is the orthogonal projection of the eye point upon the picture plane, later termed the principal vanishing point, whereas the points N are the two so-called distance points. They lie on the horizontal line through O—the horizon—at a distance from O, equalling the distance from the eye point to the picture plane (Pozzo 1707, Figure 1)

to construct perspective images of simple two-dimensional forms was to apply a distance point construction, he considered the plan-and-elevation method more concrete and more powerful than other constructions, especially for involved three-dimensional compositions.[31]

Giacomo Barozzi da Vignola's *Due regole della prospettiva* (1583) represented the next stage in the assessment of the two methods. As suggested by the title, the work was composed to present the two techniques. Like Piero, Vignola considered the distance point method more difficult to grasp but easier to carry out, because the plan-and-elevation typically involved longer and more tedious operations. To his comparison of advantages and disadvantages, Vignola added a noteworthy remark which had not occurred in earlier textbooks, namely "many have said there is only one true method." This remark suggests that in Vignola's time there was some doubt about whether both methods were correct, that is, that they resulted in a figure conforming to the definition of a perspective image.

Vignola showed that the two methods lead to identical results in the special case of the perspective image of a square. He seems to have believed that this outcome implies that the two methods in general provide identical perspective images—a conclusion that requires an argument. Most likely, Vignola thought it was evident that the plan and elevation construction was correct and that the equivalence of the results of the two methods guarantees that also the distance point construction is correct. Despite some logical shortcoming in Vignola's proof and further reasoning, his remark and contemplations are from a historical point of view extremely interesting because they document that some practitioners of perspective wondered whether commonly applied methods of perspective constructions were geometrically correct. After Guidobaldo had presented his general theory of vanishing points it became easy to prove that both methods are indeed correct.

Vignola's *Due regole* appears to be the most likely inspiration for Pozzo's *Perspectiva.* Vignola was, in the first place, an authority on building. Pozzo referred explicitly to the architect's *Regola delli cinque ordini d'architettura* (1562) as a standard work on the five orders.[32] Moreover, the two authors' treatment of perspective was largely similar, in that they both tended to use drawings rather than words to explain construction techniques. Like Piero and Vignola, Pozzo preferred the distance point construction, which was, in his opinion, a "common and easy rule."[33] Although he waited until the publication of the second volume of the *Perspectiva* before introducing the plan-and-elevation method, he recognized the value of both techniques:

> Thus, the rules [the distance point and the plan-and-elevation methods] while being good, can nevertheless at times be different, but never contradictory. On the contrary, the one throws light upon the other, as in arithmetic multiplication serves as test for division and vice versa.[34]

Pozzo also took up the question of "correctness." Like Vignola, he presented two perspective images—not of a square—but of a cupola similar to that of Sant'Ignazio, constructed by the two different techniques. He then encouraged his readers to verify that similar line segments in the two images were equal in length. His choice could indicate that he was unaware of Guidobaldo's theory of perspective, although it is also possible that he thought measuring was more convincing for his intended audience.

The Cupola of Sant'Ignazio

Although Pozzo avoided formal mathematics, his treatise nonetheless demonstrated a good grasp of the geometrical possibilities offered by the distance point technique. His construction of the cupola of Sant'Ignazio, in particular, must rank among the most elegant pre-nineteenth-century applications of perspective by a non-mathematician (Fig. 9). A design like this—involving an image projected on a horizontal plane seen from below—confronted Pozzo with a different set of

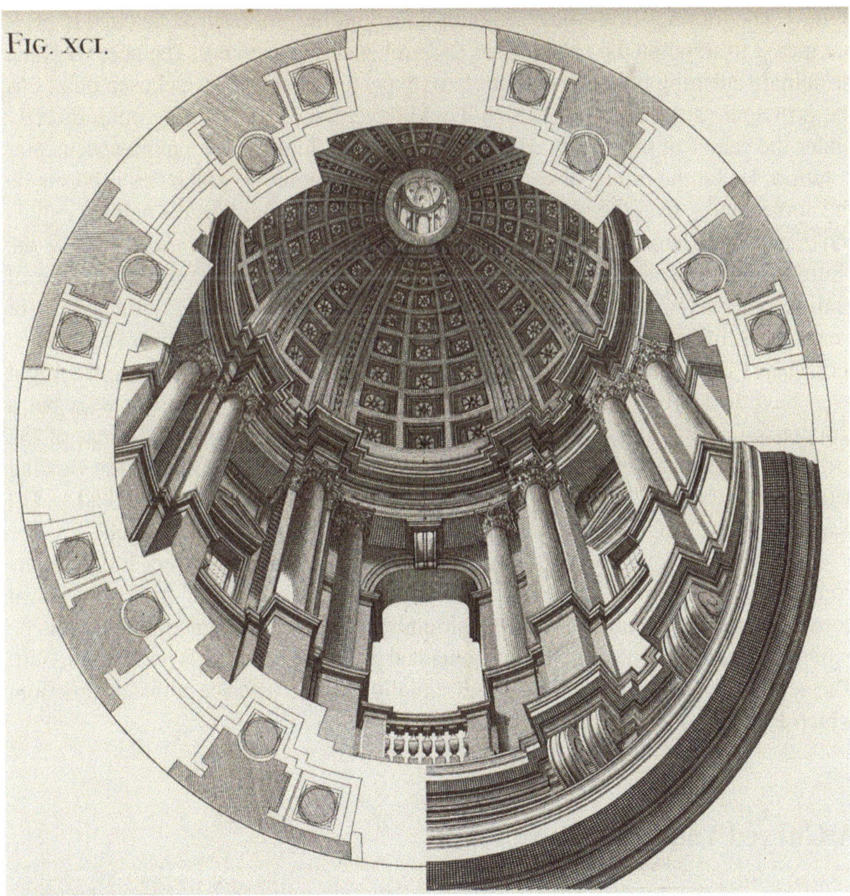

Fig. 9 Illustration of the feigned dome of Sant'Ignazio. In the figure text, Pozzo remarked that this cupola would probably last longer than the painted version (Pozzo 1707, Figure 91)

perspectival rules and an unusual challenge. On a vertical plane, vertical lines are depicted as vertical, whereas parallel horizontals (apart from those parallel to the picture plane) are depicted as converging. Contrariwise, in a horizontal plane seen from below or above, it is the vertical lines of the depicted volume that converge. The point of convergence—the principal vanishing point—is the orthogonal projection upon the ceiling of a fixed, ideal viewing point. As in the case with a vertical image, the viewer's impression of converging lines is dependent on his or her position in relation to the eye point. Viewed at some distance from that spot, the converging lines no longer look vertical. The horizontal picture plane also produces one simplification. As Pozzo himself pointed out, horizontal circles—that is, those parallel to the picture plane—are depicted as circles, not as ellipses.[35] This fact has important implications for the depiction of domes.

The perspective construction of the virtual dome of Sant'Ignazio—reproduced in the *Perspectiva*—exemplifies Pozzo's technical and graphical capability (Fig. 10). The virtual dome rises from the picture plane inside the outer circle centered on point *I*. The footings of the outer columns and window embrasures, hatched in grey, are meant to stand on the pendentives and arches of the crossing. The point *O* is the principal vanishing point. One might have expected Pozzo to have chosen point *I* as the principal vanishing point, which would have placed the viewing point directly under the center of the cupola. However, Pozzo was looking for a more spectacular solution. He let the principal vanishing point be the point *O*, which lies just outside the dome's base, over the vault of the nave. The effect is that Pozzo's painted cupola looks more and more convincing as one walks toward it from the entrance of the church. Viewed from a point directly under *O*, it gives an amazing impression of reality, whereas seen from other locations, it has an anamorphic effect. Part of Pozzo's fascination with virtual cupolas may have been that the viewer could place themselves in positions where the illusion could no longer be maintained. He might even have had something like this in mind when he wrote: "Perspective is but a Counterfeiting of the Truth."[36] Pozzo's own explanation for the placement of the principal vanishing point at *O* was so "that the Eye might be less weary'd in viewing the Work, and take in more of the Architecture, than it could have done, had … [it] be in the midst."[37] One could add that it is also less strenuous for the neck.

Pozzo's technique for obtaining the perspective image of the dome seems complicated because it involves many lines, and like most of his perspectival compositions, it requires some consideration in order to understand how he achieved his result. However, the mathematical principle behind it is not that difficult. The geometrical knowledge needed is essentially that of a distance point construction, which Pozzo had introduced beforehand.

A Curved Vault as Picture Plane

Pozzo had planned to end the first book of the *Perspectiva* with the engraving of the octagonal dome (Fig. 7), when he was subsequently asked by some friends to explain an additional problem: the "Perspective Net-Work for irregular Surfaces."[38]

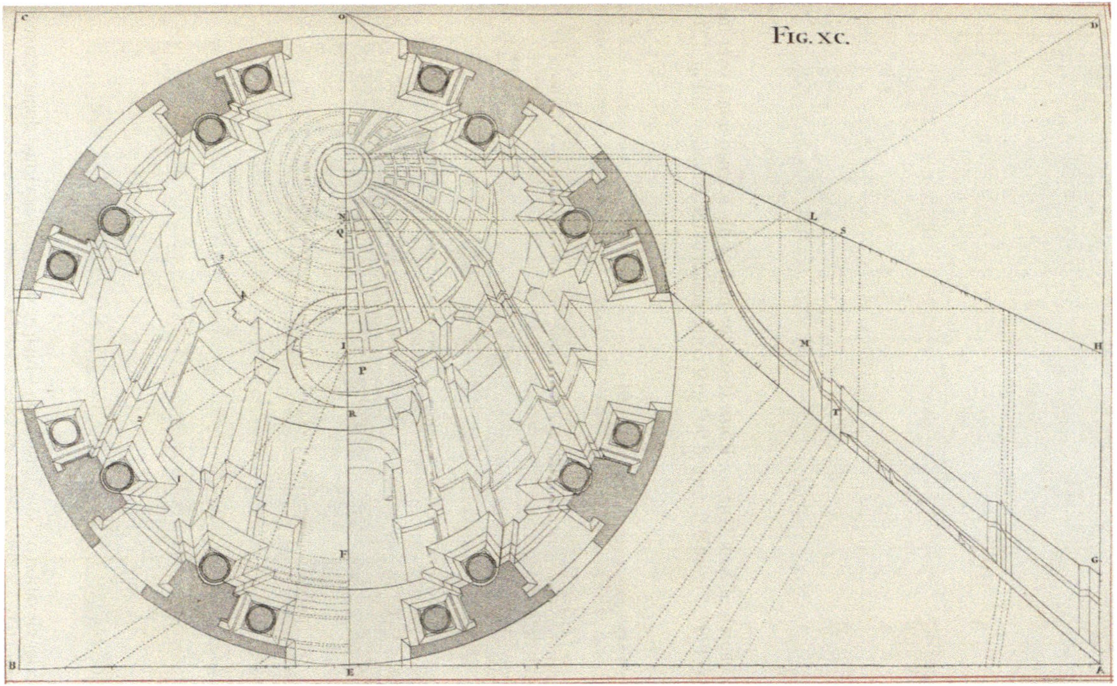

Fig. 10 Pozzo's construction of the virtual dome of Sant'Ignazio. The vertical elements converge toward the principal vanishing point *O*, while the rising horizontal sections are diminishing circles. At right is a perspectival section of half the cupola and lantern, which has its principal vanishing point in *O* and its right distance point in *D*. The section itself (which Pozzo did not include in his diagram) lies in the vertical plane through *AD* and is bounded by the two vertical lines through *A* and *H*, whose perspective images are *AG* and *HO*. The perspectival section of half of the dome can be obtained by a distance point construction. With this section at his disposal, Pozzo built up his image of the dome as a series of diminishing circles, using the perspectival section to determine their centers and radii. The cornice of the entablature, for example, is formed by a circle with center *N*. This circle is determined in the section to the right, by the point *L* on the horizontal line through *N* and the radius *LM*. For details of the construction see Andersen (2007, 392). Image from (Pozzo 1707, Figure 90)

Fig. 11 Plan of the church of Sant'Ignazio, Rome (Note the plan of his feigned dome over the crossing, from Pozzo (1707, Figure 93), turned 90° so it corresponds to the elevation in Fig. 12)

Fig. 12 Pozzo's interior elevation of Sant'Ignazio, in which he depicted his virtual dome as a real one (Pozzo 1707, Figure 94)

This theme relates to another decoration in Sant'Ignazio, namely the fresco on the barrel vaulted nave next to the virtual cupola (Figs. 3, 11, and 12). For this commission, Pozzo designed and executed a truly impressive work. *The Apotheosis of St. Ignatius*, painted between 1690 and 1694, celebrated the founder of the Jesuits and the order's missionary activities on the four continents.

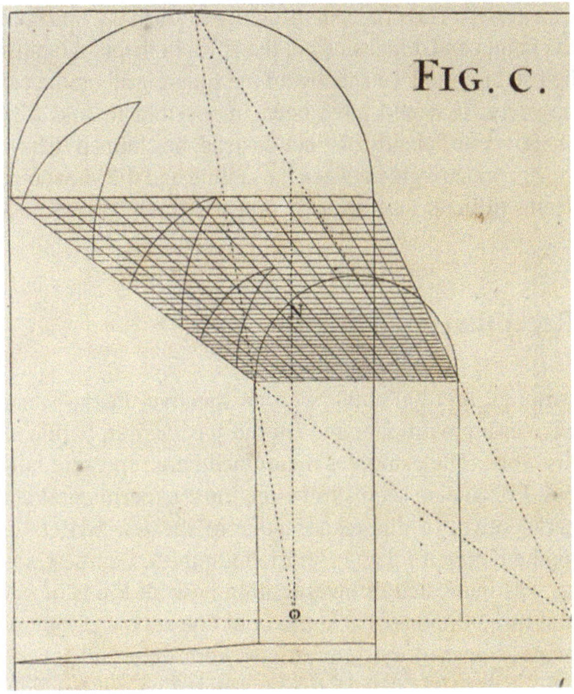

Fig. 13 Pozzo's demonstration of how to map a flat grid of squares onto a cylindrical vault. The principle of the projection is shown in perspective: The eye point is *O* and the horizontal grid of squares *N* is projected on the vault from *O*. The illustration shows projections for two points (From Pozzo (1707, Figure 100), detail)

When Pozzo received the commission to decorate this vault, his patrons restricted the placement of the principal vanishing point. Although he was allowed to select the subject matter, they did not want another eccentric eye point like the one for the cupola.[39] Instead, he placed the vanishing point centrally, organizing the entire composition around the vertical axis of the vault. Because the surface is long, the impression is very sensitive to the viewer's position and distorts easily when seen from anywhere but the eye point. From that spot, however, the viewer is rewarded with a spectacular and realistic vision of clustered columns and vast arches, rising up from the real windows of the church to frame the open sky. Airborne figures float above, scaled to the rising stages of the fictive architecture, leading the eye toward Saint Ignatius at the center of the composition.

The perspectival challenge here was the opposite to that posed by the painted dome. Where the latter required Pozzo to make a flat surface over the crossing look curved, this design required him to treat the semi-cylindrical surface of the vault as a plane. In order to work correctly as a perspectival image, in other words, the construction had to compensate for and disguise the pronounced curvature of the vault. This problem was too complicated for geometrical constructions. Instead, Pozzo used a mechanical transfer technique well-known at the time. First he set out the perspective composition on a plane surface at the full scale of a horizontal section in the vault and overlaid this drawing with a grid of squares. He then projected the vertices of this grid from the eye point to the semi-cylindrical surface (Fig. 13). The images of the vertices were then connected with curved lines, so that they formed a kind of coordinate system on the vaulted ceiling to which the flat drawing could be transferred. The practical difficulty of this task must have been considerable, given the height and dimensions of the nave. In theory a lamp placed at the eye point could be used to project a shadow of the square grid on the ceiling. As Pozzo pointed out, however, it would have been impossible to find a light strong enough to cast shadows sufficiently distinct. Instead, he claimed to have projected the vertices with string, pulled tight between the ceiling and the viewpoint, although this method, too, seems difficult to imagine in practice.[40]

Reception and Afterlife

Among pre-1800 authors on perspective, Pozzo stands out, both for his experience as a practitioner and for the spectacular, public, and large-scale character of his work. The examples mentioned here represent but a sample of his entire output. Pozzo also seems to be the most experimental and playful of the authors on perspective. In this respect, one of the few writers comparable to him was the mathematician Johann Heinrich Lambert. Lambert was not much of a painter, but he was interested in investigating how all kinds of geometrical problems in normal three-dimensional space could be solved if they were transferred to a picture plane. Some of these questions were quite advanced and esoteric: for instance, how to draw the perspective image of a painting which itself is in perspective, the perspective image of a rainbow, or that of a doubly reflected object. If Pozzo was the expert practitioner of perspective, Lambert was the expert geometrician.[41] In the second, 1774 edition of his important work on perspective, Lambert added a short survey of the literature of the field, in which he recognized the contribution by his more practical predecessor: "due to its many neat architectonic drawings for painters and architects, the work by Andrea Pozzo has always much excellence."[42]

Other readers also found the *Perspectiva* of great interest. Indeed, the first volume stayed in print throughout the eighteenth century. Pozzo's particular brand of *quadratura*, disseminated by the Jesuits and other Counter-reformation orders, became widely imitated. His designs for feigned domes, in particular, inspired creations in Italy as well as in Austria, Bohemia, Germany, Poland, and Silesia. Pozzo's influence was particularly strong among Jesuit artists.[43] Although the *Perspectiva* found many readers, it had no noticeable impact on the literature of perspective. Nor did Pozzo's showpiece, the distance-point construction of the virtual cupola, find its way into later books on the subject. Yet, this construction represents an important median between the mathematical and practical literatures on perspective. Although Pozzo showed little explicit interest in its geometrical content, his knowledge of the technique was both thorough and subtle.

Notes

I am grateful to Henk Bos, Henrik Kragh Sørensen, and Anthony Gerbino for very valuable comments on the content of the paper and to Claire Neesham and Anthony Gerbino for substantial linguistic improvements. I also want to express my gratitude to L'Accademia di Danimarca, Rome and the Max Planck Institute for the History of Science, Berlin, for housing me while, among other things, I worked on this article.

1. Pozzo (1707; repr. New York: Dover 1989), quoted from the section, *To the Lovers of Perspective*. This edition also appeared in an undated printing by J. Senex and J. Osborn. The original appeared as Pozzo (1693–1700, 2nd ed. 1702–1723). A German translation also appeared: Pozzo (1719).
2. For Pozzo's decorations see De Feo (1988) and the various contributions in Battisti (1998). For earlier biography, see the ample list of references in Frangenberg (2000, 93, n. 2).
3. General descriptions of the *Perspectiva* are found in Bösel (1996); Salviucci Insolera (1998); and Oechslin (1998). The first two editions of volume one and the first edition of volume two are used as a source material by Frangenberg (2000) and Wilberg (1970).
4. For a detailed description of Pozzo's life and *oeuvre*, see Kerber (1971). For an overview, see Hills (1993) and Kerber (1998). For a description of some of his compositions see Milman (1983, 26–27, 1986, 52–55).
5. De Feo (1988, 9).
6. Bösel (1996). For further details on this work, see Lucas (1998) and De Luca (1998).
7. Bösel (1996).
8. Felici et al. (2004).
9. On the *Jesuitenkirche* and on Pozzo's planned stay in Venice, see Bösel (1996). For his other Viennese projects, see Bösel (1998).
10. Hills (1993).
11. My thanks to Anthony Gerbino for this observation.
12. Alberti (1972, parag. 23, 59). See Andersen (2007, 18).
13. Andersen (2007, 34–79). Piero's manuscript was presumably written in the 1470s. It was first published together with a German translation in della Francesca (1899) and again in della Francesca (1942 [1974]). I have used the 1974 reprint.
14. Aguilon (1613, 637–681); Scheiner (1631); Bettini (1642, *Apiarium V*); Bourdin (1661, text to Plate 172); Kircher (1646, 161–196); Dubreuil (1642); Schott (1657, 99–169); Tacquet (1669, 158–177); Dechales (1674, vol. 2, 465–532); Deidier (1744); Rivoire (1759); Scherffer (1781, 191–225). For more on Jesuit mathematics, see Susan Klaiber's contribution (Architecture and Mathematics in Early Modern Religious Orders) in this volume.
15. Dubreuil (1642).
16. Matteo Zaccolini, *Prospettiva lineale*, unpublished manuscript from about the 1620s. See Bell (1996) and Guarini (1671, 452–462: *Tractatus XXVI de stereographia*).

17. Maignan (1648) and Niceron (1638).
18. The Dominicans were represented by Egnazio Danti's substantial reworking of Vignola (1583); the Augustinians by Bourgoing (1661), a stencilled edition of a handwritten manuscript; and the Camillians by Amato (1736).
19. Serlio (1547); Contino (1645); Sirigatti (1596); Vignola (1583); and Accolti (1625).
20. Cataneo (1567); Barca (1620, 25–27); and Viola-Zanini (1629).
21. Valenciennes (An VIII [1799/1800], *Introduction*, iv).
22. On the rising number of books by mathematicians, see Andersen (2007, 30), on the character of these texts, see pages 359–360. On Guidobaldo, see Andersen and Gamba (2008).
23. Galli-Bibiena (1740) and Piranesi (1750).
24. Kerber (1971, 267–270).
25. On this point, see also Marry (1998, [plates: 408–411], 317).
26. Pozzo (1707, 8).
27. Pozzo (1693–1700, vol. 2, Introduction), translated from 1723 edition.
28. See Dardanello (1998). For details about his treatment of stage sets, see Bjurström (1972, 103 and 108–110) and Marry (1998).
29. Pozzo (1707), quoted from text to Figure 68. For a similar opinion of Pozzo's aims, see Marry (2002, 314–316).
30. The distance point is the vanishing point of horizontal lines forming an angle of 45° with a vertical picture plane.
31. della Francesca (1942 [1974], 129).
32. Pozzo (1707), from the comment to Figure 9 in the section "For the greater Help to Beginners."
33. Pozzo (1707), quoted from the section *To the Lovers of Perspective*, c[v].
34. Pozzo (1693–1700, vol. 2, Fig. 52), translated from 1723 edition.
35. Pozzo (1693–1700, vol. 2, Fig. 78), translated from 1723 edition.
36. Pozzo (1707), quoted from the section "An Answer to the Objection made about the Point of Sight in Perspective".
37. Pozzo (1707), quoted from text to Figure 90.
38. Pozzo (1707), quoted from text to Figure 93.
39. For details of the commission, see Frangenberg (2000, 99).
40. Pozzo (1693–1700, vol. 2, Fig. 100), translated from 1723 edition.
41. Andersen (2007, 635, 686–89 and 664–71).
42. "Des Andrea Pozzo Werk … hat wegen der vielen sauberen architektonischen Zeichnungen für Maler und Baumeister immer viel Vorzügliches." Lambert (1774 [1st ed. 1759], vol. 2, 28–29). Reprinted in Lambert (1943, 309–380).
43. See Dziurla (1998); Kowalczyk (1998); Preiss (1998); and Wilberg (1998). For the Jesuit artists who continued Pozzo's style, see Bösel (1996).

Photographic Credits

Author: Figs. 2, 3
John Rylands Library, University of Manchester: Figs. 1, 4–13

References

Accolti, Pietro. 1625. *Lo inganno degl'occhi, prospettiva pratica*. Florence: P. Cecconcelli.

Aguilon, François. 1613. *Opticorum libri sex*. Antwerp: Officina Platiniana.

Alberti, Leon Battista. 1972. *On painting and On sculpture: The Latin texts of De Pictura and De Statua*. London: Phaidon.

Amato, Paolo. 1736. *La nuova practica di prospettiva*. Palermo: V. Toscano & O. Gramagnani.

Andersen, Kirsti. 2007. *The geometry of an art: The history of the mathematical theory of perspective from Alberti to Monge*. New York: Springer.

Andersen, Kirsti, and Enrico Gamba. 2008. Monte, Guidobaldo, Marchese Del. In *Complete dictionary of scientific biography*, vol. 23, ed. Noretta Koertge, 174–178. Detroit: Charles Scribner's Sons.

Barca, Pietro Antonio. 1620. *Avvertimenti e regole circa l'architettura ... prospettiva*. Milan: Malatesta.

Battisti, Alberta, ed. 1998 [2nd ed.]. *Andrea Pozzo*. Milan: Luni.

Bell, Janis Callen. 1996. Matteo Zaccolini. Oxford Art Online. www.oxfordartonline.com.

Bettini, Mario. 1642. *Apiaria universae philosophiae mathematicae*. Bologna: I .B. Ferroni.

Bjurström, Per. 1972. Baroque Theater and the Jesuits. In *Baroque art: The Jesuit contribution*, ed. Rudolf Wittkower and Irma B. Jaffe, 99–110. New York: Fordham University Press.

Bösel, Richard. 1996. Andrea Pozzo [Puteus]. Oxford Art Online. www.oxfordartonline.com.

Bösel, Richard. 1998. L'architettura sacra di Pozzo a Vienna. In *Andrea Pozzo*, ed. Alberta Battisti, 161–176. Milan: Luni.

Bourdin, Pierre. 1661. *Le cours de mathematique*. Paris: S. Benard.

Bourgoing, Charles. 1661. *La perspective affranchie, contenant la vraye et naturele pratique*. Paris: F. Follain.

Cataneo, Pietro. 1567. *L'architettura*. Venice: Figli di Aldo.

Contino, Bernardino. 1645. *La prospettiva pratica*. Venice.

Dardanello, Giuseppe. 1998. La sperimentazione degli effeti visivi in alcuni altari di Andrea Pozzo. In *Andrea Pozzo*, ed. Alberta Battisti, 121–131. Milan: Luni.

da Vignola, Giacomo Barozzi. 1583 [Facsimile reprints: Vignola: Cassa di Risparmio di Vignola, 1974 and Alburgh: Archival Facsimiles, 1987]. *Le due regole della prospettiva pratica con i commentarii del R. P. M. Egnatio Danti*. Rome: Francesco Zannetti.

De Feo, Vittorio. 1988. *Andrea Pozzo: Architettura e illusione*. Rome: Officina Edizioni.

della Francesca, Piero. 1899. *Petrus Pictor Burgensis: De prospectiva pingendi*. Strasbourg: Heitz.

della Francesca, Piero. 1942 [2nd ed. Firenze: Licosa Reprints, 1974]. *De prospectiva pingendi*. Firenze: Sansoni stampa.

De Luca, Maurizio. 1998. Gli affreschi della Galleria del Gesù a Roma: la tecnica pittorica e le nuove acquisizioni alla luce dell'ultimo restauro. In *Andrea Pozzo*, ed. Alberta Battisti, 145–161. Milan: Luni.

Dechales, Claude François Milliet. 1674. *Cursus seu mundus mathematicus*. Lyon: Anisson.

Deidier, Abbé. 1744. *Traité de perspective théorique et pratique*. Paris: C. A. Jombert.

Dubreuil, Jean. 1642. *La perspective pratique ... par un religieux de la compagnie de Jesus*. Paris: M. Tavernier.

Dziurla, Henryk. 1998. Christophorus Tausch, allievo di Andrea Pozzo. In *Andrea Pozzo*, ed. Alberta Battisti, 409–429. Milan: Luni.

Felici, Anna Candida, et al. 2004. The wall paintings in the former refectory of the Trinità dei Monti Convent in Rome. *Journal of Cultural Heritage* 5: 17–25.

Frangenberg, Thomas. 2000. Andrea Pozzo on the ceiling paintings in S. Ignazio. In *Pictorial composition from medieval to modern art*, ed. Paul Taylor and François Quiviger, 91–116. London: Warburg Institute.

Galli-Bibiena, Giuseppe. 1740. *Architettura, e prospettiva*. Augsburg: A. Pfeffel.

Guarini, Guarino. 1671. *Euclides adauctus et methodicus mathematicaque universalis*. Torino: B. Zapatae.

Hills, Helen. 1993. Andrea Pozzo. In *International dictionary of architects and architecture*, ed. Randall J. van Vynckt. Detroit: St. James.

Kerber, Bernhard. 1971. *Andrea Pozzo*. Berlin: Walter de Gruyter.
Kerber, Bernhard. 1998. Opere di Andrea Pozzo. In *Andrea Pozzo*, ed. Alberta Battisti, 465–469. Milan: Luni.
Kircher, Athanasius. 1646. *Ars magna lucis et umbrae*. Rome: L. Grignani.
Kowalczyk, Jerzy. 1998. La fortuna di Andrea Pozzo in Polonia. Altari e finte cupole. In *Andrea Pozzo*, ed. Alberta Battisti, 440–451. Milan: Luni.
Lambert, Johann Heinrich. 1774, 1st ed. 1759. *Die freye Perspective... mit Anmerkungen und Zusätzen vermehrt*, 2nd ed. Zürich: Orell, Gessner, Fuesslin.
Lambert, Johann Heinrich. 1943. *Schriften zur Perspektive*. Berlin: Lüttke.
Lucas, Thomas. 1998. La Galleria del Pozzo nella Casa Professa a Roma: una teologia riscoperta. In *Andrea Pozzo*, ed. Alberta Battisti, 141–144. Milan: Luni.
Maignan, Emmanuel. 1648. *Perspectiva horaria*. Rome: P. Rubei.
Marry, Béatrix. 1998. La scénographie chez Andrea Pozzo. In *La prospettiva: Fondamenti teorici ed esperienze figurative dall'antichità al mondo moderno*, ed. Rocco Sinisgalli, 246–248 [plates: 408–411]. Firenze: Cadmo.
Marry, Béatrix. 2002. Le traité de perspective d'Andrea Pozzo. In *Les cieux en gloire: Paradis en trompe-l'oeil pour la Rome baroque*, ed. Jean-Marc Olivesi, 312–323. Ajaccio: Musée Fesch.
Milman, Miriam. 1983. *Tromp-l'oeil Painting: The illusions of reality*. Geneva: Skira.
Milman, Miriam. 1986. *Tromp-L'œil. Painted architecture*. Geneva: Skira.
Niceron, Jean François. 1638. *La perspective curieuse*. Paris: P. Billaine.
Oechslin, Werner. 1998. Pozzo e il suo trattato. In *Andrea Pozzo*, ed. Alberta Battisti, 189–205. Milan: Luni.
Piranesi, Giovanni Battista. 1750. *Opere varie di architettura, prospettiva, grotteschi, antichità, sul gusto degli antichi Romani*. Roma: G. Bouchard.
Pozzo, Andrea. 1693–1700, 2nd ed. 1702–1723. *Perspectiva pictorum et architectorum/Prospettiva de' pittori e architetti*. Rome: Typis Joannis Jacobi Komarek Bohemi apud S. Angelum Custodem.
Pozzo, Andrea. 1707; reprint New York: Dover, 1989. *Rules and examples of perspective proper for painters and architects, etc*. London: J. Senes, etc.
Pozzo, Andrea. 1719. *Der Mahler und Baumeister Perspectiv*, ed. Johann Borbarth and Georg Konrad Bordeneer. Augsburg: Johann Friderich Probst; Jeremia Wolffs.
Preiss, Pavel. 1998. Pozzo e il pozzismo in Boemia. In *Andrea Pozzo*, ed. Alberta Battisti, 431–439. Milan: Luni.
Rivoire, Antoine. 1759. *Nouveaux principes de la perspective linéaire, traduction de deux ouvrages, l'un anglois de Docteur Brook Taylor, l'autre latin de M. Patrice Murdoch*. Lyon: J. M. Bruyset.
Salviucci Insolera, Lydia. 1998. Le prime edizioni del Trattato. In *Andrea Pozzo*, ed. Alberta Battisti, 207–213. Milan: Luni.
Scheiner, Christoph. 1631. *Pantographice ... prior epipedographicen, sive planorum, posterior stereographicen, seu solidorum*. Rome: L. Grignani.
Scherffer, Karl. 1781. *"Perspektivische Aufgaben," Beyträge zur Mathematik*. Vienna: Edler von Trattnerm.
Schott, Gaspar. 1657. *Magia universalis naturae et artis, Pars I: Optica*. Würzburg: J. G. Schönwetter.
Serlio, Sebastiano. 1547. *D'Architecttura/L'Architecture. Il secondo libro di perpettiva/Le second livre de perspective*. Paris: J. Barbé.
Sirigatti, Lorenzo. 1596. *La pratica di prospettiva*. Venice: G. F. Sanese.
Tacquet, Andreas. 1669. *Opera mathematica*. Antwerp: J. Meursium.
Valenciennes, Pierre-Henri. An VIII [1799/1800]. *Elémens de perspective pratique à l'usage des artistes*. Paris: Desenne.
Viola-Zanini, Giuseppe. 1629. *Della architettura ... con ... regole prospettiva*. Padua: F. Bolzetta.
Wilberg, Peter Vignau. 1970. *Andrea Pozzos Deckenfresko in S. Ignazio mit einem Anhang: Archivalische Quellen zu den Werken Pozzos*. Munich: Uni-Drück.
Wilberg, Peter Vignau. 1998. Le finte cupole e la loro recenzione nella Germania meridionale. In *Andrea Pozzo*, ed. Alberta Battisti, 215–223. Milan: Luni.

Part IV
Narratives for the Birth of Structural Mechanics

The three contributions in Part IV consider the origin of modern structural mechanics from the late seventeenth to the mid-eighteenth century. This subject was one of the most technically difficult for practitioners to adopt, but also the one that would have the greatest future impact on architecture. Indeed, the appearance of this "new science"—as Galileo called it—was momentous, for it portended a new way of using mathematics in the design process, one that would introduce a sharp separation between architecture and engineering that persists to the present day.

Although the essays in this section cohere chronologically and thematically, they also offer contrasting historiographical perspectives. More than our earlier sections, this one may be read as something of a conversation or debate, to some extent mirroring disciplinary divisions within the field itself. Jacques Heyman's essay, summarizing many years of research, compares the mason's traditional rules of design with the basic mathematical tools that underlie the modern science of structures, namely geometry, mechanics and analysis. His approach is that of an engineering scientist with a particular commitment to plastic, or limit-state, theory, a method of structural analysis and design that he helped to establish. Developed in the 1950s, this method proceeded from the recognition that most physical structures are, in essence, statically indeterminate. Subtle changes in settlement, imperfections in materials and in the construction process profoundly affect the internal forces of the structure and will always preclude a "true" evaluation of its conditions under loading. Elastic analysis, which seeks to model the deformation of a structure, is therefore highly sensitive to small changes in support conditions and states of self-stress. In contrast, the plastic method seeks to model the structure's behavior at the ultimate points of yield and collapse. The latter method is more structurally efficient—material is minimized—and arguably more accurate, as it does not purport to account for the actual internal stresses of the structure. It is also simpler. Whereas the elastic method relies on complex differential equations to describe the structure's deformation, plasticity theory returns to "mechanical" equations of equilibrium.

Heyman's background as an engineering scientist informs his historical perspective. Indeed, he sees the development of structural analysis in terms of its advance towards—and its detours from—the present state of knowledge. "The historian of science," Heyman has recently written, "is a fully qualified Whig historian."[1] His landmark book, *Structural Analysis: A Historical Approach* (1998), brought this avowedly "internalist" approach to bear on the history of the theory of structures, providing the first comprehensive account of the contributions of the plastic method. As Heyman shows in that book and in his contribution here, mathematicians were at the forefront of this development, which unfolded for the most part in the pages of learned journals.

In contrast to Heyman's approach, our next two contributors take what might be called a synchronic view of the same phenomena. Rather than trace the development of the subject from a known vantage in the present, Pascal Dubourg Glatigny and Filippo Camerota concentrate on the *reception* of structural mechanics in specific historical contexts. Mathematicians play a central role in their accounts, too, but Dubourg Glatigny and Camerota are concerned to see them as merely one set of actors in a wider—and often fractured—community of interest. Both authors try to recapture a sense of the novelty of this new building science, when structural mechanics did not yet have the epistemic authority it enjoyed by the mid-nineteenth century. By concentrating on the intersection of structural theory with "real world" problems of building, they show that the reception of this new building science was, in fact, messy and contentious, particularly among builders and administrators. In contrast to mathematicians, these groups worked with physical structures daily and bore ultimate responsibility for them.

Dubourg Glatigny's essay, the second in this section, looks at the expertise convened in 1742 to examine the stability of the dome of St Peter's basilica. Like the controversy over the design of the Paris Pantheon in the 1770s, the St Peter's expertise is well known, largely because of the mathematicians who were called in to provide an analysis of the structure and recommendations for repair. Their work helped to spur public engagement with the new science of structural mechanics. Dubourg Glatigny focuses on an early moment of this episode, in particular, the dispute between the ecclesiastic and humanist Giovanni Bottari and the mathematicians who were initially asked to provide advice on the restoration of dome: the Jesuit Roger Boscovich and his two Minim colleagues, François Jacquier and Thomas Le Seur, all three self-confessed Newtonians. Bottari, acting as a consultant of the Reverenda Fabbrica di San Pietro, was by no means unfamiliar with contemporary mathematics or large-scale engineering. Yet he proved to be the mathematicians' most outspoken and implacable opponent, the only participant of the congress to reject out-of-hand their analysis and recommendations. What he objected to was the process of abstraction by which the mathematicians claimed to be able to represent the physical structure of the dome as a purely geometrical object. Dubourg-Glatigny looks at this debate as contest of legitimacy and authority, exacerbated by a disciplinary and epistemological gulf between the two parties.

Filippo Camerota takes a different approach to the same theme. In the third part of his essay, he describes the influence of Galileo and his *Discorsi intorno a due nuove scienze* (1638) on the overlapping worlds of architecture, mechanical science, and engineering. The book's impact was immediate and profound. The new science of the "resistance of solids" served to undermine the Vitruvian doctrine of proportion, while that of "local", or projectile, motion called attention to the variety of mathematically definable curves that might find an application in building. In both cases, Galileo's treatise led to a re-evaluation of *firmitas* as *the* primary principle of building, elevating structural solidity over the other two Vitruvian concepts of beauty and convenience. As Camerota shows, these insights quickly shook up relations between the various professional communities of the building world. The reverberations are evident both in the discussions of Florentine academies in the 1660s and in the 1697 expertise to look into the cracks in the dome of Florence cathedral. In both cases, we see the theory of structures—backed by Galileo's authority—becoming a common source of reference in all matters of building. The charismatic Venetian monk Carlo Lodoli took this process to its logical conclusion. His teachings raised solidity from a structural value to a visual and aesthetic one.

In fact, the transformation that Camerota describes went beyond structural mechanics. As he shows in the first and second parts of his essay, Galileo was only one representative of the "new science" to alter architectural culture during the period. The work of Juan Caramuel de Lobkowitz reveals the analogous influence of Descartes. The latter's deductive method of reasoning and his conception of geometric space provided models for Lobkowitz's novel, "oblique" architectural theory. Bernardo Vittone offers a third case study. This Piedmontese architect was profoundly affected by Newtonian optics, in particular, by the idea the eye itself changes shape in viewing distant objects. For him, the perception of architecture was to be defined not by perspective geometry, but by the physiology of vision. These three cases reveal the powerful appeal of a "philosophical" mathematics for seventeenth- and eighteenth century architects.

Note

1. Heyman (2005, 3).

Reference

Heyman, Jacques. 2005. The history of the theory of structures. In *Essays in the history of the theory of structures: In honour of Jacques Heyman*, ed. Santiago Huerta, 1–8. Madrid: Instituto Juan de Herrera.

Geometry, Mechanics, and Analysis in Architecture

Jacques Heyman

Mathematics, in the form of rules of proportion, was used in architecture from very early times. However, it was not until the mid-seventeenth century that the three disciplines in the title of this paper came to be gradually introduced. Their use was crystallised into modern codes of practice for elastic design, which claim to evaluate the state of a given structure under given loading. In fact, these states cannot be observed: a real structure is subject to unknown and unknowable imperfections, which profoundly alter its behavior. Safe designs may still be made: the way forward is to use so-called plastic methods (or limit design in the US). The now established use of the word "plastic" is misleading and could be replaced more meaningfully by "equilibrium". The equations of statics, which preceded those of analysis, turn out to be the key to structural design.

The Mathematical Tools

Geometry, mechanics and analysis are three recognized mathematical disciplines, which may overlap to some extent when they are applied to the study of the influence of mathematics on architecture. The word "geometry" will cover attempts to devise shapes, rational or not, either structural (as the profile of an arch) or not (as the *entasis* of columns). "Mechanics" will be applied to investigations where values are sought for the internal forces in a structure, which could be used to assess the strength and deformation of the structure, or to compute the forces that it exerts on its environment. Finally "analysis" will be used in the technical sense to imply the use of algebraic methods, and in particular the use of calculus. For example, analytical geometry applies algebra and calculus to problems of classical, Euclidean, geometry.

J. Heyman (✉)
Emeritus, Department of Engineering, University of Cambridge, Cambridge CB2 1RD, UK
e-mail: jh10002@eng.cam.ac.uk

A. Gerbino (ed.), *Geometrical Objects: Architecture and the Mathematical Sciences 1400-1800*, Archimedes 38, DOI 10.1007/978-3-319-05998-3_8,

Hooke's master statement of 1675, concerning "The true Mathematical and Mechanichal form of all manner of Arches for Building, with the true butment necessary to each of them," evokes, implicitly or explicitly, all three of the terms. The (Latin) anagram, deciphered after Hooke's death, yields: "As hangs the flexible line, so but inverted will stand the rigid arch." The statement solves the *geometrical* problem; if a set of (arbitrary) weights is attached to a light chain, then the shape of that chain, inverted, will give the shape of the perfect (masonry) arch to carry those same loads. Moreover the *mechanics* of the hanging chain (that is, the determination of forces, not easy but possible in 1675) will yield the value of the horizontal pull necessarily applied at the ends of the chain, and so the value of the horizontal thrust of the corresponding arch. Thus the "butments" may be designed to resist that thrust.

In 1670, Hooke had stated to the Royal Society that he had solved the problem posed by his "master statement" of 1675, and he was pressed, then and later, to provide his "demonstration" (that is, the mathematical proof). He never did this, and he was in fact unable to do the *analysis*—he could not determine mathematically, by the use of classical geometry or the new calculus, the equation for the shape of the hanging chain. It was for this reason that he hid, for the time being and indeed until he died, behind the Latin anagram. He wished to find the solution himself before some more learned geometer profited from his insight. (Hooke would later be hurt, unforgivably, in this way. He had published openly his revolutionary idea that celestial bodies might somehow exert a mutual attraction, but could make no use of this mathematically. The idea was novel to Newton—once he had grasped the idea, Newton had no difficulty in deducing the inverse square law, using the language of classical geometry rather than that of calculus. Hooke was outraged that he was given no credit when the *Principia* was published in 1687.)

The three mathematical giants, James Bernoulli, Leibniz, and Huygens, were engaged in 1690, competitively and successfully, on the analytical solution to the catenary problem; they too were wary of revealing to each other the full details of their discoveries. Hooke had solved the *engineering* problem of the design of arches to carry given loads, and for this purpose the search by the three mathematicians for the equation of the catenary was irrelevant. The solution was sought for the rather limited case of uniform loading, a poor representation for many practical designs. An arch bridge, for example, is required to carry its own non-uniform weight as well as live loads. (Curiously, a parabolic distribution of load gives a reasonable basis for the calculation of an arch bridge.)

A modern example of engineering design which involves all three of the terms in the title of this paper lies in the calculation of thin reinforced-concrete shells—a structural form now somewhat out of fashion. The differential equations of the so-called membrane theory are of course analytical, and are equations of equilibrium (mechanics) written in terms of internal stress resultants; the geometry may be prescribed (for example, a hyperbolic paraboloid) or may be allowed to emerge from the mathematics. Such shells are often of constant thickness (except perhaps near the edges). It is of interest that Hooke, having determined (to his own satisfaction) the proper shape of the two-dimensional arch, sought also the shape of the perfect dome (assumed tacitly to be of constant thickness, as in the innermost of the three domes of St Paul's Cathedral). He declared it to be, again without proof, that of the cubico-parabolical conoid (that is, $y = ax^3$ rotated about the y-axis). The true shape (on the basis of certain assumptions) is very much more complex, but differs only minutely from Hooke's profile. Hooke and Wren worked closely together, and one of Wren's sketches for St Paul's shows a plot, explicitly, of Hooke's Equation.

A Historical Perspective

The rules of architecture, from before Ezekiel (600 BC) through Vitruvius and the Gothic and up to and partly including the Renaissance, were essentially rules of proportion, and thus geometrical. Notions of mechanics and analysis are entirely absent, although builders must have been well aware that structural forces perhaps required the presence of buttresses, and certainly demanded (but did not always receive) good foundations.

It seems strange to a modern engineer, used to considering problems of strength, that rules of proportion alone should lead to viable structures. In fact, the knowledge of material properties was not necessary for most masonry construction; the material used for Greek temples and Gothic cathedrals is very lightly stressed. Important elements (flying buttresses, say) are subject to about one hundredth of the crushing strength of the stone. Modern engineers work modern materials (steel or reinforced concrete) much closer to their limits. The ancient rules were directed to ensuring the overall stability of the structure. For Greek and Roman temples, for example, a limiting ratio on the height to diameter of a column ensured the stability of that column, while the restriction on the width of intercolumniation ensured that monoliths of reasonable size could be used as architraves. The ancient rules were, in general, satisfactory.

To be sure, there were anomalies. A variation of design might result in a shape of flying buttress that could not contain the required compressive forces (as at Amiens c. 1260), with resultant distress, or the overall geometry of a cathedral might be such that any small deformations imposed by a (hostile) environment could destabilize the structure (as probably at Beauvais 1284). However, a building that basically conformed to the established geometrical rules would be safe. An artist/craftsman like Inigo Jones, who had mastered Italian styles (above all, those of Palladio) and had become noted as a theatrical designer, could move on to the design of real buildings with little difficulty.

There was, of course, room for interpretation of the rules, and masons' lodges each had their own design manual. The expertises of 1399/1400 at Milan show masters from all over Europe arguing about the best way to resume the intermitted building of the cathedral. The rules under discussion were largely numerical, but there are glimpses in the recorded minutes of questions of arch thrust and of necessary buttressing. As a matter of great fascination, but of no practical consequence, the question of irrational numbers (such as the square root of 3), which had been latent since Greek times, surfaced again at Milan. Ancient and medieval building work was laid out by use of the great measure, that is, a physical rod (in England the rod, pole, or perch) on which were marked the units of length, subdivided as necessary in accordance with the prescribed rules of proportion. It was appreciated that, no matter how finely the units were subdivided, no mark could be made on the great measure to represent an irrational number. By definition (and proof by Pythagoras) an irrational number could not be expressed as a ratio, that is, proportionately. Masons' rules-of-thumb were largely formulated to avoid them.

By the middle of the seventeenth century the rules of architecture had begun to include concepts other than those of geometry, and the profession started to embrace ideas of what would now be called structural engineering. Considerations of statics were added to the geometrical rules. For example, François Blondel, in 1673, discussed the four principal problems of architecture, and these may be assigned to the first two categories in the title of this paper. The problems concerned:

1. Entasis (geometry)
2. The shape of arches (geometry)
3. Joints between voussoirs in arches (geometry, mechanics?)
4. The strength of cantilever beams (mechanics)

The creation of entasis is important visually, but has little effect on the structural behaviour of a column. The fourth problem was first studied by Galileo in 1638, and is mentioned again below.

Blondel's second and third problems relate to important "engineering" concerns in the design of arches. The masonry arch (and its three-dimensional derivatives, the cross-vault and the dome) was one of the major structural forms, and gave rise to much activity in the Académie and the Royal Society (both were founded around the time Blondel was posing his problems). In 1717, for example, Gautier (in a book on bridges) listed five questions for those "sçavans" to resolve:

1. The thickness of abutment piers for all kinds of bridges (geometry?, mechanics?)
2. The dimensions of internal piers for multi-span bridges (mechanics?, geometry?)
3. The thickness of the arch rib (geometry)
4. The shape of arches (geometry, mechanics?)
5. The dimensions of retaining walls to hold back soil (mechanics)

The problems were worked at throughout the century, and advances made without yet reaching final solutions.

Arches

Stereotomy is usually taken to denote the specification, by geometry, of the way stones are cut, so that they may fit into a coherent structural form. For the two-dimensional arch, with prescribed extrados and intrados, the voussoirs are wedge shaped, with two faces conforming to the profile of the arch, and two defining the joints between stones. Vitruvius states clearly that the joints for a circular Roman arch should be radial, directed to the centre of the circle—and the prescription survives in the term "centering" which denotes the falsework necessary to support the masonry until the keystone is in place. (The geometrical problems involved in describing voussoirs are not easy, and they become formidable when the structure is fully three-dimensional, as say a rampant skew arch.)

However, stereotomy also had implications for engineering. Radial joints may be "obvious" for a circular arch (although slightly different jointing may in fact be more rational), but how should joints be made between the voussoirs of a non-circular or pointed arch and for which the extrados and intrados are not necessarily "parallel"? A full stereotomical statement must involve something more than geometry.

La Hire addressed the problem in 1695 by using mechanics to investigate the internal forces in an arch. It was clear to La Hire that if there were no friction between two adjacent stones (that is, they were perfectly smooth) then a (compressive) force could only be passed from one stone to another if the joint between them were at right angles to that force (tensile forces were not possible—the stones would pull apart). He attacked the mechanics by constructing a force polygon (involving the weights of the voussoirs), and then the corresponding funicular polygon (that is, Hooke's inverted hanging chain) for the arch. For an arch of given shape with smooth voussoirs the funicular polygon is fixed, so that, working backwards, the force polygon can be deduced and finally the weights of the voussoirs found. Now if the springing lines are horizontal then the weights of the springing voussoirs must be infinite (Hooke's chain can never become vertical under finite loading).

La Hire realised that the assumption of frictionless joints was unproductive, and he abandoned the work until 1712. His new attack allowed for friction so that sliding at the joints was inhibited, and his analysis was directed to a more meaningful solution of the mechanics of the arch, and in particular to determining the value of the abutment thrust. He is at this point not interested in geometry, but rather solely in mechanics; he sketches a semicircular arch of constant thickness, but his method may be applied to an arch of general shape. He unlocks the statics of the arch by considering the mechanism of failure, involving hinging cracks, thus "pinning" the internal forces at a few known locations, from which the necessary equilibrium equations can be written.

Structural Design

There are many criteria that must be satisfied by a successful structure, but three are pre-eminent, and may be labelled strength, stiffness, and stability. A Greek temple, a Gothic cathedral, and a masonry arch bridge are all clearly strong—there is no failure of the material; equally, their members do not deflect in any appreciable way under the action of their own weights or of high winds—they are adequately stiff.

Their stability was assured by correct geometries—the proportional rules had been proved empirically to be correct. La Hire's work was also concerned with stability, and the introduction of mechanics yielded new information—for example, the calculation of abutment thrust for an arch.

Galileo (1638) had spotted the weakness of rules of proportion. For a material of finite strength, a limit would be reached as a structure was progressively increased in size (Galileo, before knowledge of dinosaurs, imagines the elephant to be the largest animal able to sustain its own weight). In fact, builders rarely approached the limits of their materials, but Galileo saw the value of exploring the breaking strength of a structure (implicitly made of wood, although he discusses also the behaviour of stone). Galileo's structure, a cantilever beam, is actually a device for examining the strength in bending of a prismatic rod, and his analysis should really be classified under the heading "strength of materials" rather than "theory of structures." To use modern terms, he had determined the breaking stress in tension of the rod, and wished to apply this value to the fracture in bending of the same rod. The resolution of the problem is of course brilliant, although a numerical constant that emerges was subject to much discussion over the next two centuries.

Galileo uses very little of what we think of as mathematics. Although algebraic equations were acquiring modern form, they had not yet been applied to problems of mechanics. Instead, he used the classical theory of ratios to be found in Euclid. He did not write in full the equations of equilibrium, but his results are essentially correct, and academic mathematicians continued to explore the cantilever beam for the next century along much the same lines. Conflating the beam's structural and material properties, they determined its shape not as prismatic but curved, that is, designed to fail at every section along its length (Galileo himself started this enquiry).

Parent corrected Galileo's mechanics in 1713. Implicitly (since such ideas were slow in being formulated), he assumed that bending strains would be linearly elastic through the depth of the beam and so determined a different value from Galileo's for the constant in the expression for the breaking strength. Thus began the science of the strength of materials, culminating in Navier's 1826 formulation of an elastic "philosophy" of design applied to the theory of structures as a whole. From this point, the elasticity of structures themselves was to remain explicit. Galileo had no such notion. His work was directed to the determination of fracture strength, rather than to the actual behaviour of a structure under load.

The French Panthéon

Parent's work was ignored, but the subject slowly gained ground. By 1760, the contributions of Galileo and La Hire (and others, including Euler, Musschenbroek, James Bernoulli, Vauban, and Bélidor) were included in the syllabus of the two-year course at the "university" of Mézières. Calculus was by now well established and was taught in the mathematics classes. Mézières was the school for officers of the Corps Royale du Génie, to which Coulomb was admitted in 1760; he graduated just under two years later. In 1764, he was posted to Martinique, where he stayed for nine years, engaged in the design and construction of the island's defenses. Coulomb found that what he had learned at Mézières was not sufficient to resolve the four great problems of civil engineering in the eighteenth century, all of which presented in his work in Martinique. He described the theoretical advances made during his nine years' stay to the *Académie* in 1773, on his return to Paris.

The title of his *Essai* was "On the application of rules of maximum and minimum to some statical problems, relevant to Architecture." The four topics were:

1. The strength of columns
2. The strength of beams
3. The thrust of arches
4. The thrust of soil

These echo the lists of Blondel and Gautier, but notions of geometry have almost entirely disappeared. Coulomb's tools are those of mechanics (equations of statics) and of analysis (maxima and minima determined by the use of calculus). This seems to be one of the first occasions when analysis is applied by an engineer to problems of civil engineering.

Coulomb's paper was published in the middle of the 50-year dispute over the design of the church of Sainte-Geneviève (later the Panthéon). The episode is one of the earliest instances in which the new conceptual tools were "applied" to an actual building. Although Soufflot had begun work in 1756, the project progressed slowly. By 1770, he had ceded oversight of construction to the architect Rondelet. It was in that year that Patte, another architect, wrote a *Mémoire* criticizing the design of the crossing piers on geometrical grounds. A year later, Gauthey, an engineer, refuted Patte's views, using his knowledge of mechanics to argue for the piers' soundness. These "engineering" judgments, based on considerations of statics, were clearly superior to Patte's proportion-based arguments, but they did not gain many adherents. The situation was further compounded in 1776, when defects appeared in the "inadequate" piers. As Rondelet appreciated (his views were published in a late *Mémoire* of 1798), the piers were defective because of poor construction, not poor design. However, he lacked the insights of mechanics to support his case. The *Académie* (by then the *Institut*), of which Coulomb was a member (although he appears to have played no part in the discussions), appointed several committees—of architects, of architects afforced by engineers, and again afforced by mathematicians—to look at the issue, but there was no immediate resolution of the dispute. It was not until the early nineteenth century that restoration work was put in hand, and Rondelet enlarged somewhat the four crossing piers.

Stiffness

Galileo did not discuss the deformation of his cantilever beam, and indeed, before the invention of calculus, this was hardly possible. However James Bernoulli had mastered the use of this new tool, and in 1691 he used it to hit upon an important hypothesis (which he too published in a Latin logogriph for fear of piracy), that the curvature of a uniform (elastic) strip, would be proportional at each section to the bending moment at that section. Starting from a (hypothetical) problem in mechanics, Bernoulli thus transformed the whole question of the "elastica" into a purely mathematical investigation, which culminated in the definitive analysis by Euler in 1744. The mathematics was difficult and complex, involving closed-form solutions of fourth-order differential equations in terms of new mathematical functions, and the derived shapes were clearly not representative of anything that might be useful in architecture. However, Euler reintroduced some equations of mechanics into the work, where he saw that the very smallest deformation of an initially straight strip could be maintained only in the presence of specified loading. Effectively, a straight column would stay straight under axial load until that load reached a certain limiting value, the "Euler" buckling load. This concept is of fundamental importance in all structural design that involves the possibility of buckling.

Scholium

By the middle of the eighteenth century the tool of analysis had been added to those of geometry and mechanics, and the stage was set for the development of a recognizably modern theory of structures. The description of deformation became, in Navier's hands, one of the three essential ingredients of an (elastic) theory, the other two being the equations of statics, and the specification of material properties—for example, Hooke's Law.

A statically-determinate structure may be "solved"—that is, the internal stress resultants may be found—by the use only of equations of statics. Euler had realized that there were structures (and this is the usual case) for which the equilibrium equations did not suffice: those structures that are statically indeterminate or hyperstatic. Euler anticipated Navier's schema in presenting a simple particular example (a four-legged table). Even in these cases, however, equations of equilibrium, deformation, and material properties will together provide enough information for the solution of a hyperstatic structure.

Navier presented his scheme of "indeterminate" computation by placing a rigid prop under the free end of Galileo's cantilever. When the beam is subjected to a specified transverse loading, a force is induced in the prop, but the equations of statics (*mechanics*) do not enable its value to be determined. However, the internal stress resultants (bending moments) can be calculated in terms of the unknown force. From the second-order differential equation of bending (*analysis*), the deformed shape of the beam can be found (*geometry*) still in terms of the unknown force. Finally, those calculated deformations must be such that the boundary conditions are satisfied—the deflection of the beam must be zero at the rigid prop. This condition completes the solution.

It took over a century—and arguments continue—to appreciate that the "Navier" solution is not one that can be observed in a real structure. The fault lies in the assumption of boundary conditions. In reality, the prop under Galileo's beam is not absolutely rigid: the footings under a framed building will settle slightly, the abutments of a bridge will move under the action of arch thrust. Tiny displacements of these kinds have a very large effect on the internal state of a structure. Moreover, it is impossible to specify the small movements imposed by the environment, and it is therefore impossible, as a matter of fact, to make calculations describing the "actual" state of a structure.

It is of extraordinary interest that Coulomb, in his discussion of the behaviour of the arch, does not attempt to describe the actual state. Instead, his application of the "rules of maximum and minimum" enable him to place upper and lower limits on the structural quantities—for example, on the value of the abutment thrust.

In the same way, the structural engineer may make a safe design of Galileo's propped beam, even if it is impossible to determine its actual behavior. For this simple example, modern plastic theory may be used instead of "elastic" analysis. To do this, equations describing the static state of the structure are written. For a hyperstatic structure, that is, one which cannot be described by the laws of statics alone, there is more than one possible distribution of internal stresses. However, these may be manipulated so that the maximum stress the material is capable of sustaining is nowhere exceeded. No reference need be made to the deformation of the structure or the boundary conditions. All that is needed is the use of mechanics, considering the equations of equilibrium and the limiting material properties. This gives a powerful method for the safe design of a large class of structures.

References

The historical references in this paper are well documented. Details may be found in:

Heyman, Jacques. 1982. *The masonry arch*. Chichester: Ellis Horwood.

Heyman, Jacques. 1997. *Coulomb's memoir on statics*. London: Imperial College Press.

Heyman, Jacques. 1998a. Hooke's cubico-parabolical conoid. *Notes and Records of the Royal Society of London* 52(1): 39–50.

Heyman, Jacques. 1998b, 2nd ed. 2007. *Structural analysis: A historical approach*. Cambridge: Cambridge University Press.

Epistemological Obstacles to the Analysis of Structures: Giovanni Bottari's Aversion to a Mathematical Assessment of Saint-Peter's Dome (1743)

Pascal Dubourg Glatigny

Visible faults in the dome of Saint Peter's basilica in Rome had raised fears about the structure's stability ever since its completion in 1593. The most extensively documented episode of this long history erupted in the early 1740s, a few years after Prospero Lambertini was elected Pope Benedict XIV. The debates over the causes of the cracks, the ensuing scientific analyses, and the adopted solutions are well known, due to the *Memorie istoriche della gran cupola del Tempio vaticano*, the magisterial treatise published in 1748 by Giovanni Poleni (1685–1761), the mathematician entrusted with the supervision of the restoration work.[1] One of the great points of interest of this episode was the involvement of competing protagonists and factions, including architects, master carpenters, and natural philosophers. Each of these groups benefited from varying degrees of credibility. Beyond the technical issues concerning the dome's structure, the debate raised important questions about the social and intellectual legitimacy conferred by different forms of expertise.

Church officials also took part, both as consultants to and members of the Reverenda Fabbrica di San Pietro, the administrative body in charge of building, decorating, and maintaining the holy temple.[2] Among this group was Giovanni Bottari (1689–1775), one of the few figures involved in the discussions who was neither a scientist nor a builder or architect. Although he had no official position in the curia, Bottari was an influential figure of the Roman intellectual scene.[3] A man-of-letters and connoisseur, Bottari took part in the controversy from its earliest days as a member of the *congresso*, the special commission convened by the pope to explore the problem in January 1743. In his *Memorie istoriche*, Poleni mentions Bottari's name infrequently and only incidentally, reporting neither his statements nor his position. Bottari, however, claimed to speak with the voice of public opinion

P. Dubourg Glatigny (✉)
Centre Alexandre Koyré, 27 Rue Damesme, 75013 Paris, France
e-mail: pascal.dubourg@cnrs.fr

A. Gerbino (ed.), *Geometrical Objects: Architecture and the Mathematical Sciences 1400-1800*, Archimedes 38, DOI 10.1007/978-3-319-05998-3_9,

(*la voce commune*), and he was the only participant who publicly opposed any form of restoration. Indeed, he regarded the building improvement as unnecessary and extremely risky.

A series of official documents and private letters, preserved in Venice, Rome, and the Vatican, provides evidence of the peculiar outsider's role that Monsignor Bottari played in this story. He greatly distrusted both architects and mathematicians, regarding both groups as opportunists. According to him, the architects were motivated primarily by the prospect of gaining new work—indeed, a prominent commission in the most important church of the Christendom—while the scientists seemed eager to grasp any opportunity to strengthen their influence on society. Paradoxically, it was the mathematicians to whom Bottari objected most vigorously. Their approach served to reduce the Vatican Temple to a mere mathematical object, denying its historical, spiritual, and aesthetic significance.

The *Parere di tre mattematici*: The Building as a Mathematical Object

The controversy developed very quickly toward the end of summer 1742, when rumors about a possible collapse of the dome first began circulating. In October, the papal physician (*archiatra pontificio*) Antonio Leprotti wrote to his friend Bottari to learn more about the worrying gossip (*le strepitose ciarle*) spreading through town.[4] The chatter must have been widespread and at least partly credible. On 21 November 1742, following a site visit in late September, the Pope decided to commission an expertise by "the most eminent mathematicians," fathers François Jacquier (1711–1788) and Thomas Le Seur (1703–1770) of the Minim order. The Jesuit Roger Boscovich (1711–1787) joined the two monks soon after. The French mathematicians had a peculiar position in the Roman scientific landscape. As editors of the most recent edition of the *Principia*, published in Geneva, they were well-known Newtonians.[5]

This "external" consultation was supposed to put an end to growing uncertainty, the result of several inconclusive investigations conducted over the years by the Fabbrica's own architects. The papal commission strived to be clear therefore about the subject, scope, and limits of the review, which was intended to be definitive. The document underlines the need to establish an opinion "not so much on the present state of the damage observed on the dome, but rather on its restoration, so that the architects can implement the solutions considered to be the most necessary."[6] Faced with a great diversity of opinion among the architects and the master builders, the Fabbrica called on the mathematicians with the hope of settling the argument.

To publicize its results, the commission published an official report in early 1743. The *Parere di tre mattematici* is a short booklet organized into three parts.[7] After a brief description of the dome, the first part presents the damage observed by the mathematicians themselves. It lists meticulously the various cracks and attempts to establish, as far as possible, when they appeared by comparing "the present state of the dome with an earlier one." The latter effort was only partially successful, due to the lack of archival evidence. The second part tries to determine if the cracks were due to a structural instability or to accidental causes, such as wind loading or ground settlement. The last part is dedicated to the presentation of the "system", a theoretical model of the structure used to elaborate the proposed solution. The authors carefully correlate the system to the present state of the building and calculate the interaction of weights and forces occurring within it. In conclusion, they recommended the placement of iron rings around the dome and the buttresses of the drum. Despite the scientific analysis and extensive calculations, the solution was traditional. Iron chains and rings were used widely to ensure the stability of domes. In fact, two such devices had already been incorporated into the dome of St Peter's during its construction.

The explanation of the "general system" is accompanied by certain hypotheses, with the restrictions Newton gave to this word.[8] They serve to determine the centre of gravity and its practical consequences, as well as the method of calculating the thickness of the pillars. Architects were used to treating these issues with practical rules-of-thumb, but here they are treated in a novel way. The system conceives the dome as an organic complex of abstract elements, while eliminating indeterminate causes, that is, factors that cannot be calculated. Such unknowns might include the state of the foundations, the firmness of the ground, or the effects of natural phenomena, including wind, lightning, or earthquakes. To elaborate a solution grounded on reason, the mathematicians try to isolate their object, as a natural philosopher working in his cabinet would isolate a phenomenon in an experiment. In this connection, their use of Musschenbroek's conclusions on the resistance of materials is particularly relevant.[9] Excerpting his value for the resistance of iron, they applied it directly to the existing iron ring around the dome. Possible variables, however, that might influence this force—the regularity of the ring's dimensions, the quality of the screws fixing the different elements together, or its present state of conservation—are not considered. In this analysis, the dome is transformed into an object of purely theoretical knowledge. Indeed, most of the authors' supporting references are to recent work by the French Academy of Sciences, in particular to the essays of Philippe de La Hire.[10] The publication's sole engraving illustrates the damaged cupola in elevation and section, juxtaposed with force diagrams of the different theories discussed in the text (Fig. 1). Architects were no strangers to problems of stability, but the authors were clear that their approach would be different. In the preamble to their study, the mathematicians affirmed that this "special situation is one that requires the theories of mathematicians more than practice."[11] Architects' expertise, in particular, was implicitly excluded from the demonstration.

The mathematicians' conclusions were printed in early January 1743 and distributed widely among scholars and amateurs. On 22 January, the Pope convened a congress at the Quirinal to agree on a definitive solution. During this meeting, Boscovich demonstrated the system, showing the audience the network of cracks then visible. For this, he used the model of the dome constructed by Michelangelo between 1558 and 1561, on which Luigi Vanvitelli (1700–1773), official architect of the Fabbrica, had drawn the positions and shapes of the faults (Figs. 2 and 3).

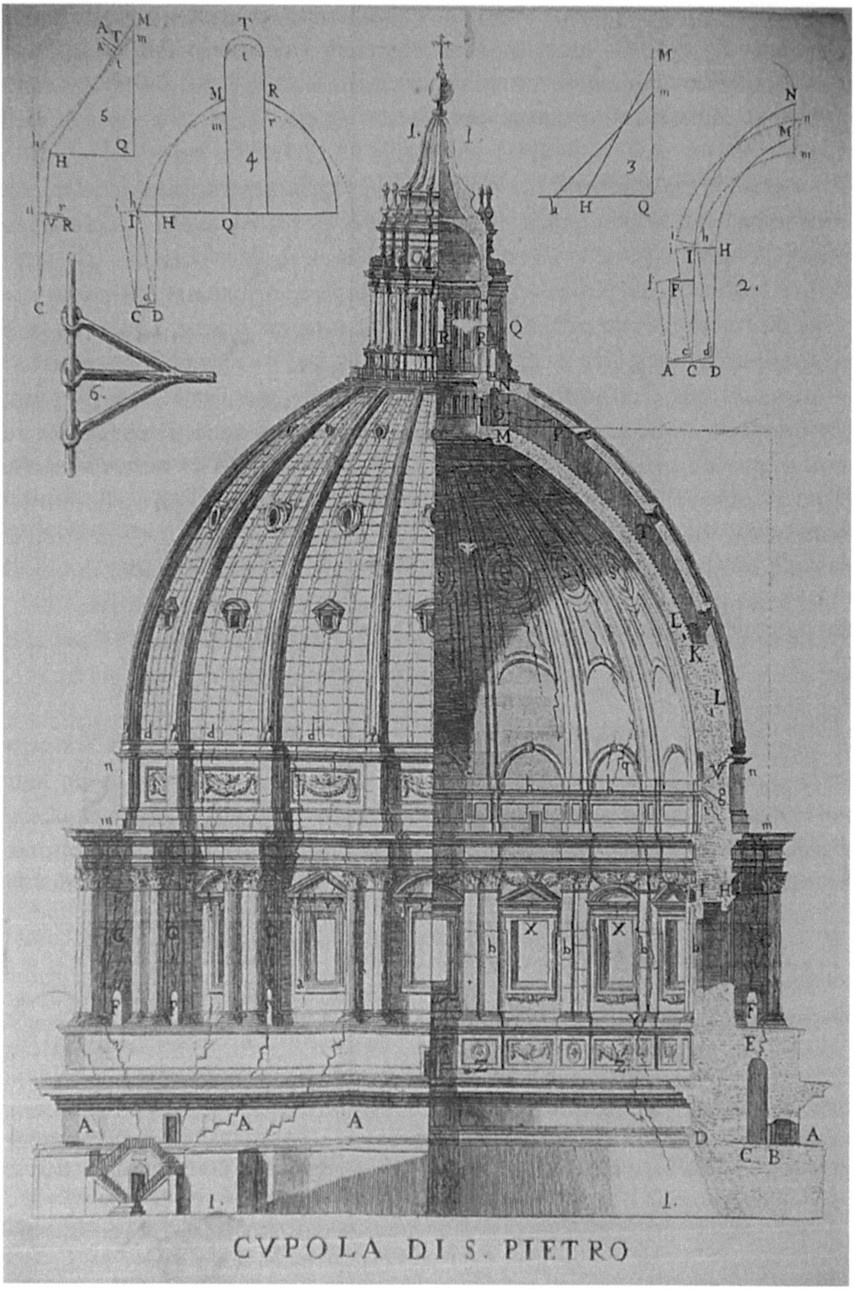

Fig. 1 Elevation and section of the damaged cupola of St Peter's, juxtaposed with force diagrams (From Jacquier et al. [1742])

Fig. 2 Wooden model of Michelangelo's dome of St Peter's basilica, 1558–61, with later additions by Giacomo della Porta

The audience was selected by the administrator of the Fabbrica, Francesco Olivieri.[12] The participants were composed of three different groups. First were the architects and builders from inside and outside the Fabbrica, including Luigi Vanvitelli, Ferdinando Fuga (1699–1782), and Nicola Salvi (1697–1751). The second group consisted of other mathematicians like our three authors. They included Diego Revillas (1690–1746) and Michelangelo Giacomelli (1695–1774). There were, finally, two amateur non-specialists, the marquis Girolamo Theodoli (1677–1766) and Giovanni Bottari. Curiously, other than Olivieri, none of the cardinals of the

Fig. 3 Detail of the wooden model of Michelangelo's dome, showing painted "cracks" in the structure by Luigi Vanvitelli

Fabbrica was present. This was contrary to the normal custom of the institution, which required any work in or on the Basilica to be approved by a committee of its members.

"Everyone acknowledges the damage" was the official conclusion of the congress. The few expressed reservations were mostly trivial. Revillas, for one, doubted whether the cracks in the crossing piers were related to those observed in the dome. The former had given rise to a controversy in 1680 over Bernini's intervention in the crossing.[13] Likewise, concerning the prescribed solution, all architects agreed with the three mathematicians: that new rings should be installed. Among the architects, only Filippo Barigioni claimed to have difficulties with the proposed solution, but only "from the attic and above." His reservations appear to contest the placement of rings, rather than the solution itself. The relative unanimity among the participants, however, was more apparent than real. Bottari was the sole member of the commission to take the unusual step of withholding his opinion. He announced that he would offer one only after a site visit and would let it be known in written form.

Giovanni Bottari and the Nature of Architecture

How and why Bottari was included in the commission is hard to establish. Bottari was a scholar from Florence, famous for his work on grammar and literature. Also trained in geometry, he had edited an edition of Galileo's works, as well as a compendium of Apollonius' *Conics* by the Pisan professor Guido Grandi.[14] Close to Bartolomeo Corsini, nephew of then-pope Clement XII, Bottari followed his patron to Rome in 1730. In the early 1730s, he was asked to consult on several hydraulic projects, particularly in the river Po area. It was during this period that Bottari may have met the Bolognese mathematician Eustachio Manfredi, who would also later be involved in the dome controversy. In 1732, Bottari accompanied Manfredi to the Tiber in Perugia to examine the possibility of making a portion of the river navigable.[15] This experience may have given Bottari some legitimacy in the fields of engineering and architecture. The election of Benedict XIV furthered Bottari's opportunities for patronage. He was nominated to several academies, in particular, that of Saint Luke, becoming an "academician of honour" in 1738.[16]

Bottari had a polemical mind, involving himself continuously in cultural, doctrinal, and theological controversies. He had a great interest in the arts, Tuscan artists, in particular; his name is still famous for his edited collections of artists' letters and unpublished documents.[17] His own views on art were transmitted mainly through the *Dialoghi sopra le tre arti del disegno* published in 1754, in which he proposed a fictitious dialogue between the biographer Giovanni Pietro Bellori and the painter Carlo Maratta. Bottari apparently started it when he arrived in Rome in the 1730s, intending to defend the prestige of Tuscan artistic values in the Holy City. A hypercritical book, first published anonymously, it is tinged with a strong feeling of *campanilismo* and disparages many contemporary Roman painters. It also remains a cornerstone of the historiography of painting conservation.[18] More important for our purpose, however, is the large part of the second dialogue dedicated to architecture, specifically to the question of domes and vaults. It is here that Bottari makes precise reference to Saint Peter's dome.

The discussion concerns Michelangelo's intervention in the design following the death of Antonio Sangallo the Younger and focuses, in particular, on the sculptor's battles with the "Sangallo sect", who accused him of having ruined the building. To this charge—evidently still a live issue in the early eighteenth century—Bottari countered vigorously: "Michelangelo surpassed all the Greeks and made something look rather like a divine miracle than any human artifact."[19] In the mouth of Maratta, Bottari later qualifies the dome as "an admirable machine whose excellence is quite obvious to anybody."[20] This conception of the dome contrasts emphatically with that of the three mathematicians. Bottari saw the work primarily as the product of human art and ingenuity and as the legacy of a "divine"—not to mention Tuscan—artist. It was the pre-eminent building of the modern age, surpassing not only the architecture of contemporaries like Sangallo but that of the ancients themselves. Bottari's humanist notion of excellence as a criterion of quality in the arts extends even to its structural character. The dome is emphatically not a *sistema*, but a *macchina*—in the ancient sense of a structure excelling by its size and stability.

Bottari's defense of Michelangelo also reveals a general suspicion of the professional ambition of Roman architects, an admonition he would later make repeatedly in his assessment of the various proposals to restore the dome. In the *Dialoghi*, this sentiment is particularly apparent in his criticism of Luigi Vanvitelli, whom the author chastised for having spoiled Michelangelo's work at Santa Maria degli Angeli.

To Bottari, Vanvitelli's work manifested nothing less than "the depravity in which this century has fallen in architecture."[21] Vanvitelli was, of course, the acting architect of the Fabbrica of Saint Peter's, responsible for the restoration work on the dome in the 1740s.

Bottari's position here must not be misunderstood. In no sense was he rejecting the role of mathematics in architecture. In the *Dialoghi* themselves, he emphasized the necessity of "universal rules… to teach how to calculate the strength of the arches and of the vaults, the resistance of the wall structure, how to balance the forces, which can be only learned from geometrical doctrine, mechanics, and similar learning."[22] Nor did Bottari have difficulty with mathematicians. He was particularly friendly with Poleni, with whom he corresponded both before the controversy broke out and after.[23] Bottari's position, rather, is that mathematics—indeed, science in general—must be absorbed by *ingegnium*. This intellectual capacity is not only linked with grace and beauty, it also relies on historical knowledge. In the same way that an architect chooses a particular element according to a detailed knowledge of its historical use and from the analysis of a particular situation, a suitable mathematical analysis must respond to the history and use of its object. Indeed, this was a case where the analysis had to be particularly sensitive, given the building's symbolic importance and institutional context.

After the meeting of 22 January 1743, Bottari reserved the right to express his views in a written statement. His declaration is known to us through a manuscript copy kept in the Vatican library.[24] In this document, he first described the *Parere* of the three mathematicians. He characterized it as consisting not of three parts, but of four: "two of facts and two of speculation, the last two based on the foundations of the first." Where the authors had described the visible damage to the dome in one section, Bottari was careful to distinguish the more recent cracks from the older, historically verified ones. He insisted that "the system of damages" suggested in the last part constituted a construction logically based on the facts established in the first. His main criticism was that this part of the work was imprecise and incomplete, rendering the conclusions false. The description of the cracks had not been elucidated in enough detail, nor did it provide a comprehensive picture of the situation. Furthermore, the vocabulary used by the three mathematicians was vague and unspecific. Bottari acknowledged that the mathematician's statement had been carefully established after many visits on the scene, but he still considered the results far from satisfying. The investigators, he pointed out, had measured only the length of the cracks, not their width or depth. Some facts, moreover, had been taken from another report, without verifying them or quoting the source. The authors had relied on this account for the measured incline of the leaning walls and piers of the drum. "Blind belief is the enemy of geometry," wrote Bottari, castigating the authors for failing to make their own direct visual investigation. A particularly important indicator was the dovetailed marble revetments (*marmi a coda di rondine*) placed in different places on the most visible cracks. These had been recently installed only to be broken by continued movement of the structure and therefore constituted one of the mathematicians' main arguments for an urgent intervention. Bottari, however, considered it impossible to deduce anything from their assertions. The authors had failed to report the dates when the marble plates were fitted, nor the exact variation between broken and unbroken pieces, or the number of the plates involved.

Bottari's Jansenist sympathies may have prejudiced him against the three mathematicians—members of both the Minims and the Jesuits—but religious differences are not sufficient to explain his position. Nor was it simply a question of method and insufficient precision. His reticence was also linked to the idea of the "system" and to a lack of diachronic and historical perspective in the analysis of the building. He recognized the system as an ingenuous proposition, but was very severe on its content. The principal difficulty was the lack of firm historical data about the state of the dome in the seventeenth century. Earlier site reports, for instance, those made by Mattia dei Rossi or others from 1680, were insufficiently precise to allow any conclusions to be drawn about subsequent deterioration. Nor did the commission's apparent unanimity deter him. As the three authors had themselves admitted, the solutions proposed were traditional: "more from architects than from geometers." As Bottari suggested in his discourse about the restoration of the works of art, architects and artists remain on the side of the project, while scholars control the reception and the *a posteriori* judgment. This demarcation was to be even more carefully observed in such a highly symbolic historic monument. The judgment of the committee's technical experts therefore meant little to him.

The mathematicians realized immediately that they faced a dangerous enemy. They even had to apologize to the physician Leprotti, promising him that they would do nothing to offend Bottari.[25] He made the mathematicians all the more defensive by spreading rumors that their calculations were false. As Boscovich complained to Poleni three days after the meeting, "a voice is increasing in Rome that all of our calculations are wrong, that everything relies on false suppositions."[26] Boscovich hoped to gain Poleni's support as an undisputed authority, far from the curia with its scheming and back-stabbing. He defended his work in a series of letters: "We have made the calculations several times," he wrote just after the congress, "and verified the data by numerous measures because we found many inaccuracies in the previous drawings." Boscovich also defended the principles followed in the redaction of the booklet, saying "we could only allude to the more difficult things and had to summarize even the most elementary ones in order to be understood in a country where very few people are acquainted with geometry and even fewer with calculation."[27] Such an excuse would hardly have satisfied his Roman audience.

Bottari was the only participant of the congress to reject the presentation of the three mathematicians in its entirety. This refusal was aimed not only at their intellectual background, their "Newtonianism". Bottari's criticism on the quality and reliability of their inquiry disguised a deeper opposition. That is to say, he used their own tools and arguments to contest their entire approach to the question. As we have seen in the *Dialoghi* and in his statement following the congress, Bottari preferred to qualify the dome with the word "machine". He understood this term in its Renaissance sense as a complex of material elements connected one with the other in order to serve a specific function, whether structural, productive, or demonstrative.[28] This "machine", moreover, had a history. The design and construction of the dome belonged to a long tradition of handicraft and erudition brought to a culmination by Michelangelo. A product of his *ingegnium*, it surpassed the works of both the Greeks and the moderns, to become something like a "divine, almighty miracle." The word machine subsumes all of these meanings. "System", on the other hand, is the word the three mathematicians endorse. They use "system" both for the dome and equally for the analysis of its present state. A system is a complex of *abstract* elements, not necessarily linked to a material structure. A system can be studied independently from its context, its historicity, its function.

In the end, Bottari was not able to prevent the Fabbrica from engaging restoration work. The dome, it was believed, could collapse at anytime. This fear was not ungrounded; some of the cracks were wide enough to step through. Nor was Bottari able to impose an alternative solution to the abhorred metal rings. Although he claimed Michelangelo rejected their use, he was perfectly aware that a couple of rings had already been incorporated into the structure. Bottari, however, did achieve two of his goals. The first was to exclude the three mathematicians from the supervision of the restoration work; they were soon replaced by Giovanni Poleni. The second was to compel Poleni and the Fabbrica to gather a special team of collaborators, responsible for compiling documentation about the dome for the official history, published in 1748. The enormous amount of technical and historical material that they gathered became the basis of Poleni's *Memorie istoriche della gran cupola*, still the major source for this controversy. Although Poleni's analysis—and proposed solution—were similar to those of the three mathematicians, he presented them in a way that did not neglect, but rather complemented the demonstrative historical method that Bottari advocated. Poleni, too, recognized the very peculiar character of the "Vatican Temple"—in both its symbolic role and as the product of a very protracted and unusual decision-making process—and adapted his analysis to it. Subsequent studies have downplayed the broader scholarly context of Poleni's mathematical analysis, but it was integral to the conception of the treatise and points to a hitherto unknown aspect of its background and origin.

Notes

1. Poleni (1748).
2. On the Reverenda Fabbrica di San Pietro, see Marconi (2004, 19–36); Basso (1987) and Sabene (2012).
3. The bibliography on Bottari is very scarce. No monograph has yet been dedicated to this important figure of the roman *Settecento*. In the meantime, see Pignatelli and Petrucci (1971) and Consoli (2004, 143–50).
4. Letter from Leprotti to Bottari (20/10/1742), Carteggio Bottari 1660 (32E21), Biblioteca Corsiniana, Rome, fol. 62r–63r.
5. Newton (1739–1742).
6. Arm. 50, B, 17, Archivio della Reverenda Fabbrica di San Pietro (ARFSP), Vatican City, fol 997r.

7. Jacquier, Leseur, and Boscovich (1742). This study has been presented and discussed elsewhere. See, most recently and with references to earlier work: Dubourg Glatigny and Le Blanc (2005), 189–218.
8. Chaudhury (1962).
9. Jacquier, Le Seur, and Boscovich (1742, 27–28).
10. See, in particular, de La Hire (1695, 1712). Also see Couplet (1712, 1730).
11. Jacquier, Le Seur, and Boscovich (1742, 4).
12. The manuscript of Olivieri's notification to participants is kept in Cicognara V-3849, int. 2, #1, Biblioteca Apostolica Vaticana. The list of invited speakers is in Arm. 50, B, 17, ARFSP, fol. 998r.
13. On this episode, see Dubourg Glatigny (2009) and Marder (2008).
14. See Galileo (1718) and Apollonius (1722).
15. Bottari's account was later published in Gambarini (1746).
16. Libro dei decreti dell'insigne Accademia di S. Luca dalli 22 luglio 1726 a li 12 Maggio 1738 (vol. 49), Archive of the Accademia di San Luca, Rome, fol. 183 v.
17. Bottari (1754–1773).
18. See also Bottari's important re-edition, in 1730, of Raffaello Borghini's *Il Riposo*. On this work, see Procacci (1955, 229–49).
19. Bottari (1865, 31).
20. Bottari (1865, 60).
21. Bottari (1865, 35).
22. Bottari (1865, 84–85).
23. Bottari wrote to Poleni about Vitruvius in 1741 and kept him posted on the dome controversy, at least until 1744. Several letters from Poleni to Bottari are kept in Carteggio Bottari, 32G33, Biblioteca Corsiniana, Rome, fol. 4r, 8r, 10r, 24r.
24. Cicognara V-3849, int. 2, #3, Biblioteca Apostolica Vaticana.
25. Carteggio Bottari, 1660 (32E21), Biblioteca Corsiniana, Rome, fol. 66r.
26. Boscovich to Poleni, 25 January 1743, Mss. Italiani, cl. 10, n° 304 (6544), Biblioteca Marciana, Venice.
27. Boscovich to Poleni, 22 January 1743, Mss. Italiani, cl. 10, n° 304 (6544), Biblioteca Marciana, Venice.
28. Popplow (2007).

Photographic Credits

Private collection: Fig. 1
Fabbrica di S. Pietro in Vaticano: Figs. 2, 3

References

Apollonius. 1722. *Compendio delle sezioni coniche*. Florence: Tartini e Franchi.

Basso, Michele. 1987. *I privilegi e le consuetudini della Rev.da Fabbrica di San Pietro in Vaticano (sec. XVI-XX)*. Rome: Typis polyglottis vaticanis.

Borghini, Raffaello. 1730. *Il riposo*, ed. Antonio Maria Biscioni and Giovanni Gaetano Bottari. Florence: M. Nestenus, F. Moücke.
Bottari, Giovanni Gaetano. 1754. *Dialoghi sopra le tre arti del disegno*. Lucca: F.M. Benedini.
Bottari, Giovanni Gaetano. 1754–1773. *Raccolta di lettere sulla pittura, scultura ed architettura*. Rome: Niccolò e Marco Pagliarini.
Bottari, Giovanni Gaetano. 1865. *Dialoghi sopra le tre arti del disegno*. Parma: Pietro Fiaccadori.
Chaudhury, Pravas Jivan. 1962. Newton and hypothesis. *Philosophy and Phenomenological Research* 22(3): 344–353.
Consoli, Gian Paolo. 2004. Giovanni Gaetano Bottari e Francesco Algarotti: per una genealogia del neoclassicismo. In *Ferdinando Sanfelice: Napoli e l'Europa*, ed. Alfonso Gambardella, 143–150. Naples: Edizioni Scientifiche italiane.
Couplet, Pierre. 1712. De la poussée des voûtes. *Histoire de l'Académie royal des sciences, avec les mémoires de mathématique et de physique*: Part II, Mémoires, 79–117. Paris: Imprimérie royale, 1729.
Couplet, Pierre. 1730. Seconde partie de l'examen des voûtes. *Histoire de l'Académie royal des sciences, avec les mémoires de mathématique et de physique*: Part II, Mémoires, 117–141. Paris: Imprimérie royale, 1730.
de La Hire, Philippe. 1695. *Traité de mechanique*. Paris: Jean Anisson.
de La Hire, Philippe. 1712. Sur la construction des voûtes dans les edifices. *Histoire de l'Académie royal des sciences, avec les mémoires de mathématique et de physique*: Part II, Mémoires, 69–77. Paris: Imprimerie royale, 1731.
Dubourg Glatigny, Pascal. 2009. Bernin disputé: science, art et architecture dans la Rome de 1680. In *Conflicting duties: Science, medicine and religion in Rome, 1550–1750*, ed. Maria Pia Donato and Jill Kraye, 89–108. London: Warburg Institute.
Dubourg Glatigny, Pascal, and Marianne Le Blanc. 2005. Architecture et expertise mathématique: La contribution des Minimes Jacquier et Le Seur aux polémiques de 1742 sur la coupole de Saint-Pierre de Rome. *Mélanges de l'Ecole française de Rome, Italie et Méditerranée* 117(1): 189–218.
Galilei, Galileo. 1718. *Opere di Galileo Galilei Nobile Fiorentino*, ed. Tommaso Buonaventuri, Guido Grandi, and Benedetto Bresciani. Florence: Gio. Gaetano Tartini e Santi Franchi.
Gambarini, B. 1746. *Delle ragioni, e de'rimedi delle inondazioni del Tevere, della somma difficoltà d'introdurre una felice e stabile navigazione da Ponte Nuovo sotto Perugia, fino alla foce della Nera nel Tevere e del modo di renderlo navigabile dentro Roma*. Rome: Antonio de'Rossi.
Jacquier, François, Thomas Leseur, and Ruggero Giuseppe Boscovich. [1742]. *Parere di tre mattematici sopra i danni, che si sono trovati nella cupola di S. Pietro sul fine dell'anno [1742] dato per ordine di Nostro Signore Papa Benedetto XIV*.
Marconi, Nicoletta. 2004. *Edificando Roma Barocca*. Rome: Edimond.
Marder, Tod A. 2008. A finger bath in rosewater: Cracks in Bernini's reputation. In *Sankt Peter in Rom 1506–2006*, ed. Georg Satzinger and Sebastian Schütze, 427–434. Munich: Hirmmer.
Newton, Isaac. 1739–1742. *Philosophiae naturalis principia mathematica*. Geneva: Barillot et fils.
Pignatelli, G., and A. Petrucci. 1971. Giovanni Gaetano Bottari. Dizionario Biografico degli Italiani. http://www.treccani.it/biografie.
Poleni, Giovanni. 1748. *Memorie istoriche della gran cupola del tempio Vaticano*. Padova: Stamperia del Seminario.
Popplow, Marcus. 2007. Setting the world machine in motion: The meaning of the *Machina Mundi* in the middle ages and the early modern period. In *Mechanics and cosmology in the medieval and early modern period*, ed. Massimo Bucciantini, Michele Camerota, and Sophie Roux, 45–70. Florence: L.S. Olschki.
Procacci, Ugo. 1955. Di uno scritto di Giovanni Bottari sulla conservazione e il restauro delle opere d'arte. *Rivista d'arte* 30: 229–249.
Sabene, Renata. 2012. *La fabbrica di San Pietro in Vaticano*. Rome: Gangemi.

A Scientific Concept of Beauty in Architecture: Vitruvius Meets Descartes, Galileo, and Newton

Filippo Camerota

The artistic and architectural theories of the seventeenth and eighteenth centuries were strongly marked by a critical comparison between the greatness of the ancients and the inventions of the moderns. In some cases, the parallel assumed the sterile form of a purely academic debate, but in others it gave rise to entirely original reasoning and theoretical elaboration. This is the case, for example, of the critical revision of Vitruvius's concepts in light of the extraordinary developments in modern science. A number of architects and architectural theorists believed they could reinvigorate architecture by re-elaborating the ancient theory of proportions in light of the recent achievements in optics, mechanics and projective geometry. Three cases are, I believe, particularly eloquent: the oblique architecture of Juan Caramuel de Lobkowitz in relation to Descartes' philosophical thought, the optical-perceptivity theory of Bernardo Vittone in relation to Newton's optical discoveries, and the rationalism, or functionalism, of Carlo Lodoli in relation to Galileo's studics in mechanics.

Caramuel and Descartes

Caramuel's theoretical contribution to architecture consists of the presentation of an *ars nova*, a branch of mathematics, which he termed *architectura obliqua* (Fig. 1).[1] He presented his theory in two publications: his fundamental mathematical work, the *Mathesis biceps* and his more specific *Architectura civil recta y obliqua*. *Recta* and *obliqua*, Caramuel explains, do not stand in the Ciceronian sense for "good" and "bad", "right" and "wrong", but for "done according to the authority of the learned" and "done according to what is dictated from time to time by reason."

F. Camerota (✉)
Museo Galileo, Piazza dei Giudici 1, 50122 Florence, Italy
e-mail: f.camerota@museogalileo.it

A. Gerbino (ed.), *Geometrical Objects: Architecture and the Mathematical Sciences 1400-1800*, Archimedes 38, DOI 10.1007/978-3-319-05998-3_10,

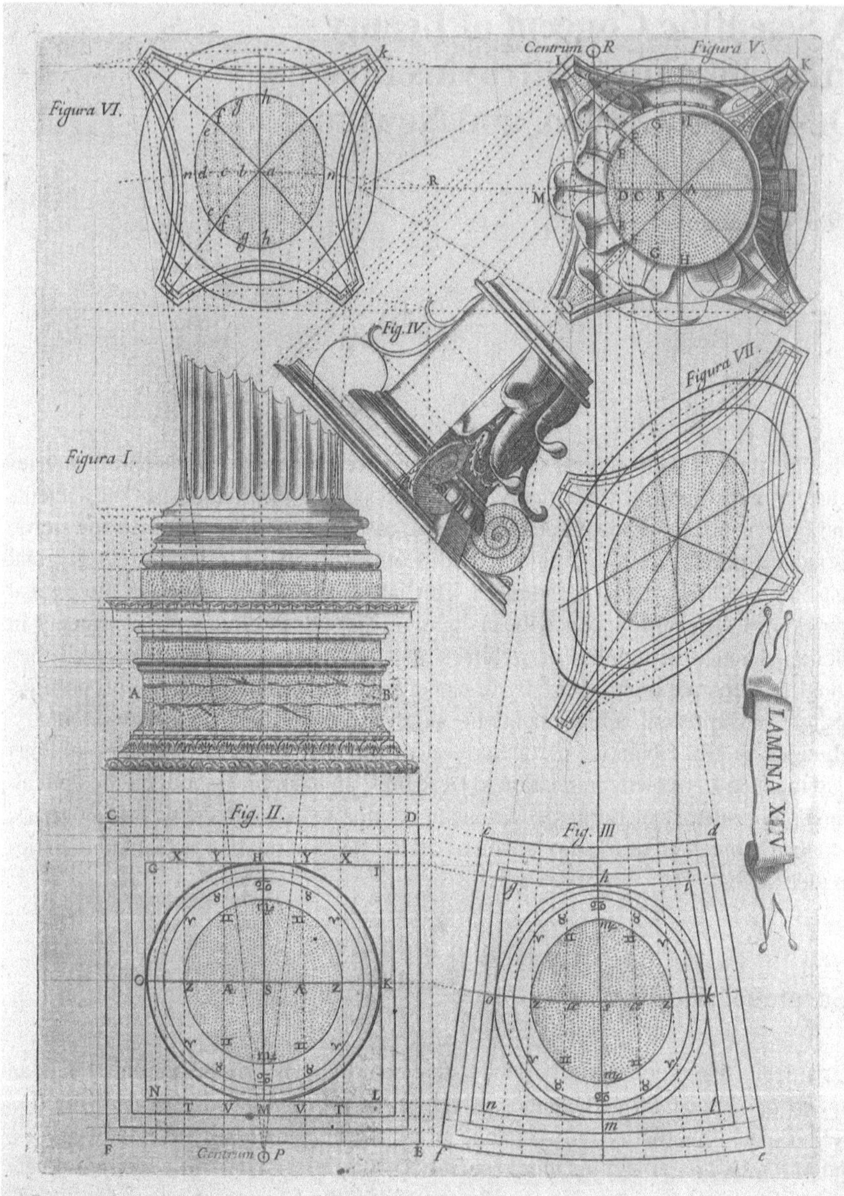

Fig. 1 Oblique deformation of a Corinthian column (From Lobkowitz 1678, Part III, plate XLV)

Thus, in treating of architecture, he specifies, we can either follow the authority of Vitruvius or, instead, "positively ignore what others say and follow only what is dictated to us by reason."[2]

The reason appealed to by Caramuel is mathematical reason, founded on geometry and on the exterior senses, in particular, sight. The space in which architecture takes shape is, in fact, a geometric space, empirically conceived in the Cartesian sense as a measurable entity in its three dimensions and defined by the presence of a body. Adopting for his own purposes the concept expounded by Descartes in his *Principles of Philosophy*, Caramuel argues that without body there is no space, and that both are indissolubly linked because the one manifests the other.[3] "Intrinsic space", as Descartes defined his concept of material space, was particularly well suited to the idea of architecture as materialized geometry. And this theoretical suitability was favoured by an illuminating, practical corroboration found on building sites and in particular in stonecutters' yards. Caramuel, in fact, saw his theoretical hypothesis confirmed not in the abstract world of geometry, but in the art of stonecutting (Fig. 2).[4]

The geometric identity between space and body prevents the form from being conceived outside its geometrical structure. Architecture for Caramuel is therefore "materialized geometry". The genesis of the form conforms to a series of principles that may be summed up as follows. According to Caramuel's philosophic probabilism, nothing can be taken for granted, or assumed to be immutable, and therefore the authority of the ancients can be challenged. Second, everything must be consistent with logical reasoning, even at the cost of deforming the morphological structure of the architectural elements; architecture must be the visual expression of geometric reasoning, and all the stages of this reasoning must be perceived. Third, each line that intervenes in the resolution of the geometric problem must be visualised in architectonic form. So, if the plane is inclined, as in staircases, and the vertical lines intersect only with oblique lines, the "straight" (Vitruvian) order must give way to the "oblique" order. Obliquity is thus understood as the inescapable transformation from a "straight" initial state (Fig. 3).

The concept of geometric transformation, after the great period of Renaissance perspective, had been further refined by the more recent developments of anamorphosis and the projective geometry of Desargues. Following the same process of transformation undergone by painted objects in passing from the "straight" form of the orthogonal projection to the "oblique" form of the perspective projection, Caramuel explains how the oblique delineations are derived from the straight ones (Fig. 4).[5] The cases in which oblique transformation was necessary were all those in which the initial condition of "orthogonality" was lacking. These cases were especially encountered in octagonal, circular and elliptical plans, and in the ornament of staircases. Many examples cited by Caramuel are buildings in Rome: the baptistery of San Giovanni in Laterano, the colonnade of the Piazza San Pietro and the Scala Regia in the Vatican, and the church of Sant'Andrea al Quirinale.[6] Caramuel's brief sojourn in Rome during the very years in which work was beginning on the Piazza San Pietro, from 1655 to 1657, was undoubtedly a crucial experience for the development of his concept of architectural obliquity (Fig. 5).

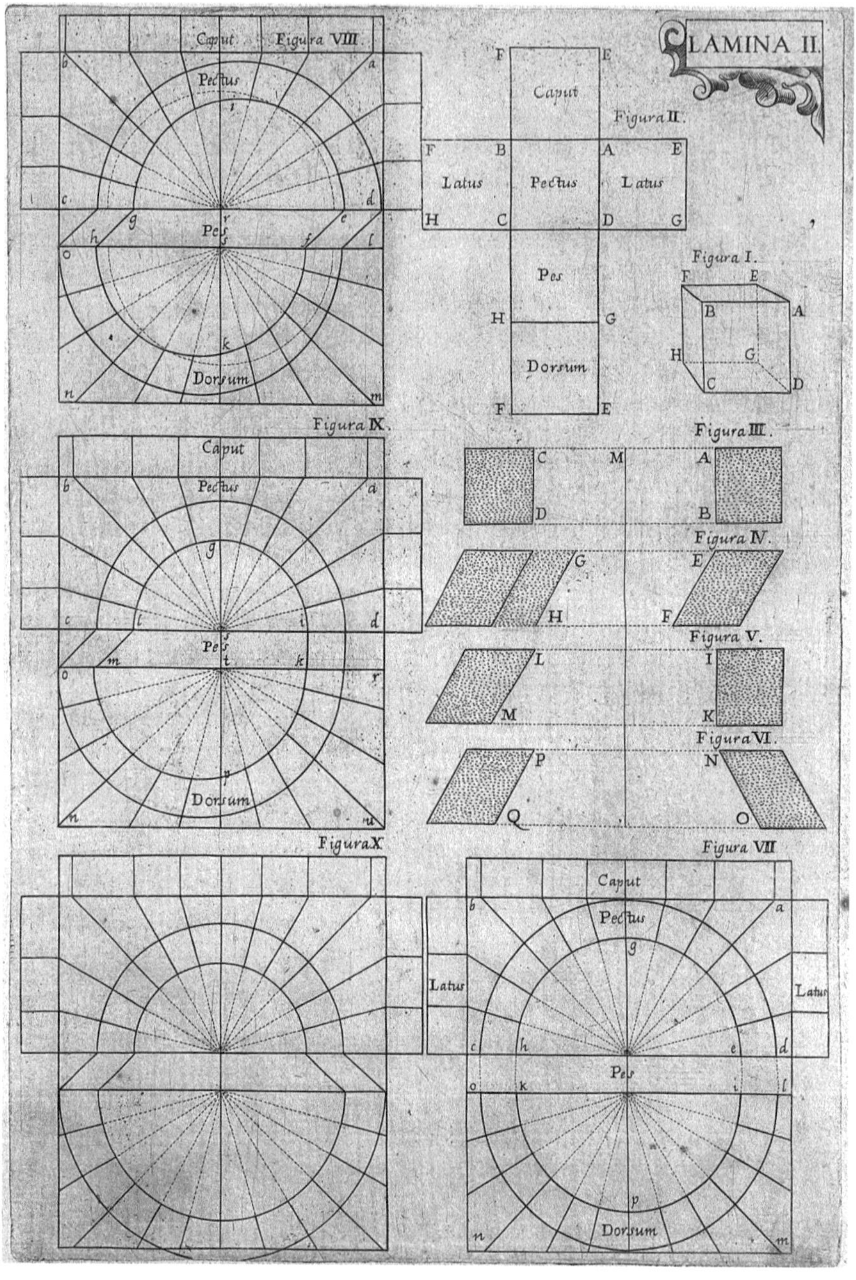

Fig. 2 Oblique arches (From Lobkowitz 1678, Part IV, Plate II)

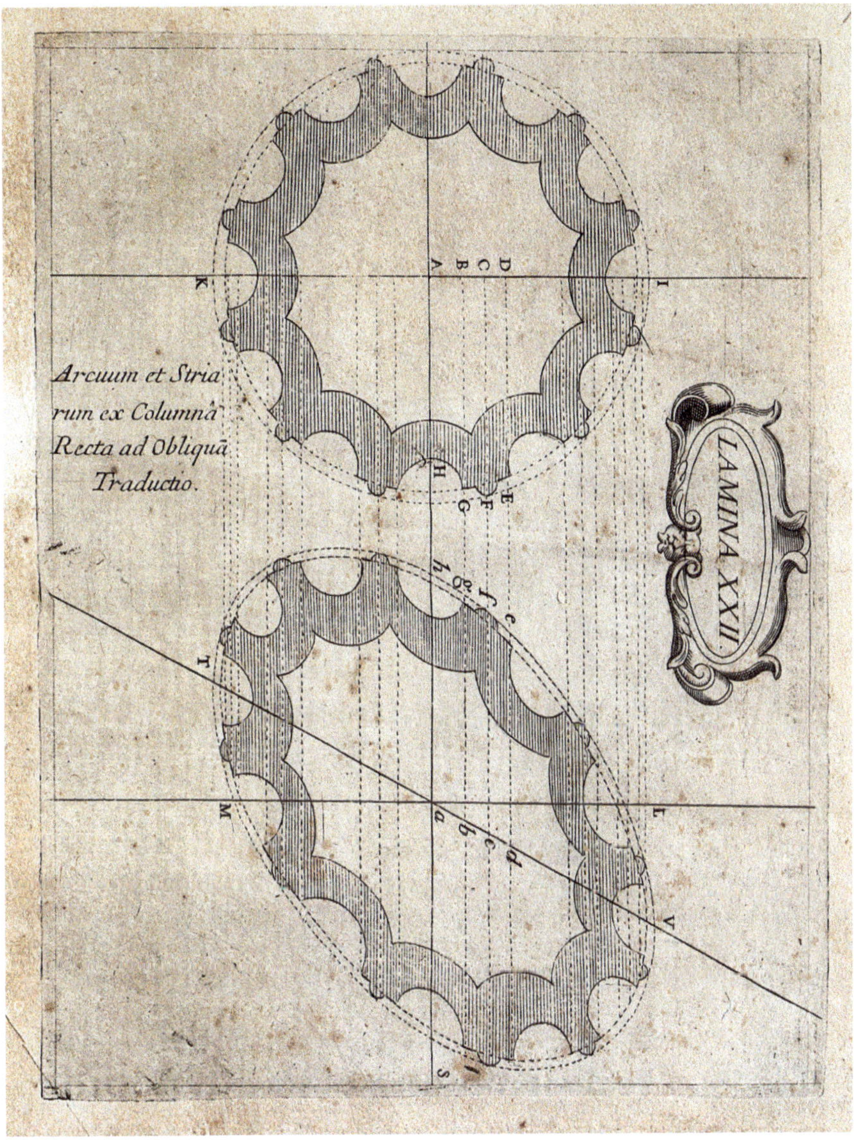

Fig. 3 Oblique deformation of a fluted column shaft (From Lobkowitz 1678, Part IV, Plate XXII)

In Rome, Caramuel had occasion to enter into contact with artists and scientists who had long been trying to fuse together their respective skills. Rome was the capital of the Baroque, but it was also the capital of science, and the activity of several artists was strongly related with that of the three important scientific centers

Fig. 4 Oblique deformation of the ornaments on a staircase (From Lobkowitz 1678, Part IV, Plate XX)

of the city: the Accademia dei Lincei, the convent of the Minims at Trinità dei Monti, and the Collegio Romano of the Jesuits.

In a chapter on "astronomical architecture", Caramuel expounded the idea of a building as a mathematical instrument (Fig. 6). This overlap between architecture and science was inspired not only by the famous Danish observatory of Ticho Brahe, but also by a number of Roman corollaries.[7] At least two cases can be considered in connection with Caramuel's astronomical architecture: the allegorical engraving dedicated by Orazio Busini to Urban VIII, with a cosmological version of Palazzo Barberini (Fig. 7), and the project for Villa Pamphilj, as visual expression of the mathematical sciences that sprang from the extraordinary collaboration between Francesco Borromini and Emmanuel Maignan.[8] Aside from astronomy, the new inventions of baroque architecture offered extraordinary occasions for experimenting with the principles of obliquity. Bernini's elliptical church of Sant'Andrea al Quirinale (begun in 1658) is praised by Caramuel as an exemplary case of *architectura obliqua*.[9] By contrast, the colonnade of the Piazza San Pietro was, in his view, a shockingly squandered opportunity.[10]

Caramuel's ideas were supported by the creativity of many baroque architects, first among them Borromini who, in his own way, adopted obliquity as a stylistic mark. But criticism was not lacking, and it did not come from the classicists alone. The most clearly articulated attack on Caramuel's theory came from the only

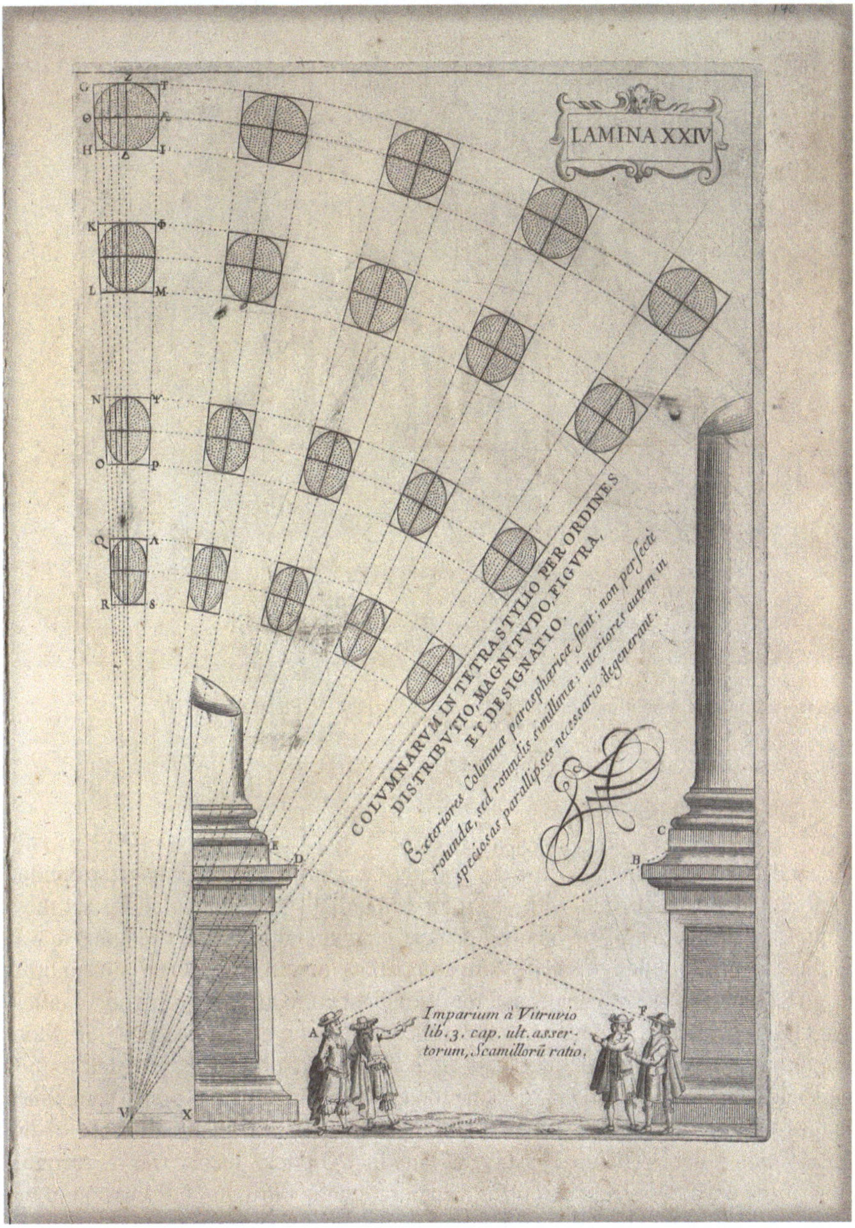

Fig. 5 Elliptical tetrastyle portico (From Lobkowitz 1678, Part IV, Plate XXIV)

Baroque architect with a comparable mastery not only in scientific theory, but also in construction and on-site practice, namely Guarino Guarini.

On the scientific level, Guarini compared Caramuel's treatise to the similarly encyclopedic *Cursus* of Milliet de Chales, but on the practical level he denounced

Fig. 6 Astronomical building (From Lobkowitz 1678, Part IV, Plate XXXVIII)

the total absence of building experience in it.[11] Guarini could not see how the oblique logic could be reconciled with any proportional system, nor could he see how oblique elements resting on inclined planes could guarantee static strength. He was willing to admit obliquity only as ornamentation for stairways. Guarini also criticized Caramuel's use of perspective for the optical correction of proportional ratios. For Guarini, it was not enough to adopt schemes of geometric transformation. More important was to consider problems of perception. Subtle apparent deformities depended basically on two factors: "the force of our imagination," which sometimes judges proportional ratios erroneously and "the site", that is, the distance of the observation point.[12] The novel aspect of Guarini's 'optical' theory was its recognition of the psychological component. Elaborating on Vitruvius's discussion of the phenomenon, he believed that corrections could sometimes be avoided because our imagination, in the same way in which it can itself be deceived, manages to correct the deception of the eye.[13]

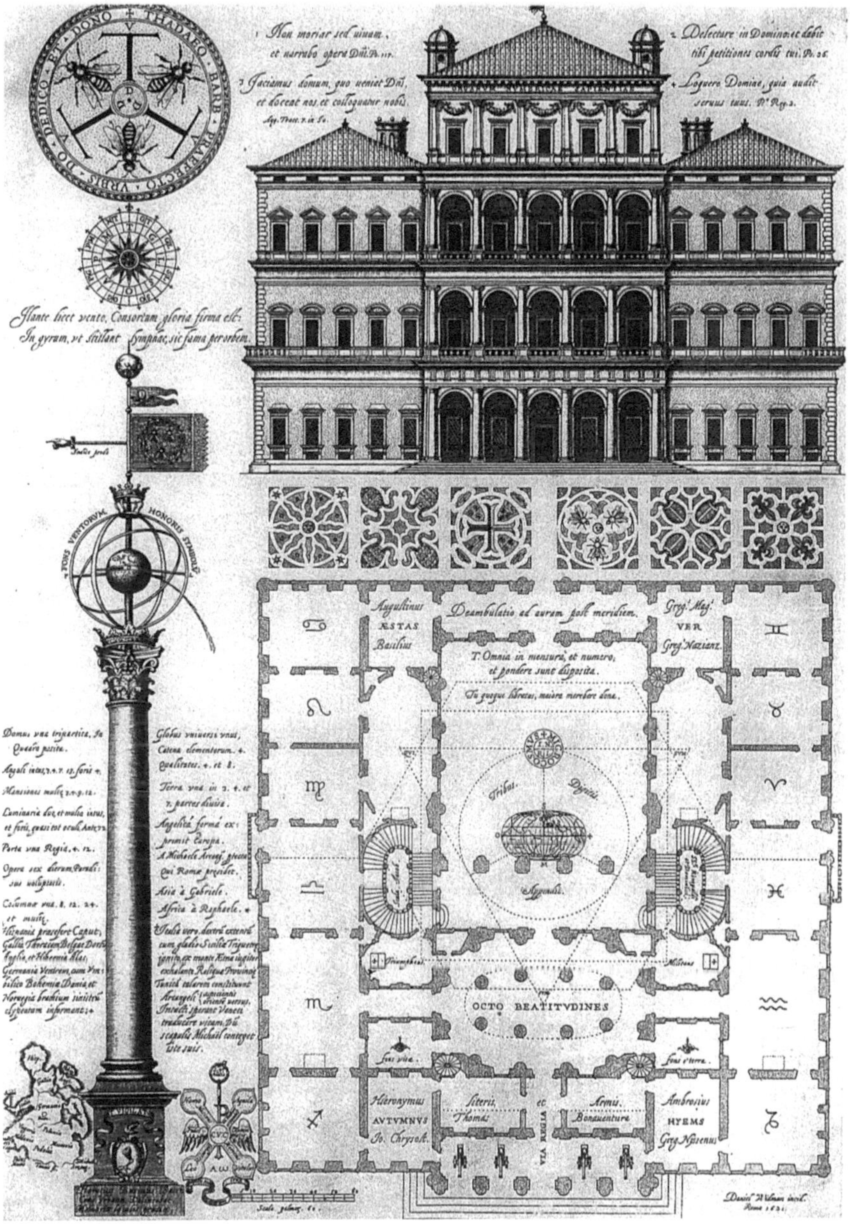

Fig. 7 Orazio Busini, *Allegorical design of Palazzo Barberini*, engraving by Daniel Widman, Rome, 1631

Vittone and Newton

These premises form the basis for the optical-perceptivity theory of Bernardo Vittone, architect, editor of Guarini's treatise on civil architecture, and author himself of two treatises on architecture. His approach to architecture was powerfully sensorial. Guarini's dictate that architecture had first of all to satisfy the senses was taken literally by Vittone, who developed his master's theory of optical corrections still further, singularly superimposing on Euclidean geometry the nature, at times inexplicable, of appearances. Inspired by the recent discoveries of Isaac Newton, Vittone explained in new terms the ancient Vitruvian question of visual appearances and the relevant proportional corrections. The optical effect became so predominant in his architecture as to lead him to an unexpected fusion of classicist control over proportions with the Baroque spectacle of illusionism. Appearance became for him an aspect as important as statics, and with felicitous results he managed to transform the stereotomic structure and geometric nature of Guarini's architecture into a visual spectacle (Fig. 8).

In his preface to the second book of the *Istruzioni elementari*, published in Lugano in 1760, Vittone specified the role of perspective and of the projection of shadows as regulating principles of the architectural concept. To this aspect of architecture, Vittone dedicated a specific theoretical category, calling it *adaptation*,

Fig. 8 Bernardo Vittone, dome of the Capella della Visitazione, Vallinotto, Carignano, 1738–39

Fig. 9 Bernardo Vittone, dome with mirrors and prisms at San Gaetano in Nice, France

which served to identify the elements that concurred in the perception of a building in relation to two basic factors: "the temperament of the light, and the state of the eye."[14] Light was a basic element of architecture. Spectacular baroque decorative schemes continued to inspire the creativity of architects, but the scenographic effect had to be controlled, Vittone believed, "in light of the Physical and Mathematical principles" recently established by Newton with his discovery of the prismatic colors of light. The importance of Newton for Vittone lay in his having "perfected our senses," allowing a more direct, almost corporeal, relationship with this impalpable "substance".[15]

In the architecture of Vittone, light is filtered, reflected, and even refracted, as in the case of San Gaetano in Nice, where mirrors and glass prisms are inserted in the furnishings of the church, perhaps in the attempt to enrich with elusive prismatic colors the already opulent decoration in colored stucco (Fig. 9). The most surprising aspect is that the architectural form seems actually modeled in such a way as to exalt the qualities of the light. The curved surfaces (concave, convex, and spheroid) display baroque spatiality as a visual manifestation of the wave-like nature of light and sound, almost as if to confirm a vision of architecture as a material expression of the physical and mathematical principles of space.[16]

Consideration of the "state of the eye" was also vitally important. Vittone showed himself well versed in recent optical research when he described the eye as "changeable in figure, and in position."[17] The changeability to which he referred was that of the crystalline lens, which in relation to the distance of an object, as Newton had explained, changes *shape*, becoming more or less convex, and *position*,

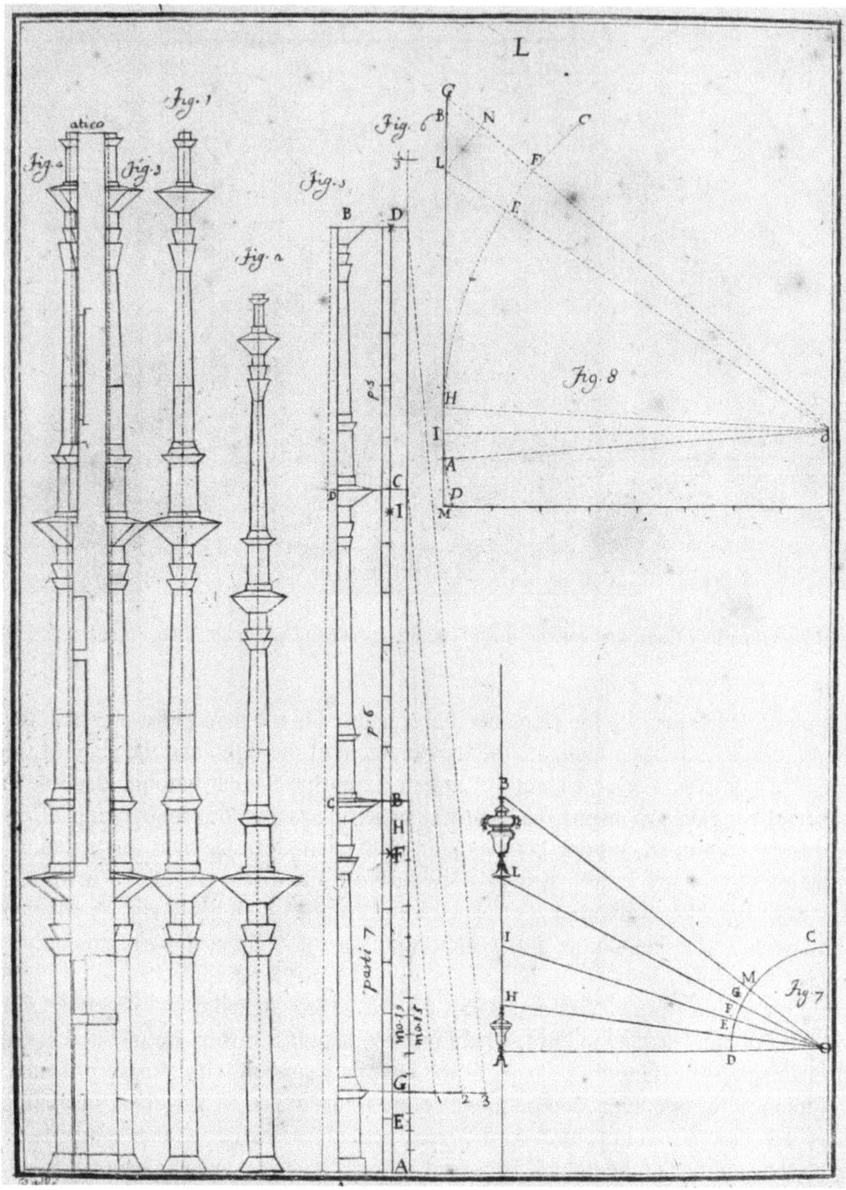

Fig. 10 Geometrical scheme for optical refinements in architecture (From Vittone 1760, Plate 50)

approaching or moving away from the retina. Vittone's knowledge of Newton's optics may have derived from the explanatory work of Francesco Algarotti, *Il newtonianismo per le dame*, a book Vittone had in his own library. Algarotti had stated, in regard to the crystalline lens: "Some others said that the retina being immobile, the crystalline humour approaches it, and moves away from it, or that the crystalline humour only changes in shape, becoming more convex for nearby objects and less so for more distant ones, and there were even those who claimed that both things occurred at the same time."[18]

The attention dedicated to the incidence of light and to the shape and position of the crystalline lens reveals a laudable attempt to update the timeworn Vitruvian precepts. With Vittone, the optical science of architecture acquired a quality that was truly optical and not mixed with perspective geometry. The geometric-proportional relationship between the distance to an object and its apparent diminution, as construed by the theorists of perspective, was no longer admissible. By changing shape and position, the crystalline lens could by itself change the amplitude of the visual angle and with it the object's apparent size. Vittone thus admitted that the perception of distant objects was determined in nature by other rules, different from those elaborated by geometric perspective (Fig. 10).[19]

Vittone fully grasped the scientific developments of his century and even perceived in them a logic that would ultimately lead to neo-classicism. For him however, "rationalism" did not mean deprivation of ornamentation and austerity of forms, but a logical process combining theoretical knowledge and practical experimentation, open to any possible innovation. Vittone may have become familiar with the rationalist hypothesis of architecture through the same "vulgarizer" who had introduced him to Newton's principles. In 1757, Francesco Algarotti had published the *Saggio sopra l'architettura*, in which he expounded the theoretical principles of the Franciscan monk Carlo Lodoli.

Lodoli and Galileo

The theoretical concepts of Carlo Lodoli were disseminated by two of his pupils, Francesco Algarotti and Andrea Memmo. Lodoli was a *maestro*, in the ancient sense of the term. He handed down his ideas orally, and was known, with some justification, as the "architect Socrates". Lodoli professed a concept of architecture "founded on the true reason of things," that is, on knowledge of the properties of the materials that determined the function and, consequently, the form of any architectural element. These principles are explicitly evoked in the motto surrounding the engraved portrait by Antonio Longhi (Fig. 11): "Experience and reason must be conjoined, and let function be the representation," a motto that stressed the supremacy of function over ornamentation.[20] This emphasis on the properties of materials and on the figure of the architect as philosopher shows a clear Galilean derivation. Lodoli taught in Venice, in the monastery of San Francesco della Vigna, only a few steps from the Arsenal, the naval boatyard that had inspired Galileo's famous *Discorsi intorno a due nuove scienze attinenti alla meccanica e ai movimenti locali* (1638).

For architects and theorists, Galileo's *Discorsi* called attention to the Vitruvian category least dependent on the trends of fashion and culture, that is, *firmitas*. The

Fig. 11 Antonio Longhi, *Portrait of Fra Carlo Lodoli* (From Andrea Memmo 1786, title-page)

Fig. 12 Cantilever beam (From Galilei 1638, 114)

static strength of buildings and the construction of worksite machines became the subjects of "speculative intellects" which, while finding nourishment in the technical culture of craftsmen, sought the reason for things in the truths of mathematics. In particular, the *Discorsi* rendered ineffective the ancient theory of proportions, which had dominated Vitruvian mechanics throughout the Renaissance.[21] A large machine, for instance, could not simply be composed of elements proportionately larger than those of a small machine, but had to be constructed of stronger materials. The "resistance of solid bodies to being broken" was one of Galileo's "two new sciences" (Fig. 12).[22] The theory of proportions that governed every other aspect of the architectural project, *venustas* in particular, was no longer applicable to the statics of buildings, because each material had its own degree of strength.

Accordingly, the form and dimensions of a structural element had to be determined on the basis of its material properties. The need to reduce the weight of a cantilever beam without impairing its strength had led Galileo to hypothesize an optimal section of parabolic shape. François Blondel would make "Galileo's problem" (*Galileus promotus de resistentia solidorum*) central to architectural debate in the Parisian royal academies.[23]

The parabola, which Galileo improperly identified as the curve formed by a hanging chain, was a subject of discussion also for the other "new science", the one that dealt with the study of local movements (Fig. 13). The law of the acceleration of free-falling objects bore implications of great importance, especially in the military field, where ballistics had for nearly a century been included among the mathematical sciences. The parabola was an "ancient" curve. However, when associated with the movement of a body, it fell within the context of other new curves elaborated by seventeenth-century mathematicians: the "cycloid" and the "sinusoid" (Fig. 14).

The architectural implications of these studies in geometry were immediate. According to Carlo Dati, Galileo had invented the cycloidal curve to adapt it to the design of arches for bridges, "and in particular to the design of the new bridge in Pisa, when it was proposed that it be made of one arch alone, [Galileo] saying, that this line provided a curve for a bridge of most beautiful form."[24] It was not long before these curves began to influence architects. The sinusoid curve made its first appearance in simplified form in some works by Francesco Borromini, such as the Filomarino altar in Naples and the façade of S. Carlino alle Quattro Fontane in Rome. Blondel proposed the ellipse and the parabola for the section of beams, the entasis of columns, the shape of rampant arches, and the design of the joints of these arches. Later on, Robert Hooke was to deem the catenary curve the ideal one for the statics of cupolas, a suggestion taken up by Giovanni Poleni (Fig. 15).

Around 1660, Galileo's discoveries were subjected to rigorous examination by two scientific academies in Florence. The first and most famous was the Accademia del Cimento founded by Leopoldo de'Medici. The second, almost unknown, was a private academy founded by Abbot Ottavio della Vacchia with the aim of mathematically solving and experimentally verifying problems relevant to technology, architecture, and the figurative arts. In the records of the sessions of this Academy, whose activity lasted only two years, we find problems of statuary, theatrical scenography, perspective, architectural technology (such as the construction of vaults, bridges and trusses), problems of hydrostatics relevant to the channeling of rivers, problems of gnomonics, and of military architecture (Fig. 16).[25]

Outstanding among its members was Cosimo Noferi, a disciple of Galileo and author of twelve mathematical treatises, one of them entirely dedicated to architecture. To his readers, Noferi declared himself to be nothing less than an architect-philosopher: "I do not merely give the rule for what I propose to make, but in regard to it I hold a session, and a discussion, and I prove it either by geometrical reasons or natural reasons, and so these discussions are not those of a simple practitioner but of a speculative philosopher, such as I will prove that the expert Architect

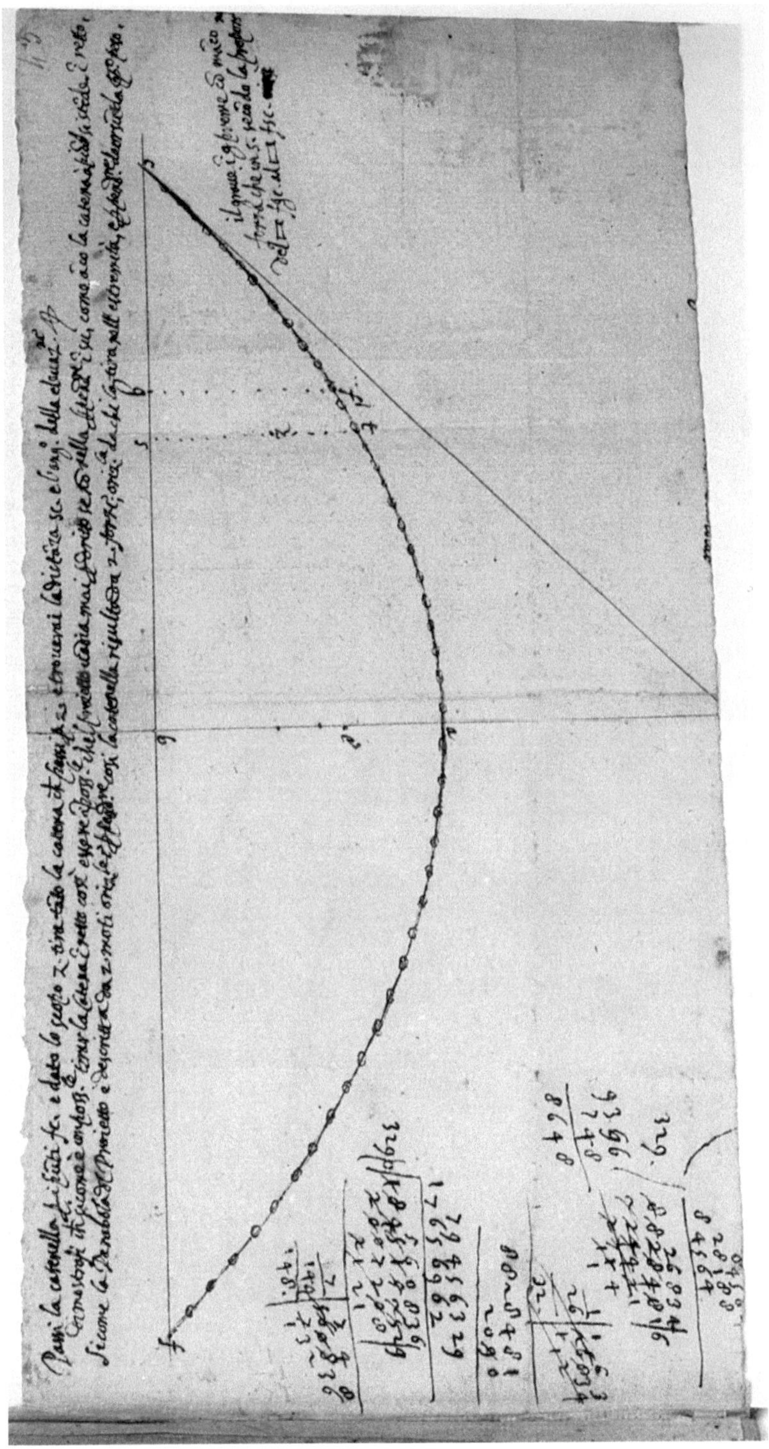

Fig. 13 The "catenella" (small chain) (From Galileo Galilei, Notes on motion and mechanics, Biblioteca Nazionale Centrale di Firenze, Ms. Gal. 72, fol. 43)

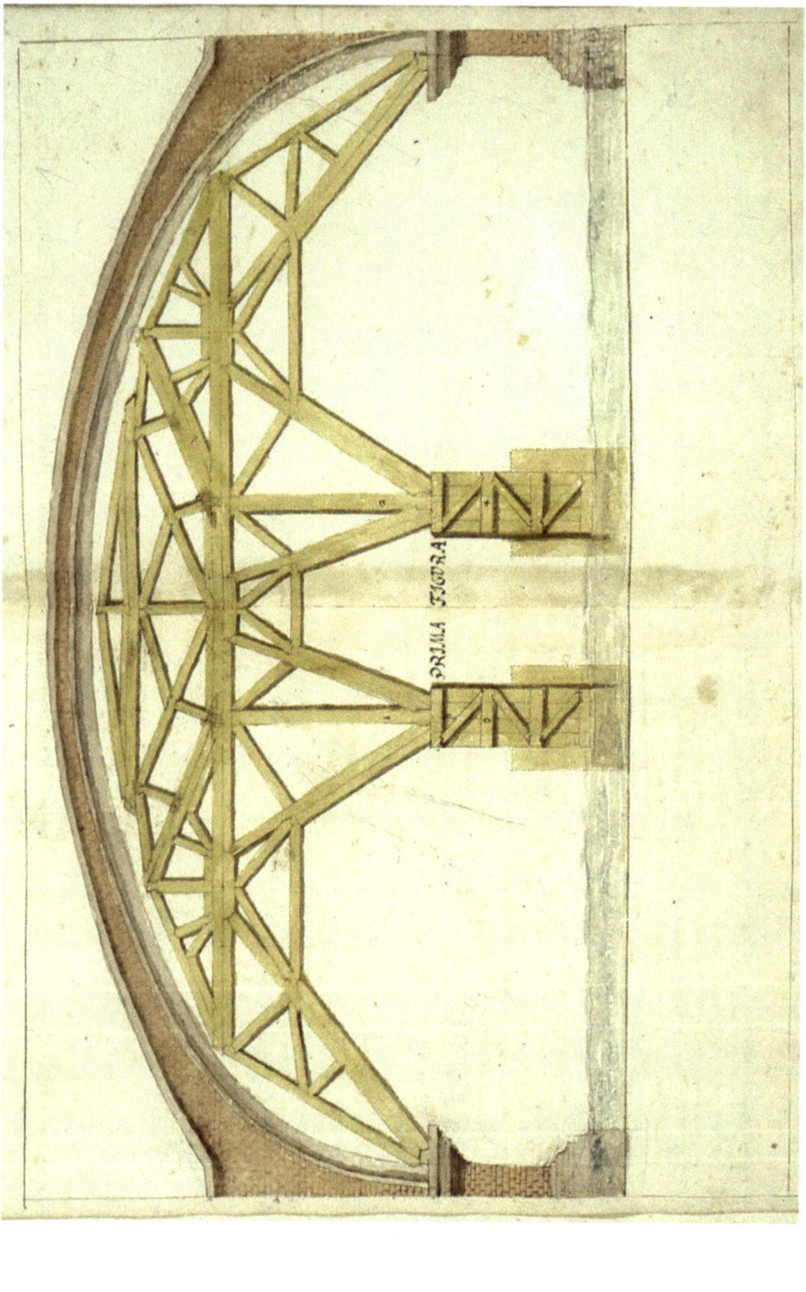

Fig. 14 Cycloidal arch for a bridge (From Cosimo Noferi, "Tomo secondo della travagliata architettura," Biblioteca Nazionale Centrale di Firenze, Ms. Gal. 123, cc. 32v–33r)

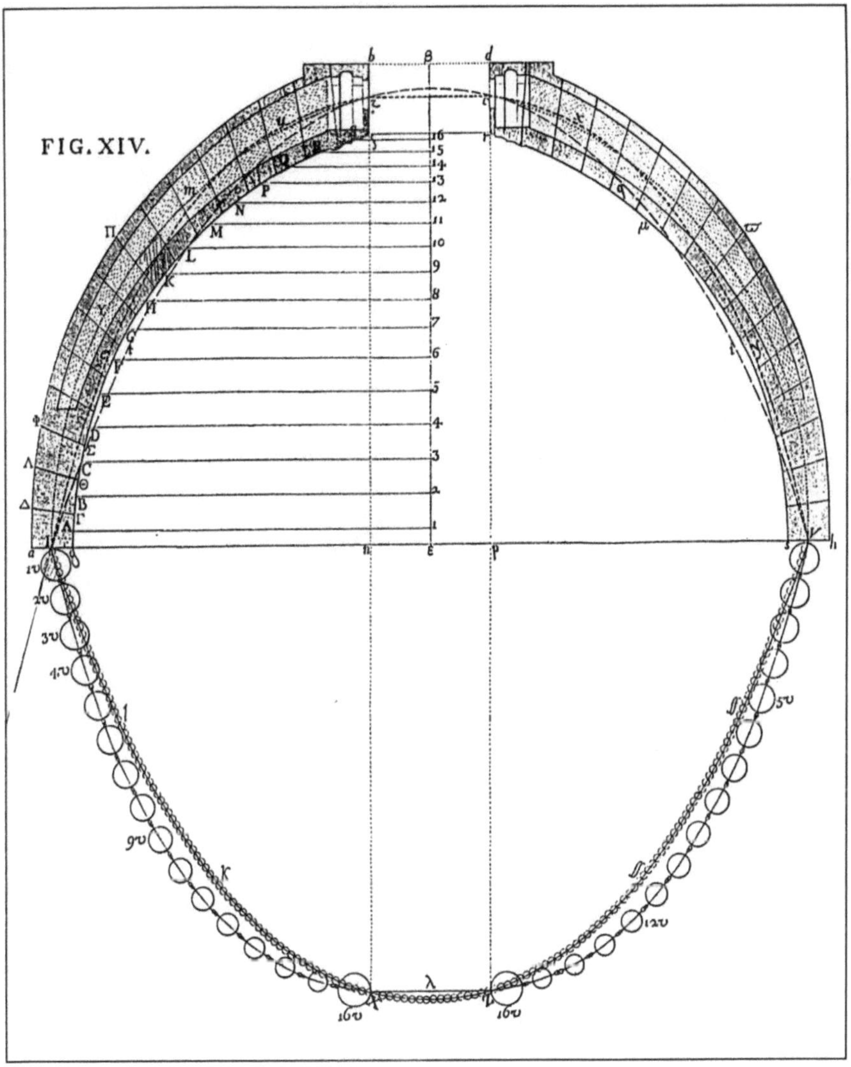

Fig. 15 Section of Saint Peter's dome with inscribed catenary curve (From Poleni 1748, Fig. XIV)

should be."[26] The foremost quality of the architect, according to Noferi, was a perfect knowledge of the "science of quantities," that is, the "Mathematical professions."[27] For Noferi, the architect had to don the guise of the philosopher, to hold "sessions and discourse," to demonstrate "by geometrical reasons or natural reasons," and to give proof of his own "speculative" capacity. In Florence, Noferi found an exemplary historical example of the architect's speculative mind: the

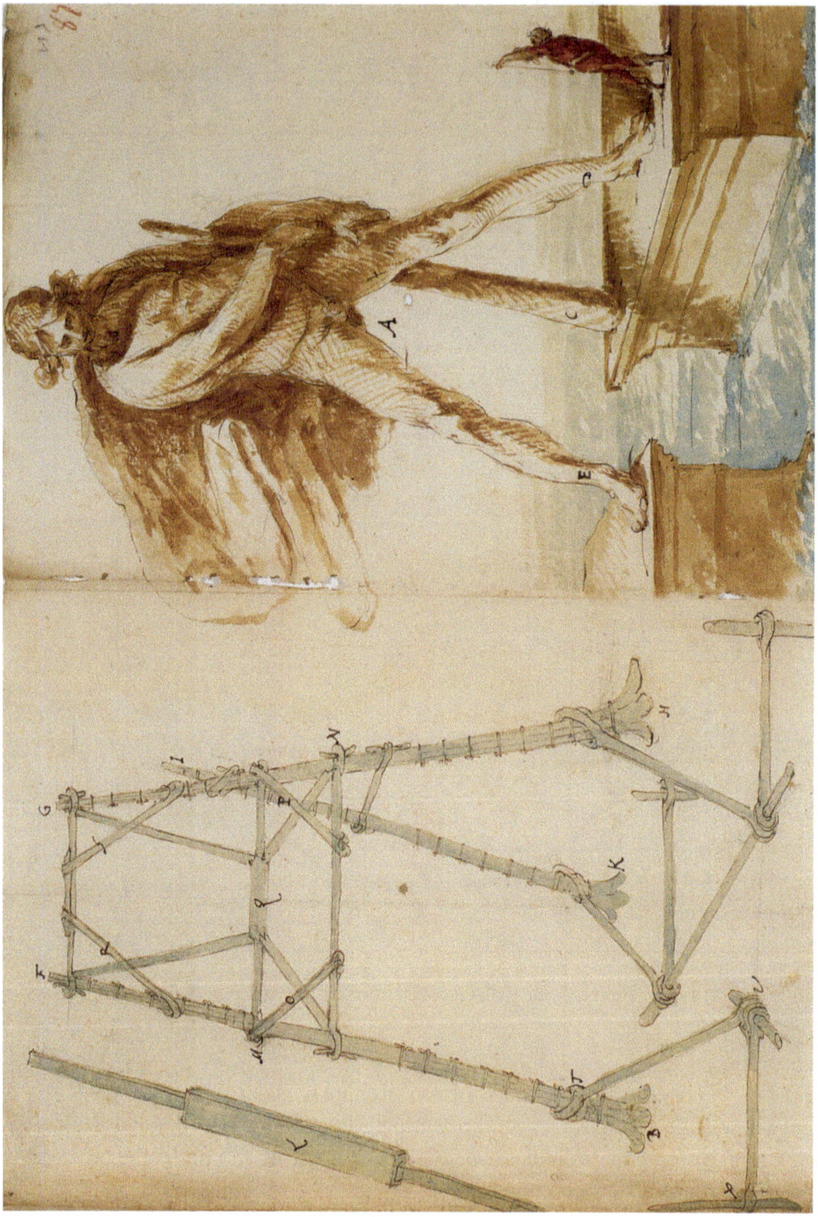

Fig. 16 Cosimo Noferi, *La fabbrica del Colosso Rodiano*, in "La risoluzione di più problemi stati proposti nel Accademia del Sig.r Abate Ottavio della Vacchia l'anno 1662," Biblioteca Nazionale Centrale di Firenze, Fondo Nazionale II-46, cc. 86v–87r

dome of Santa Maria del Fiore: "after many opinions given by expert Architects, only Filippo [Brunelleschi] was the one who could boast of it." Brunelleschi alone had been able to:

> understand and penetrate with true and demonstrative reasons what force is exerted and where by an arch vaulted in that manner, how it should be fastened, how connected, what order should be observed in joining the tiles, in making the necessary vents as defense against the insults of the winds, how to load it reasonably, to arrange the channels, the drains for rainwater, having in mind the manner of the scaffolding, of the reinforcement, of how the work should be carried out.[28]

All these things, Noferi continued, "would make any speculative and philosophizing mathematician, or any simple builder, sweat." To be sure, Brunelleschi was hardly the kind of philosopher that Noferi described. Yet, his description nonetheless reveals the kinds of problems that architecture was beginning to open up to scholars.

These problems became all too real toward the end of the century, when the cracks in the dome of Florence cathedral made "philosophizing mathematicians" sweat indeed. The project for consolidating the great mass of brickwork was assigned to a number of mathematicians of the Galilean school, among them Vincenzo Viviani, and a group of architects, including Giovanni Battista Foggini and Giovanni Battista Nelli (Fig. 17).[29] The problem at stake was one of statics. The general opinion was that the cracks were caused by insufficient buttressing and that installation of chains around the drum was necessary to contain the external thrust. In this respect, Viviani and the other members of the commission were backed up by the authoritative judgment of the papal architect Carlo Fontana, who had recommended a similar expedient for the Vatican dome. Viviani had even found a mathematical demonstration of the function of the chains in a text by Evangelista Torricelli on the effects of ring-braces on the columns of the Uffizi and of the Palazzo Medici-Riccardi.[30]

This unanimity, however, collapsed when the architect Alessandro Cecchini cast doubt on the proposed solution, suggesting that the effect might even be harmful. Cecchini maintained that the cracks were not caused by the excessive weight of the lantern, but by the yielding of the foundations, which had in the meantime fully settled. Cecchini also held that the natural action of domes was not to spread outward, but rather to collapse inward from the center. Chains were thus not only useless, but constituted an additional weight that would only complicate things still further. Cecchini also discounted the experience acquired with reinforcement rings applied to smaller domes, since the ratio between forces and resistance would vary in proportion to the dome's size and was therefore not applicable to Brunelleschi's much larger work: "and anyone who does not believe this proposition with me," concluded Cecchini, "is obliged to believe it with Galileo."[31] By calling into cause Galileo, Cecchini had practically ended the dispute. The installation of the chains was interrupted in 1697, and the members of the commission postponed a final decision to the results of further investigation. Apart from intriguing mathematical questions raised on both sides, this debate had a number of important consequences. In the first place, it exposed the limitations of practical experience, which could no

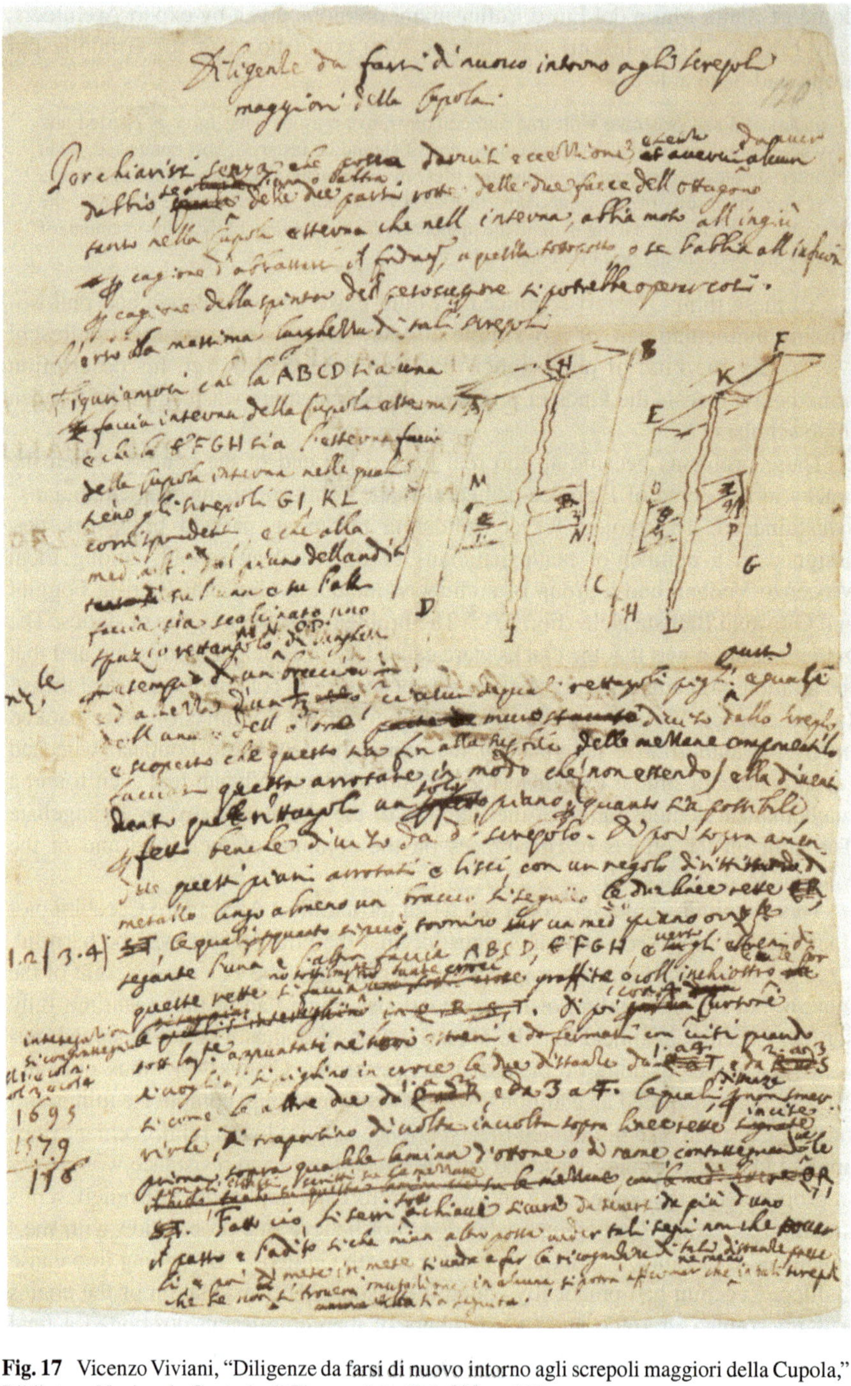

Fig. 17 Vicenzo Viviani, "Diligenze da farsi di nuovo intorno agli screpoli maggiori della Cupola," Biblioteca Nazionale Centrale di Firenze, Ms. Gal. 222, c. 120r

longer count on the universality of the proportional laws. The fact that mathematicians like Viviani were included on the commission may be taken as an implicit recognition of the new situation. However, the episode also severely tested the technical expertise and credibility of the "philosophizing" architects. The confrontation with a real object—especially one as complex as the dome of Florence cathedral—highlighted the distance between the new theory and the traditional world of practice.

The "new science" of the *resistentia solidorum* called then for a true science of building that would make it possible to control every aspect of architecture, especially "the construction of Arches, and of Cupolas." According to Giovanni Poleni, the Paduan mathematician who directed the restoration of the dome of St. Peter's in the mid-eighteenth century, these structural elements "must without doubt be deemed the most difficult thing in the particular Mechanics of Architecture."[32] This science had produced important theoretical results, such as the *Exercitationes* of Bernardino Baldi (1621) on the "mechanical problems" of Aristotle, the *De resistentia solidorum* by Alessandro Marchetti (1669), and the *Trattato delle resistenze* by Vincenzo Viviani (1718), but it also begat important practical results, expounded mainly by the French academicians. After François Blondel, it was Philippe de La Hire who took up this baton with his *Traité de Méchanique* (1695), a book dedicated expressly to problems of construction. La Hire would later publish an important advance, "Sur la construction des voûtes," in which he proposed a real model of calculation to determine the dynamics of failure in arches and vaults.[33]

As the "new science" progressed, it is not surprising that it also began to influence the Vitruvian category most dependent on questions of taste, that is, *venustas*. Carlo Lodoli exemplified the new figure of the architect-philosopher and represents the most extreme attempt to re-establish the canons of beauty in Galilean terms.

Notes

1. The principles of this "*Arte Nueva*" are expounded by de Lobkowitz (1678). See Ferrero (1965); Oechslin (1970); and Marino (1973).
2. "Dos caminos tenemos de hallar la verdad en questiones oscuras; El uno, que es mas trillado y conocido sigue la autoridad de gente docta; y el otro que es mas subtil y delicato haze lo que le dicta la razon […] Luego estos dos caminos son los que podremos seguir trattando de Architectura. Pero (valgame Dios!) que podre revolver si sigo la autoridad de gente docta. Vitruvio no se ajusta en todo a lo que dixeron los Antiguos, de los Modernos unos le alaban, otros le corrigen, otros le vituperan […] Luego sera mejor ignorar positivamente lo que dixeron otros, y seguir solo lo que nos dicta la razon. Y que razon es esta? Se funda en los sentidos exteriores y principalmente en la vista, potenzia que se halucina facilmente, lo que haze buena vista en los ojos de uno, parece mal en los de otro, de donde viene a ser, que es tan dudoso lo que resuelve la razon como lo que definio la autoridad." de Lobkowitz (1678, vol. 2, 43–44): Trat. V, Art. IV, Nota II.

3. "Digo pues, que el lugar Intrinseco o el espacio es una extension, que tiene longitud, latitud, y profundidad." de Lobkowitz (1678, vol. 3, 66): (Trat. IX, Art. II: "De el Espacio o Lugar Intrinseco". See also Descartes (1974, vol. 8, 45).
4. de Lobkowitz (1678, vol. 2, 17–18): Trat. VI, Art. XI: "De las Escaleras". Also see Trat. IX, Art. IV, Lamina I: "Saca Ellipses de Circulos, y Cuerpos Ovales de Globos. Propone los primeros Fundamentos de la Dotrina Obliqua; y enseña, como se han de hazer rexas y ventanas en lugares inclinados, que se llaman *viajes* en nuestra Lengua Castellana." On Caramuel's architectural training, see Correa (1984 [facsimile reprint]).
5. de Lobkowitz (1678, vol. 2, 6): Trat. VI, Art. IV: "Como de las Rectas nacen las Delineaciones Obliquas".
6. de Lobkowitz (1678). On the Lateran baptistery, see Trat. VI, Art. V, "De los yerros, que tienen las Colunas de la Capilla, en que se bautizo Constantino," 2: 8–9. On the Scala Regia, see Trat. VI, Art. XII, "De los Balaustres y Colunas Obliquas, con que se suelen adornar las Escaleras," 2: 19. For Bernini's piazza, see Trat. VIII, Art. III, "De algunos Edificios de Roma", Seccion VIII, "De el Templo de San Pedro de Roma," 2: 51. Again on Bernini's elliptical piazza, the equestrian statue of Constantine, the façade of the basilica, and other interventions, see Seccion X, "De los errores que en el Templo de S. Pedro han cometido los Ingenieros y Architectos," 2: 52–53. Again on the Lateran Baptistery, see Trat. IX, Art. III, Lamina LVII, "De la Capilla, en que S. Sylvestro Papa bautizo al Emperador Constantino," 2: 102. On the Piazza San Pietro and Sant'Andrea al Quirinale, see Trat. IX, Art. IV, Lamina XXIV, "De como han de ser estas mismas colunas, si formaren un Portal de tres Naves," 2: 108.
7. de Lobkowitz (1678, vol. 2, 64–65): Trat. VII, "De algunas Artes y Ciencias, que accompañan y adornan a la Architectura," and Art. IV, "De la Astronomia"; de Lobkowitz (1678, vol. 2, 109): Trat. IX, Art. IV, Lamina XXXVIII–XXXIX, "De la Architectura Astronomica".
8. On Busini's engraving, see Scott (1995) and Fagiolo (2000, 222–27). On the Villa Pamphilj, see Camerota (2000a, b).
9. de Lobkowitz (1678, vol. 2, 108): Trat. IX, Art. IV, Lamina XXIV: "y en voz tacita alabando a quien las delineo y labro".
10. "… la Ellipse Tetrastylia, que en Roma delante de S. Pedro erigio Alexandro VII, … Tiene tantos errores, como piedras." de Lobkowitz (1678, vol. 2, 108): Trat. IX, Lam. XXIV. Also see Trat. VI: "Architectura obliqua," Trat. VIII: "Architectura pratica," Art. III: "De algunos edificios de Roma," and Sect. X: "De los errores que en el Templo de S. Pedro han comettido los Ingenieros y Architectos."
11. Guarini (1737, 71): Trat. II, Cap. VIII. See Ferrero (1966, 37–53). Also see Oechslin (1970).
12. Guarini (1737, 157: Trat. III, Cap. XXI).
13. On this subject, see Robison (1991).
14. Vittone (1760, vol. 2, 242). On Vittone's work, see Portoghesi (1966) and Accademia delle scienze di Torino (1972).
15. Vittone (1760, vol. 2, 243).
16. On the "scientitic" interest in light informing the illumination of Baroque churches, see Connors (1992).

17. Vittone (1760, vol. 2, 243).
18. Algarotti (1737, 103). The book is registered with the number 701 in the inventory compiled in Vittone's house immediately after his death. See Portoghesi (1966, 250).
19. Vittone (1760, vol. 2, 398).
20. "Devonsi unir e fabbrica e ragione e sia funzion la rappresentazione." Memmo (1786, title-page). See Cellauro (2006).
21. See di Pasquale (1996).
22. See Camerota (2004, 544–560).
23. Blondel (1673). See also Becchi (2004) and Gerbino (2010).
24. ([Dati], 1663, 4). Also see Pascal ([1658]).
25. "La risoluzione di più problemi stati proposti nel Accademia del Sig.r Abate Ottavio della Vacchia l'anno 1662 con i nomi di chi propose et di chi ha risoluto," Fondo Nazionale II-46, Biblioteca Nazionale Centrale, Florence. The manuscript has been recently published in Schlimme (2006, 155–263).
26. Cosimo Noferi, "Tomo primo della travagliata architettura," Ms. Gal. 122, Biblioteca Nazionale Centrale, Florence, "Proemio," fol. 2r. On this treatise, see also Pellicanò (2004).
27. Noferi, "Tomo primo," fol. 6r.
28. Noferi, "Tomo primo," fol. 4r.
29. On this topic, see Galluzzi (1977); Righini (1978); Barbo and Teodoro (1983); and Fanelli (2004).
30. See Torricelli (1919, vol. 2, 243–46), cited in Galluzzi (1977, 95).
31. See Nelli (1753, 88).
32. Poleni (1748, cols. 30–31).
33. de La Hire (1712).

Photographic Credits

Museo Galileo / Istituto e Museo di Storia della Scienza, Florence: Figs. 1–12, 15
Biblioteca Nazionale Centrale di Firenze: Figs. 13, 14, 16, 17

References

[Dati, Carlo]. 1663. *Lettera a Filateti di Timauro Antiate. Della vera storia della cicloide, e della famossissima esperienza dell'argento vivo*. Firenze: all'Insegna della Stella.

Algarotti, Francesco. 1737. *Il Newtonianismo per le dame, ovvero dialoghi sopra la luce e i colori*. Milan.

Baldi, Bernardino. 1621. *In mechanica Aristotelis problemata exercitationes*. Mainz: I. Albini.

Barbo, Luciano, and Francesco di Teodoro. 1983. Le lesioni della cupola di Santa Maria del Fiore: una proposta di datazione. *Bollettino degli Ingegneri* 31(9): 13–18.

Becchi, Antonio. 2004. *Q.XVI. Leonardo, Galileo e il caso Baldi: Magonza, 26 marzo 1621*. Venice: Marsilio.

Blondel, François. 1673. *Résolution des quatre principaux problèmes d'architecture*. Paris: Imprimerie royale.
Camerota, Filippo. 2000a. Le bizzarrie dell'ingegno: architettura e scienza per villa Pamphili. In *Francesco Borromini: atti del Convegno internazionale: Roma, 13–15 gennaio 2000*, ed. Christoph Luitpold Frommel and Elisabeth Sladek, 297–311. Milan: Electa.
Camerota, Filippo. 2000b. Architecture and science in Baroque Rome: The mathematical ornaments of Villa Pamphilj. *Nuncius* 15(2): 611–638.
Camerota, Michele. 2004. *Galileo Galilei e la cultura scientifica nell'età della controriforma*. Rome: Salerno Editrice.
Cellauro, Louis. 2006. Carlo Lodoli and architecture: Career and theory of an eighteenth-century pioneer of modernism. *Architectura: Zeitschrift für Geschichte der Baukunst* 36(1): 25–59.
Connors, Joseph. 1992. Virtuoso architecture in Cassiano's Rome. In *Cassiano del Pozzo's paper museum*, ed. Ian Jenkins, 23–40. London: Olivetti.
Correa, Antonio Bonet. 1984 [facsimile reprint]. Estudio preliminar. In *Juan Caramuel de Lobkowitz, Architectura civil recta y obliqua*. Madrid: Ediciones Turres.
de La Hire, Philippe. 1712. Sur la construction des voûtes dans les edifices. *Histoire de l'Académie royal des sciences, avec les mémoires de mathématique et de physique*: Part II, Mémoires, 69–77. Paris: Imprimerie royale, 1731.
de Lobkowitz, Juan Caramuel. 1678. *Architectura civil recta y obliqua considerada y dibuxada en el Templo de Jerusalem*. Vigevano: Camillo Corrado.
Descartes, Réné. 1974. *Oeuvres*, 2nd ed. Paris: Vrin.
di Pasquale, Salvatore. 1996. *L'arte del costruire: Tra conoscenza e scienza*. Venezia: Marsilio.
Fagiolo, Marcello. 2000. Il segno del giovane Borromini nella 'Città del Sole' di Urbano VIII (1624–1631). In *Francesco Borromini: atti del Convegno internazionale: Roma, 13–15 gennaio 2000*, ed. Christoph Luitpold Frommel and Elisabeth Sladek, 215–232. Milan: Electa.
Fanelli, Giovanni. 2004. *La cupola del Brunelleschi. Storia e futuro di una grande struttura*. Florence: Mandragora.
Ferrero, Daria De Bernardi. 1965. Il conte I. Caramuel de Lobkowitz, vescovo di Vigevano, Architetto e Teorico della Architettura. *Palladio* 15: 91–110.
Ferrero, Daria De Bernardi. 1966. Guarini trattatista e l'architettura obliqua del Caramuel. In *I "Disegni d'architettura civile e ecclesiastica" di Guarino Guarini, e l'arte del maestro*, 37–53. Turin Albra.
Galilei, Galileo. 1638. *Discorsi intorno a due nuove scienze attinenti alla meccanica e ai movimenti locali*. Leiden: Elsevier.
Galluzzi, Paolo. 1977. Le colonne 'fesse' degli Uffizi e gli 'screpoli' della Cupola. Il contributo di Vincenzo Viviani al dibattito sulla stabilità della Cupola del Brunelleschi (1694–1697). *Annali dell'Istituto e Museo di Storia della Scienza di Firenze* 2(1): 71–111.
Gerbino, Anthony. 2010. *François Blondel: Architecture, erudition, and the scientific revolution*. London/New York: Routledge.
Guarini, Guarino. 1737. *Architettura civile*. Turin: Gianfrancesco Mairesse.
Marino, Angela Guidoni. 1973. Il colonnato di piazza S. Pietro: dall'Architettura obliqua di Caramuel al 'classicismo' berniniano. *Palladio* 23: 82–120.
Memmo, Andrea. 1786. *Elementi dell'architettura lodoliana o sia L'arte del fabbricare con solidità scientifica e con eleganza non capricciosa*. Roma: Stamperia Pagliarini.
Nelli, Giovanni Battista Clemente. 1753. *Discorsi di architettura. E due ragionamenti sopra le cupole di A. Cecchini*. Florence.
Oechslin, Werner. 1970. Osservazioni su Guarino Guarini e Juan Caramuel de Lobkowitz. In *Guarino Guarini e l'Internazionalità del Barocco*, vol. 1, 573–595. Turin: Accademia delle Scienze.
Pascal, Blaise. [1658]. *Histoire de la roulette, appellée autrement la trochoïde, ou la cycloïde, où l'on rapporte par quels degrez on est arrivé à la connoissance de la nature de cette ligne*.
Pellicanò, Antonio. 2004. Cosimo de Noferi e Galileo Galilei. Un inedito trattato di architettura nella Firenze del 1650. *Rivista dell'Ordine degli architetti, pianificatori, paesaggisti e conservatori della Provincia di Reggio Calabria* 3: 68–73.

Poleni, Giovanni. 1748. *Memorie istoriche della gran cupola del tempio Vaticano.* Padova: Stamperia del Seminario.
Portoghesi, Paolo. 1966. *Bernardo Vittone, un architetto tra illuminismo e Rococò.* Rome: Edizioni dell'Elefante.
Righini, Guglielmo. 1978. Considerazioni su teoremi del Viviani a proposito degli 'screpoli' della cupola brunelleschiana. *Annali dell'Istituto e Museo di Storia della Scienza di Firenze* 3(1): 105–108.
Robison, Elwin C. 1991. Optics and mathematics in the domed churches of Guarino Guarini. *Journal of the Society of Architectural Historians* 50(4): 384–401.
Schlimme, Hermann (ed.). 2006. *Practice and science in early modern Italian building: Towards an epistemic history of architecture.* Milan: Electa.
Scott, John Beldon. 1995. Patronage and the visual encomium during the pontificate of urban VIII: The ideal Palazzo Barberini in a dedicatory Print. *Memoirs of the American Academy in Rome* 40: 197–236.
Torino, Accademia delle scienze di (ed.). 1972. *Bernardo Vittone e la disputa fra classicismo e barocco nel Settecento: Atti del Convegno internazionale... 21–24 settembre 1970.* Turin: Accademia Delle Scienze.
Torricelli, Evangelista. 1919. *Opere.* Faenza: G. Montanari.
Vittone, Bernardo. 1760. *Istruzioni Elementari per indirizzo dè giovani allo studio dell'Architettura Civile.* Lugano: Agnelli.
Viviani, Vincenzo. w.d. [Florence, 1718]. *Trattato delle resistenze principiato da V. Viviani per illustrare l'opere del Galileo.*

Part V
Architecture and Mathematical Practice in the Enlightenment

Part V rounds out the volume with two essays on new forms of mathematical practice in eighteenth-century architecture. The period represents both a watershed in the evolution of the subject and, for that reason, a fitting endpoint for the volume. We see in the late eighteenth century a weakening of the issues that animated early modern architects—the authority of Vitruvius, in particular. At the same time, new themes appeared that point to a transformation of architectural practice itself in light of contemporary science. The previous essays, in Part IV, showed both of these processes at work in the early history of structural mechanics. To take one example: Galileo's theories on the resistance of solids—that the strength of a form was dependent on its own size and weight—had immediate and irreversible effects on the Vitruvian doctrine of proportions. The two essays that follow explore similar transformations in different areas of architectural thought and practice.

Jeanne Kisacky reports on the efforts of Enlightenment medics and health reformers to improve the design of hospitals and prisons by exhausting "unhealthy" air. In some respects, the subject is well-known to historians. It formed part of Michel Foucault's interest in the modern disciplinary society and the basis of many subsequent studies inspired by his work. Kisacky, however, concentrates on a specific and relatively obscure aspect of this phenomenon: the associated attempts to quantify both the amount of air "consumed" by bodies and its subsequent movement through physical spaces. The behavior and properties of air, in this view, were mathematically definable. Subject to measurement by volume and by rate of flow or chemical transformation, these intangible properties could then be used to determine the form and dimensions of the ideal hospital. Both medical men and architects—in various degrees of collaboration—responded to this theory with a number of imaginative schemes. Although few of these projects were realized, they nevertheless represented something new: an architecture that reflected a mathematical-chemical-medical understanding of the building's "use". In redefining the building as an instrument or a "machine" of medical reform, they were also changing the nature of architecture itself.

A similar transformation is at work in the volume's final essay. David Yeomans, Jason Kelly, and Frank Salmon delve into the sketches and field notes compiled by James Stuart and Nicholas Revet as they prepared their groundbreaking serial publication, *The Antiquities of Athens* (1762–1830). In many ways, this project was traditional. Architects since the early Renaissance had sought to measure accurately the remains of ancient buildings. The qualities that made Stuart and Revet's approach novel were the same ones that tied it to the intellectual culture of the Enlightenment. In the first place, they insisted on the primacy of Greek as opposed to Roman architecture, thereby framing their work as part of a broader "philosophical" discourse of travel and discovery. Moreover, they sought an unprecedented level of precision in their measurements. The direct encounter with unfamiliar ancient monuments and at such a level of detail had unforeseen consequences. It revealed lapses and inconsistencies in Vitruvius's account of ancient design methods, while highlighting the difficulty of reconstructing those methods from the buildings themselves.

The authors' emphasis on Vitruvius brings us full circle. Bernard Cache's essay (Proportion and Continuous Variation in Vitruvius's *De Architectura*) served to place the ancient author at the beginning of a "mathematical" tradition of architecture that survived—on and off—for almost two millennia. Stuart and Revet, too, saw Vitruvius as central to their own practice. Despite protestations to the contrary, they continually sought to interpret, and in some cases complete, their findings in light of the ancient author's prescriptions. Yet, those attempts were mostly unsuccessful. If anything, their work revealed a fundamental incompatibility between the actual dimensions of ancient monuments and Vitruvius's written instructions. It was paradoxically the mathematical culture of the time that did the most to undermine the authority of the ancient text. The values of minute precision and strict empirical observation that Stuart and Revet brought to their work revealed—above all—Vitruvius's contingency: his was not the "true" voice of ancient architecture, but merely an incidental and fortuitous survival.

Stuart and Revet's work also signaled a change in the relationship between architecture and mathematics and, in some ways, an end to the tradition that our contributors have been exploring. A complete account of this phenomenon would go beyond the scope of this volume, but it may be worthwhile to mention some of the factors that likely contributed to it. The first would include a decline in the persuasive and pedagogical power of "mixed" mathematics: the idea that the various mathematical sciences were linked not only by the study of geometrical quantity but by their potential practical application. Architecture could be a form of mathematical practice and mathematics could be architectural as long as both disciplines saw themselves as part of a shared disciplinary culture, with similar methods and aims. A second, related transformation involved the rise of a new concept: technology. The early modern period had privileged the notion of "art" as a form of practical rationality, simultaneously linked to both nature and science. This role would slowly be supplanted by "technology", a term more closely linked to the values of industrial and scientific progress. Although structural engineering might play a valuable role in a truly technological endeavor, design, not to mention other more mundane aspects of architectural practice, would not. A third blow—perhaps more clearly evident in Stuart and Revet's case—was the increasing separation of contemporary mathematics from antiquarian and classical scholarship. For much of the early modern period, architecture and mathematics were joined not only as mathematical disciplines, but as privileged—and eminently recoverable—areas of ancient knowledge and culture. Indeed, both fields were emblematic of Greek and Roman civilization and, perforce, of Renaissance humanism itself. Stuart and Revet's work showed how these two disciplines, if not yet separate, might come into conflict.

For much of the early modern period, these overlapping intellectual frameworks acted as a bridge not only between two different disciplines, but different professions. They helped to link the work of scholars and practitioners, the world of the library with the building site, theory with practice. Although architecture and mathematics remained closely conjoined into the nineteenth century, their relationship would no longer be defined in the same way.

Breathing Room: Calculating an Architecture of Air

Jeanne Kisacky

How do the findings of science—the precise mathematical measurements and calculations of phenomena that reveal the deep truths of the natural world—get translated into real life, into real practices, and, for the purposes of this paper, into architectural theory and practice? The answer, far from straightforward now in an era of rapid change, was equally difficult at a time when architects were concerned about tradition as much as innovation.

By correlating pneumatic research in the 1700s to architectural designs specifically intended to promote a healthier internal air, this paper tries to trace how scientific findings became practical knowledge. The pneumatic research sought to quantify internal volumetric requirements and to outline ideal patterns of air movement in the creation of healthy spaces. Its practical application, however, posed a particular challenge for architects, who typically paid little attention to room occupancy and air flow. How architects dealt with (or ignored) this challenge illuminates the larger historical issue of how innovation is disseminated from initial laboratory-based mathematical findings to later empirically-processed practical changes.[1] This work focuses on prison and hospital design. Those building types were the subject of intense discussion and experimentation, particularly over their air quality. They have, moreover, received considerable historical scrutiny.[2]

Architecture and Air

In the late 1600s, Robert Boyle put living creatures into a pneumatic engine (an airtight metal sphere), pumped the air out of the sphere to create a vacuum, and then measured the time it took for them to die.[3] For architecture, which in essence

J. Kisacky (✉)
Independent Scholar, Ithaca, NY, USA
e-mail: jskisacky1@verizon.net

A. Gerbino (ed.), *Geometrical Objects: Architecture and the Mathematical Sciences 1400-1800*, Archimedes 38, DOI 10.1007/978-3-319-05998-3_11,

enclosed spaces, these experiments generated a pressing question: what did the enclosure have to do with the deaths? The question was particularly resonant in a period that associated increased incidence of certain diseases with foul environments, particularly foul air. Deadly plagues had troubled several European cities for centuries. Incidents in London in 1666 and Marseilles in 1720 were particularly serious. These plagues were frequently blamed on the pestilential atmosphere that developed in the congested and fetid cities. The tendency of ladies to faint at the theatre and other places of congregation was likewise blamed on their poor internal air.[4]

A longer-lived and more concentrated malaise seemed to plague the habitants of specific kinds of environments. The everyday prevalence of "ship", "jail", and "hospital" fevers in their eponymous environments revealed that specific kinds of buildings created specific kinds of disease. So-called Black Assizes (when whole courtrooms full of magistrates and spectators grew ill after the trials, most notoriously in 1577 at Oxford and 1750 at Newgate), caused intense concern about the danger of jails as breeding grounds for diseases that could escape into the larger environment. These episodes suggested there was a connection between building form and illness; examining and remedying that connection would involve both architects and doctors.

To study this connection, doctors and researchers examined rates of disease incidence. Early mortality and morbidity statistics, which counted the ratio of deaths or illness within a given localized population group, typically correlated disease incidence or death to physical surroundings, not medical practice or interpersonal interaction.[5] This correlation proved significant at both the scale of the city and at the scale of individual buildings. Urban areas had higher mortality than rural areas. Dark, dank, close rooms (usually corresponding with impoverished residents) consistently registered higher mortality rates than bright, airy rooms (usually correlated to privileged inhabitants).

Doctors since Hippocrates had focused on air as a vehicle for the spread, if not the cause, of disease. Their inferences from these statistics were straightforward. Buildings contained bad air; that bad air could cause disease. This connection was particularly noticeable in cities, where numerous densely occupied buildings were presumed to affect air flow and quality. The mathematically inclined physician and polymath Dr. John Arbuthnot believed that "the Air of Cities is unfriendly to Infants and Children," because it took animals time to develop a tolerance of the "artificial air" of urban areas.[6]

With life and death literally at stake, the aerial performance of a building thus held potentially drastic consequences for the architects and their clientele. Unfortunately, even basic questions of air flow and quality within enclosed spaces were not answered by traditional architectural expertise. Architectural treatises and works to the 1800s typically did include at least some discussion of air, but that knowledge was still largely based on Vitruvius (Books 1 and 5). At best, these works linked the layout of towns to winds, vapors, and air quality. They included no discussion of how much interior air space a person needed to subsist, how to induce specific patterns of air flow in rooms, or even how to ventilate rooms.

With smoky chimneys and cold drafts topping the list of building user complaints in Northern Europe, the most common air-related topic of architectural concern in the eighteenth-century was chimney and fireplace design.[7] This literature, too, was far from adequate: chimneys continued to smoke.

A healthy architecture thus needed answers to two questions: how enclosed air became corrupted and how buildings could mitigate those corruptive effects. Science could answer the first question; whether architects or doctors could answer the second was less clear.

The Science

Air was a hot topic of study in the 1700s. This research, however, often relied on mathematical calculations and experimental procedures unfamiliar to architects. Its results were disseminated in formulas and figures: the weight of the air, the proportional composition of different gases, the proportional effects of respiration on atmospheric air, and the physical behavior of air in motion. It was unclear how to apply the new laboratory knowledge to architecture.

Pneumatic research made graphically clear that the root cause of bad air in enclosed spaces was the breath of the occupants. The reason for this effect, however, was unknown. Did people add poisons to the air when they breathed, or did they use up some vital substance? This question had long been debated, and in the eighteenth century it spurred a number of experiments. The most significant of these attempted to quantify either the typical tidal volume of a breath or the proportional alteration of the composition of air before and after breathing. In the 1720s, for example, Stephen Hales breathed into a sealed bladder until he could barely inhale or exhale and nearly passed out (Fig. 1). He calculated a 1/13th reduction of air volume in the bladder after less than a minute. He also surmised that exhaling added quantities of unknown, and so unmeasurable, deleterious and potentially deadly substances to the air. Believing that all scientific knowledge was the result of measurement, Hales used complicated calculations of the surface area of the alveoli to estimate average volumetric lung capacity. He then multiplied that volume by the normal number of inhalations per minute (20), to get the total amount of air exposed to one person's lungs in a given time.[8] Every person, every minute of the day, spoiled roughly a gallon of air. This figure stood as an authoritative measure for decades.

Both Boyle and Hales had considered air a uniform fluid. The problem grew more complicated in the last third of the eighteenth century. The chemical and physical experiments of Joseph Black, Joseph Priestley, and Antoine Lavoisier, among others, revealed air to be a compound mixture of different gases, each with different effects and behaviors (Fig. 2). Their work shifted attention away from total respired air volume, focusing it instead on the individual gases—their effects on life and the effects of life upon them. Oxygen clearly supported life, nitrogen did not. Carbon dioxide actively harmed life. Respired air held an increased proportion of carbon

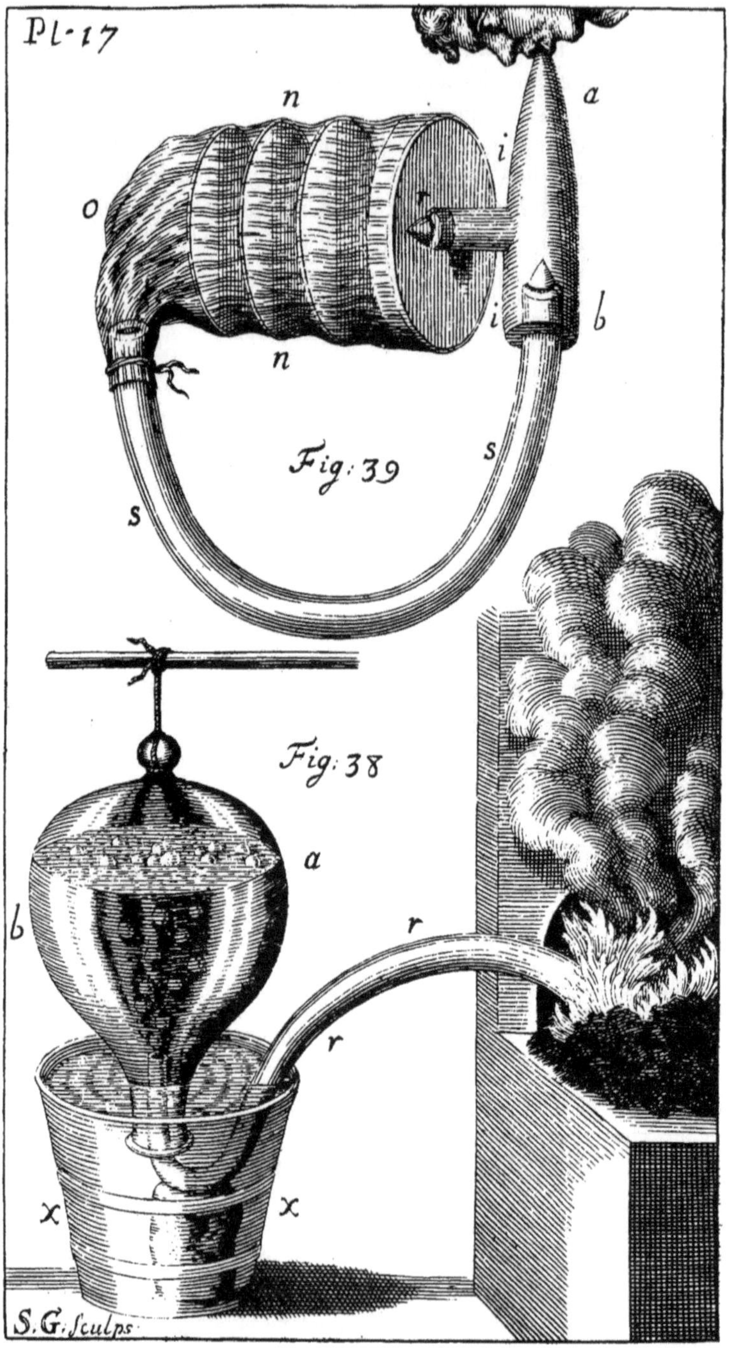

Fig. 1 Breathing experiment with sealed bladder (*top illustration*) (From Hales 1727)

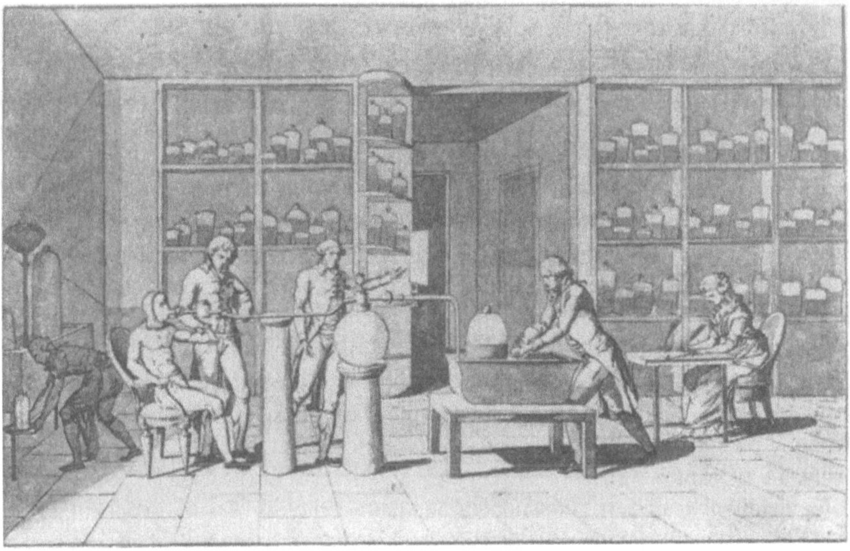

Fig. 2 Breathing experiment on respiration and transpiration (Reconstructed by Grimaux 1888)

dioxide and a reduced proportion of oxygen. People literally created 'bad' air.[9] This also implied that breathing in the air another person exhaled was inadvisable. Steady refinements in the scientific understanding of air's composition, however, did not alter the main question: if living beings were constantly using up the good air and increasing the bad air, how did they not suffocate en masse, whether inside or out?

This question was answered, in part, by the discovery of the beneficial effects of "nature" on air. If researchers were in agreement that breathing made air impure, they were in equal agreement that natural air circulation made it pure.[10] By the late 1700s, Joseph Priestley and Jan van Ingenhousz proved that plants consumed bad air (CO_2) and generated dephlogisticated air (oxygen). Enclosed spaces, which cut off the interior air from the larger atmosphere and its purifying forces, were the trouble. The best answer to noxious vapours and already-breathed air in enclosed spaces was thus ventilation—moving air between outside and inside.

Pneumatic science, however, tended to complicate this seemingly straightforward solution. The physical and chemical properties of air seemed to contradict each other in explaining its movement after it left the mouth. Physically, a by-product of respiration was heat, and warmed air inevitably rose. Chemically, the specific gravity of nitrogen was lighter than oxygen, while carbon dioxide was much heavier; theoretically this would create vertically stratified layers of gases. Respired nitrogen, by both its lightness and warmth would rise to the ceiling. Oxygen might remain in the middle or, if heated, rise. It was unclear if the heavier carbon dioxide would fall to the floor or rise because of its added heat.[11]

Architects or Physician-Architects?

These pneumatic researches put air circulation theoretically at the heart of architectural design, but left uncertain exactly who—whether architects or physicians—were to translate the science into practical architecture. As learned gentlemen, many architects were likely aware of the revolutionary results of pneumatic science. In 1798, Benjamin Henry Latrobe, trained in Europe but practicing in America, mused on the cause of yellow fever. His journal includes details of some of Joseph Priestley's experiments and the use of the word "azot" (coined by Lavoisier) to refer to one of the atmospheric gases.[12] Latrobe, however, is an exception. There is a relative silence on the matter of air and ventilation from the vast majority of non-specialist architects. Few mention individual researchers or experiments, nor are there any pneumatically inspired architectural treatises. Retranslations of canonical texts did not incorporate the new pneumatic findings. Although some do reveal some of the eighteenth-century fervor for air, they do not pose new solutions to the problem.[13]

In this relative architectural silence, the main promoters of a new practical pneumatic approach to architecture consisted of doctors, researchers, and inventors. They were not just filling a void; some physicians and inventors saw it as an opportunity to expand already fluid professional boundaries. They did so by discrediting architects as well as by establishing counter-expertise. Many aggressively denounced the positively destructive effects of traditional architectural design on the health of its occupants. Hospitals were seen to be the most egregious offenders. Even architectural critics such as Antoine Laugier decried their "misplaced splendor". The doctors and researchers were even more critical. In 1759, the agronomist, researcher-engineer, and naval official, Henri Louis Duhamel du Monceau, stated that in the majority of hospitals "the design of architects has not been to procure the renewal of air." Given his numerous preceding statements of the importance of fresh air to health, that was an indictment. In 1774, the physician, Antoine Petit stated that architectural knowledge alone was not sufficient for the design of a building as difficult as a hospital, which required some knowledge of the effect of exterior agents (air, water, exhalations) on the health of the patients. By 1787 Jean-Baptiste Leroy commented caustically that the prevalence of architectural books on palaces and theaters and the paucity of books on hospital construction showed "that men always preferred things of beauty and even of frivolity to those which offer only a sad object of utility." The design of hospitals, he insisted, required knowledge of medicine and physics as well as of building design and construction.[14]

While challenging the architects' traditional knowledge base for designing hospitals, the doctors also invariably promoted new medical and scientific grounds for hospital design. These physicians did not intend to take over all aspects of architecture, but only to determine rules for "healthy" buildings. They clearly expected that future rules of hospital architectural design, if promoted by doctors, would be implemented by architects.[15] These competing experts—researchers, doctors, architects—waged their war over the theory and practice of ventilation. For any building design, there were four possible ventilation strategies in the late eighteenth century—what I will call *chemical* (altering the composition of the air), *spatial* (sizing a room in proportion to the number of its occupants), *architectural* (making a room porous enough with numerous windows and doors not to count as confined), and *mechanical* (installing ducts and blowers or heat sources to induce air movement regardless of room size, shape, and features). Each came with its adherents and detractors; each promised to solve the problem of bad interior air.

Chemical Ventilation

Chemical ventilation sidestepped the problem of architecture's role in ventilation and placed doctors and researchers in the catbird's seat. It operated on the simple expectation that deleterious substances in air (which smelled rotten or foul) might be chemically corrected by strong-smelling, typically antiseptic substances. Smelling salts, nosegays and fumigants had been in use for centuries; most simply covered one smell with another less offensive one.

Personal ventilators promised to allow individuals to exist briefly in deathly gases and airs. In the 1660s, Sir Christopher Wren proposed an instrument "for straining the Breath from fuliginous Vapours." In 1727, Stephen Hales proposed a cloth filter and later a four-compartmented bladder impregnated with vinegar or sal tartar as a means of removing noxious vapours from air before breathing. Hales suggested their efficacy for miners in combating the mine damps. By 1759, in a curious parody of modern surgery, Duhamel du Monceau suggested surgeons wear a cloth mask to keep from breathing the bad air in a ship's ward, while in 1777 John Howard sniffed vinegar as a preventive before his visits to the jails.[16] Personal ventilators did not, of course, solve the larger problem of meliorating the air for large numbers of people in a given space.

Room-scale fumigations were also common. In 1742, the physician John Huxham suggested camphor and vinegar vapours. Hales proposed burning brimstone and hanging cloths dipped in vinegar between the decks of a ship. Duhamel du Monceau suggested fumigation by saltpeter, gunpowder, sulfur, or even perfumes. The *Gentleman's Magazine* recorded a letter promoting a specific commercial composition developed by a painter that would "destroy any noxious, pestiferous quality, either in the air or goods." In 1784, Dr. Thomas Day suggested that a fumigant should target only the infectious quality of the air. He believed it was fixed air (nitrogen) that was the culprit, and suggested lime water as a fumigant since he believed it absorbed the fixed air out of the atmosphere. He even developed an interior 'rain machine' to shower the lime water across every inch of air space in a compromised room and so purify all parts of the air (Fig. 3). The main problem with fumigation, however, was that most substances required evacuation of the room of its occupants. This was not possible in the most problematic environments such as ships, jails, and hospitals.[17]

Some commentators suggested simply increasing the supply of oxygen in rooms. Joseph Priestley speculated that in crowded rooms oxygen "might be brought into a room in casks; or a laboratory might be constructed for generating the air, and throwing it into the room as fast as it should be produced." This expedient would have reduced "breathing room" to a commodity, measured in casks of oxygen delivered per

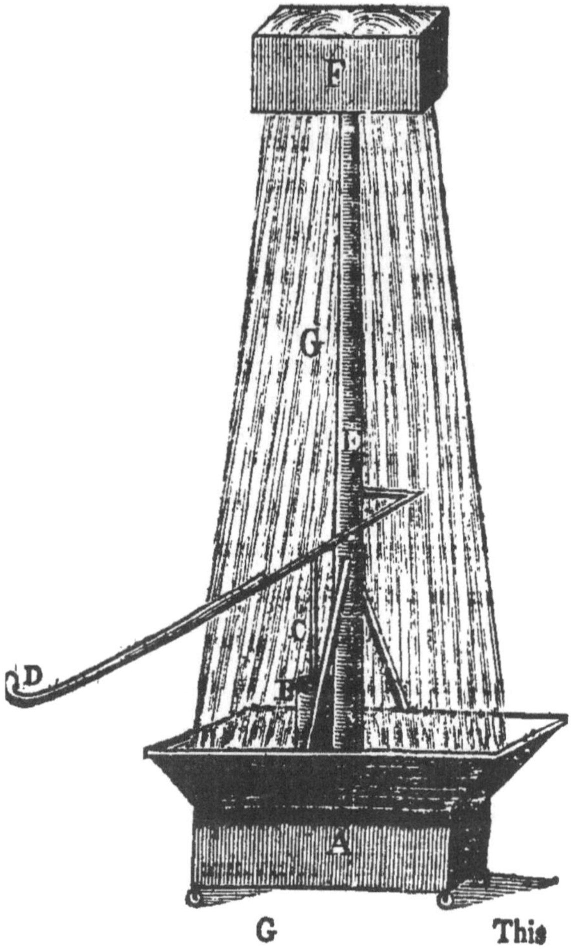

Fig. 3 Air-purifying machine (From Day 1784)

hour, but it was never tried. Jan von Ingenhousz took a simpler approach: he recommended putting oxygen-generating plants inside and around houses as air purifiers.[18]

All of these solutions were independent of architecture. Most were merely correctives designed for sporadic use. Chemical ventilation was principally the purview of researchers and perhaps some physicians. It had little to do with building design and architects had little to add to such suggestions.

Spatial Ventilation

Spatial and architectural ventilation, on the other hand, resulted directly from the physical building; the former type was dependent on the size of rooms, the latter on the openings in and the shape of rooms to permit air flow. The two approaches were interdependent; spatial ventilation relied on at least some air flow for air exchange, and architectural ventilation relied on having enough open space within a room to allow air flow. The pneumatic researches into air revolutionized both aspects of ventilation, particularly in terms of quantifying how much air space and flow was needed.

Stephen Hales' early estimate that a person "spoiled" a gallon of air every minute of their life stood as the first and longest-lasting scientific estimate. In simplistic terms, this implied each person required a gallon of fresh air every minute to maintain their existence. Other cubic estimates followed. In 1733, John Arbuthnot estimated that a person could not survive for 20 min in a tun (40 cubic feet) of air without air exchange. Based on normal breathing rates by an average person, Lavoisier calculated that five gallons of air per person per hour were needed. The influential 1786 Hospital Commission Report of the Parisian Academy of Science used Lavoisier's calculations to estimate that a man consumed 108 cubic Feet of air in one and a half hours.[19]

Although these rates were typically measured in spatial terms—cubic feet—how they related to room size was unclear. According to traditional architectural knowledge, room size was the product of harmonious mathematical proportions. In plan, Vitruvius allowed any rectangular proportion from a square to a double square to be pleasing. Beyond that, a room became a mere gallery. Laugier stressed room proportions in plan that resolved into whole number relations (1:1, 1:2, etc.) as the most pleasing. Rooms with proportions based on geometrical and arithmetical series were also admirable. With some allowance for the client's social standing (lower ceilings for the middling classes and higher ceilings for the wealthy) and the requirements of varying climates (higher ceilings in warmer climes, lower ceilings in colder climes), room heights were determined in harmonious proportions to the dimensions of the room's breadth and width. The height was typically not to exceed the breadth. The longer and narrower the room, the more displeasing a very high ceiling would be.[20]

This proportional system for room design left out considerations of occupancy (how many people would be in a given space) and cubic volume (how much air space would be available to each person in a given space). Eighteenth-century medicine and science pushed for spatial proportions based on these latter requirements. Doctors were the earliest to try to interpret how the volumetric figures of air usage in breathing would be best arranged into rooms. Their first approach was simply to translate these figures directly into architectural dimensions, establishing a minimum air volume per person.

With each inhabitant of a room understood as a point source of aerial corruption, the necessary volumetric air "space" around a person required specific shapes. Clearly, a very tall ceiling would provide a large air volume, but would also crowd occupants together at floor level. With medical opinions suggesting that the infected air from a patient spread anywhere from a couple feet to ten or more feet, crowding was dangerous. The literary physician, John Aikin pointed out the flaws of such room shapes in hospital wards: "If we conceive for a moment in imagination, that [a large ward] was partitioned into as many separate divisions as there are patients, we shall be sensible how narrow a space is allotted to each."[21] What mattered was the placement of enough air "space" between people to minimize the interaction of their breath. In this approach, a large room was really a number of smaller aerial compartments, each centered on an occupant. Floor area per person was as important as room height.

This consideration entailed an architectural system that dictated where people and objects were placed within a space. Room proportions, in other words, had to relate to both the number and dispersal of people in a room. The Academy of Sciences' 1786 Report on Hospitals, for example, explained the need for a clear aisle 3 feet wide on each side of a patient. The precise amount of space would meet three requirements: utility (access to care for the patient from either side); freedom of air flow; and disease prevention through distancing one focus of infection from another. Making the shift in architectural knowledge even clearer, the hospital physician and reformer, Jacques Tenon proposed that a ward's ceiling height be proportioned to the inhabiting patients' malady. Feverish patients, who breathed more quickly and gave off more heat and effluvia, needed more than the minimum air volume. Fever wards were thus to receive higher ceilings.[22] In Tenon's view, room size and shape were ideally related to the placement, number, and physiological characteristics of its intended occupants.

These requirements suggested a new system for determining room size, one that could not easily be worked into the existing aesthetic proportional system. Laugier, for example, appeared to be aware of these current medical discussions, but without knowing exactly what to do with them. Separately from his discussion of pleasing room proportions, he admitted that ceilings should be at least 6 feet in height, so that the inhabitants could breathe freely and maintain a considerable volume of air around themselves. Double that minimum would give a more sufficient air volume for health; triple would be even better.[23] He did not, however, attempt to integrate the varying spatial requirements of health and aesthetics. Laugier might have felt out of his depth here. While architects were capable of calculating a room's spatial volume based on expectations of occupancy, they could not authoritatively calculate the volumetric requirements of breathing, nor could they overtly control occupancy.

With little engagement on the level of design, spatial ventilation strategies first manifested not in new building forms but in new—typically unenforceable—regulatory requirements for room sizes and shapes, which promoted fewer people in larger rooms. It also made room size the product of room occupancy. Lavoisier, for example, calculated the maximum occupancy of a theater based not on the number of seats, but on the ratio of air-consuming spectators to total spatial volume. John Aikin stated that large, crowded hospital wards were impossible to make healthy and suggested that every hospital should as much as possible remove the convalescent patients from the wards to minimize the aerial load on the spaces. The regulations of Haslar Hospital required a few empty wards in the fever wing to allow all the wards to be emptied and aerated in rotation.[24]

Along these lines, Joseph Priestley added yet another extra-architectural wrinkle to spatial ventilation: time of occupancy. According to Priestley large, lofty dining rooms were preferable to small and low ones, but no matter how large, if enough people stayed long enough in the room they would contaminate the air. Smaller dining rooms, although containing less air, would be healthier, because they would get stuffy sooner and be opened to allow a flow of air. The doors and windows, being proportionally larger compared to the rest of the room size, would more effectually ventilate the space. In the end, Priestley advised the best recourse for the problems of large dining rooms was to have a provision for exhaust ventilation in the ceiling, or to vacate the room after a certain length of time, "to have the dinner in one room, and the dessert in another."[25]

Architectural Ventilation: Studying Air Flow

Architects proved largely immune to spatial ventilation strategies that placed occupancy and temporal limits on architectural design. They were more receptive to new ideas about architectural ventilation. As French architect Bernard Poyet pointed out, "even the largest and best held apartments contract a foul and infectious odor when one neglects to open them."[26] Air space without air flow was insufficient.

Windows and doors were traditionally located not for air flow, but for utility or aesthetics. Vitruvius, for example, prompted the placement of windows and doors for the efficient circulation of people and goods and as architectural embellishments that enhanced the beauty of the façade. Other architectural treatises also considered window and door placement and sizing in relationship to a pleasing exterior façade. Soane, for example, felt it necessary to point out that all the windows on one floor should be the same height and width.[27] Windows were often placed not with reference to an individual room's interior needs, but to the symmetrical requirements of the entire façade. This practice often left rooms with irregular window locations and compromised air flow.

The size of openings was also aesthetically, not pneumatically, determined. Proportional maximums stated that windows were not to be wider, say, than one-quarter of the breadth of the room or narrower than one fifth of its breadth. Local deviations were common. In England, windows were thought better not to extend to the floor; in France, they often did so. An architect's basic understanding of air flow was simple: smaller, fewer openings reduced air flow and larger, more numerous openings increased it. William Chambers recommended smaller, fewer openings in colder climates, where the preservation of heat was critical in the winter months. The actual pattern of air movement inside a room was a consequence, not a determinant of form. As Lavoisier put it, air "circulation exists more or less in all rooms, often even despite the architect who has directed the construction."[28]

In contrast, late eighteenth-century doctors and researchers wanted to prevent one person breathing in air another person had exhaled. In theory, this required moving a specific quantity of air in controlled, often complicated patterns within a room. In 1780, in a report on prisons, Lavoisier described a building design to induce different atmospheric gases to move in different patterns. Large openings at the highest point of the ceiling would remove the great quantity of nitrogen and the warmed carbon dioxide exhaled by the inhabitants. Smaller, lower openings would allow any carbon dioxide that had fallen to the floor to escape but would also supply the incoming cooler, oxygen-rich fresh air.[29] Extending the windows to the ceiling prevented the buildup of foul air, while shaping the ceiling like a funnel promoted exhaust flow. Aesthetics were not considered. This was an architecture designed around air, but it remained speculation.

As this example suggests, practical advances in architectural ventilation were impeded to a certain extent by a rudimentary understanding of air flow. After undertaking experiments on air flow and designing a wind device to siphon air through chimneys, researcher and physician Jean-Baptiste Leroy promoted the placement of windows on opposite walls simply because "[air] entered on one side and left by the other" (Fig. 4). Similarly, Jacques Tenon pointed out that any intermediate partition in a room obstructed air.[30] Both facts were well known in architectural practice on equally basic terms.

Invisibility made conclusions about air flow particularly difficult. Early anemometers could provide the measurements of relative air velocity at a given point but could not reveal the overall pattern of flow in a room. In the late 1700s, there was a brief hope that eudiometers—chemical devices believed to reveal the purity or impurity of air—could be used to correlate architectural details to air quality and thus create an architectural primer for healthy design.[31] However, the early eudiometers were not sensitive enough to register the fine differences in air composition in most locations. Doctors in hospital wards and sick rooms had the advantage of regular experience with fumigations, which often used smoke or visible vapors—the dissipation could reveal air circulation in a room. There were, however, too many variables between sites—the effect of exterior wind and weather, fireplaces, and open or shut windows—to establish sure conclusions from this procedure.

Empirical research into building environmental performance was spotty and often flawed, but the eighteenth century did see several notable attempts to answer some basic questions about air flow: did specific existing building forms promote air flow? Should bad air be exhausted from the ceiling or from the floor, and could specific building forms be correlated to better or worse health?

The Hôtel Dieu of Lyon was the site of one attempt to answer these questions. This grand building had an even grander dome. The architect, Jacques Germain Soufflot, justified this feature as more than an opulent marker of the hospital's central altar. It was designed and built for ventilation. Soufflot had visited hospitals in Italy and had come to his conclusions for how to make a healthier infirmary based on his observations there. In Soufflot's design, the heated, contaminated air from the patients was to rise and be sucked out of the dome's heights. Indeed, Soufflot, among others, believed that the volume of foul air coming through the dome would be enough to kill birds.[32]

In 1782, Dr. Maret of Dijon decided to test this thesis. A partisan of the downward flow of vitiated air, he hung some bird cages in the dome. The birds were still doing fine after 40 days. He hung several pieces of meat in the dome and at the patient's bed level. The meat at the floor level corrupted in less than a day, while the meat in the dome was still unpolluted after 5 days of exposure.[33] Maret took this as proof that the air at ground level was much more contaminated than the air in the

Fig. 4 View of Jean-Baptiste Leroy's siphon ventilator (From Leroy 1780)

height of the dome. Soufflot defended his position and continued to hold that the dome was favorable to the salubrity of the wards, but there were no further experiments to support either position.

In 1785, Lavoisier initiated a similar attempt to end the uncertainty over whether air flowed upwards or downwards. He took samples of air from ceiling and floor level of a notoriously foul hospital ward and of a theater. He then tested the proportional mixture of gases in the samples. The nitrogen content in all samples remained the same. All samples registered a reduction of oxygen, but the ceiling air experienced a greater loss than the floor. Similarly, all samples registered an increase in carbon dioxide, but the ceiling air registered the greatest. Despite Maret's more graphic experiments, Lavoisier gave more credence to the partisans of upward flow.[34]

Jacques Tenon took a different approach to the same problem. He calculated detailed mortality rates for every ward in the Hotel-Dieu and then correlated those rates to the architecture of the wards.[35] He recorded the location of windows, the proximity of service spaces, the placement, number, and size of beds, the ceiling height, and the room's total spatial volume. The conclusion—rooms with windows in only one wall, with low ceilings, with little floor space between beds (or between patients), and rooms adjacent to noxious facilities—had higher mortality rates than those with windows in more than one wall, higher ceilings, more open floor space, and windows onto open areas. This was like a new architectural primer, unlike and unrelated to Vitruvius, but equally capable of serving as the basis of building form.

Architectural Ventilation: Projects

These methods of studying air flow were supplemented by another, more frequent source of experience: remedial small-scale alterations to existing rooms with poor ventilation. Such interventions were the closest thing to effective practical experiments on air flow in interiors. They were also often where scientific and architectural knowledge interacted. Sir Christopher Wren's late-seventeenth-century insertion of pyramidal air exhaust ducts into the ceiling of the House of Commons, where speakers had been known to faint during speeches, was an early example.[36] In this case, Wren stood as both experimental researcher and architect, whereas these alterations were more commonly collaborations between the two groups. The doctors explained what was wrong, and the architects decided what could or could not be feasibly altered. Duhamel du Monceau was involved in just such a project. In 1759, he recommended carving several new exhaust holes in the ceiling of the St. Landry ward of the Hôtel-Dieu in Paris. The hospital architect, M. Ducret, confirmed it was structurally feasible and oversaw the work.[37]

Conjectural building designs organized around architectural ventilation also appeared in the late 1700s. Some were designed entirely by doctors or researchers, while others were designed by doctors but drawn up and embellished by architects in working collaborations. A few, however, were architect-driven, and it is this latter category which reveals that the new scientific knowledge was exerting some influence on the profession. In this interaction, conflicting sources of knowledge about good building design began to blur the lines between the two professions, but the two groups still appeared to need each other. Doctors and researchers were reluctant to propound on details of architectural construction or management, while architects were reluctant to express positive figures for air space or wind flow.

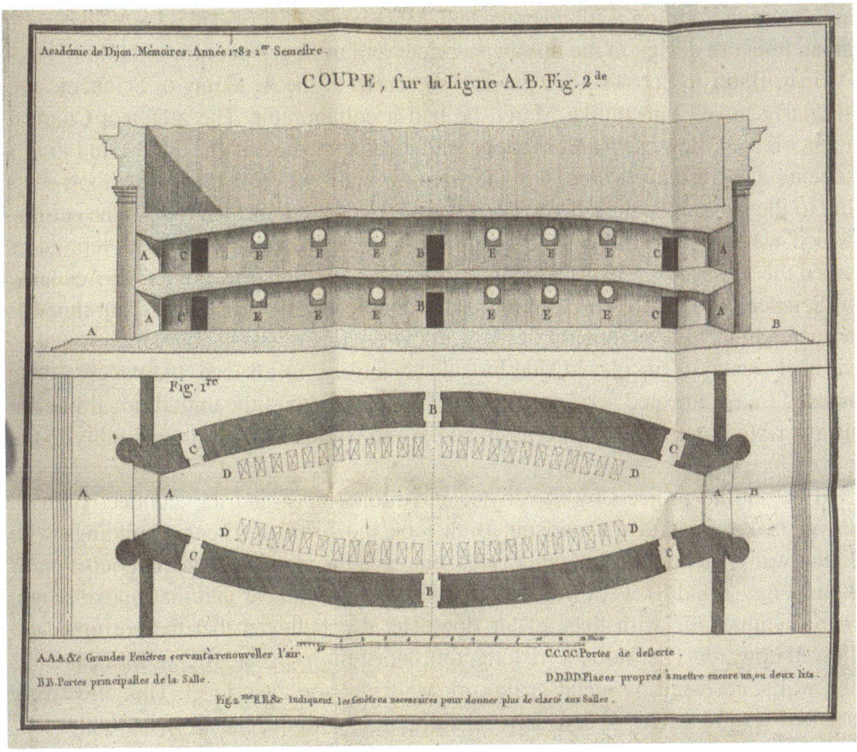

Fig. 5 Ideal hospital design (From M. Maret 1782)

A number of ideal hospital projects appeared after the devastating fire at the Hotel-Dieu in 1772. They provide unique perspective, not only onto the perceived importance of air in health, but also onto the relations between architectural and scientific practitioners in the period.[38] Dr Maret's 1782 project for an ideal hospital ward was a building designed around a complete (if inaccurate) expectation of how air moved and acted. According to Maret, bad air lingered near the patients' bodies; this made floor-level ventilation critical and reduced the value of high ceilings. To prevent the bad air from contaminating the fresh, hospitals needed a complete exchange of air. Maret also believed air flow to be linear and to be constantly reflected perpendicularly from the surfaces it encountered. His resultant ward design was geometrically structured to permit the complete exchange of air (Fig. 5). The ward was elliptical in plan and in ceiling section so that the moving cones of air would all eventually bounce back to the center axis. The two truncated ends of the plan were set with wide portals. Whenever the air seemed tainted, the doors could be opened and a flow of fresh air would sweep along the axis from one end to the other, completely flushing the bad air with it, thus "guaranteeing" a complete, controllable exchange of air. This egg-shaped building was completely outside of architectural tradition, with no precedent. Maret aptly observed that it would require an architect to design in the missing services and to figure out how to build it.[39]

Jean-Baptiste Leroy's 1773 project, presented to the Academy of Sciences, was similarly novel, but, unlike Maret, he had a collaborator. The architect Charles-Francois Viel drew the plans, sections, and details for the design (Fig. 6). In Leroy's scheme, the fresh air flowed in to the ward through 'wells' in the raised floor, passing to the patients in their beds. Their warmed, respired air then rose to the ceiling, which was shaped to direct the air upwards to an exhaust duct. Wind ventilators aided the upward flow by sucking the air out at the ceiling high point. The Academy of Sciences committee on Hospitals criticized the length of his wards, but this was a misreading.[40] Leroy thought of the ward as a series of contiguous small rooms—vertical 'tents' of air placed side by side for utility. In essence, the patients were housed in a connected series of "chimneys". Viel clothed the untraditional interior as a peripteral temple, perhaps as an attempt to lend architectural credibility to the novel design.

Two similar projects by the physician Antoine Petit and the architect Bernard Poyet make a useful comparison. Both took the form of circular buildings with radial wards and permeable outer rings, but each promoted a different course of air flow (Figs. 7 and 8). Petit believed that air flow was linear and that lines of sight were sympathetic with lines of air flow. He also believed that the corrupted air flowed upwards. His wards were long, four-story-high spaces. Niches for beds lined the walls, accessed by metal grillework walkways on each level. This effectively stacked several stories of beds into one tall space, with the bad air rising to the ceiling (Fig. 9). The outer end walls of each wing were closed; the interior end walls opened to the large funnel-shaped central dome, which encouraged the upward air flow. Petit expected the air to enter through the numerous windows placed along the long sides of each wing and then be sucked down the length of the ward to the funnel dome. Since there were windows between each bed, this placed moving 'walls' of inflowing fresh air between each patient. Petit even channeled the building's chimney flues through the walls of the funnel dome to provide extra heat and extra updraft. The air would be continually pulled through the wards, regardless of the external wind force.[41]

Wind flow also determined the layout of Bernard Poyet's proposed hospital. Each ward was to be 26 ft high and 30 ft wide, with an open middle aisle of 12 ft (Fig. 10). While the floor area was generous, it was determined as much by utilitarian needs (bed size, aisle width) as by cubic volume. Poyet simply doubled the floor area per patient available in the existing Hôtel-Dieu.[42] His expectation of air flow was vastly different than Petit's. According to Poyet, the flow of water along the river encouraged the flow of air. The old Hôtel-Dieu straddled the Seine, and, just as the water flowed past the building without entering, so, theoretically, did the air. In contrast, Poyet sited his hospital on an island in the middle of the river, maximizing its participation in the downstream air current. The long, finger-like wards were open at the short ends to allow these winds to flow the length of the wards and into the open central court. As winds traveled along the river for hundreds of miles, Poyet reasoned, there was reason to believe they would also do so inside a building

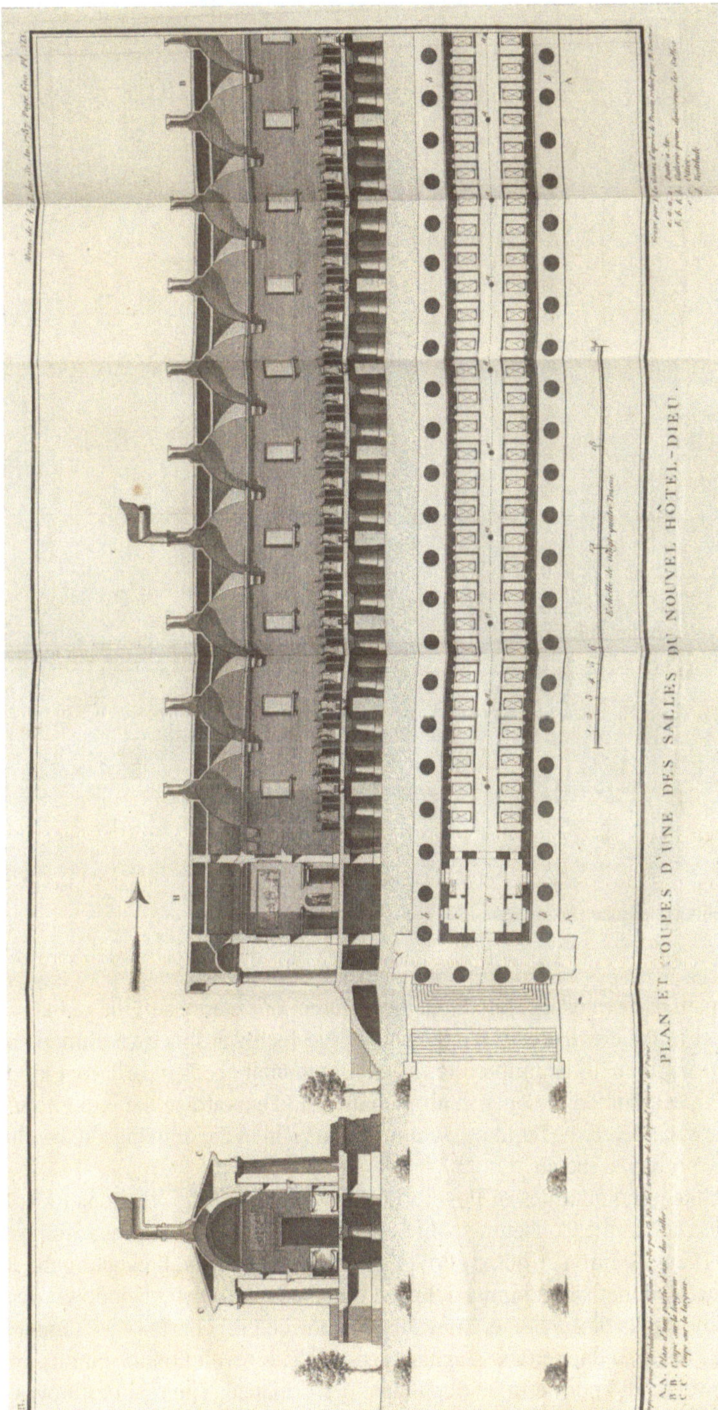

Fig. 6 Ideal hospital design (From Leroy 1787)

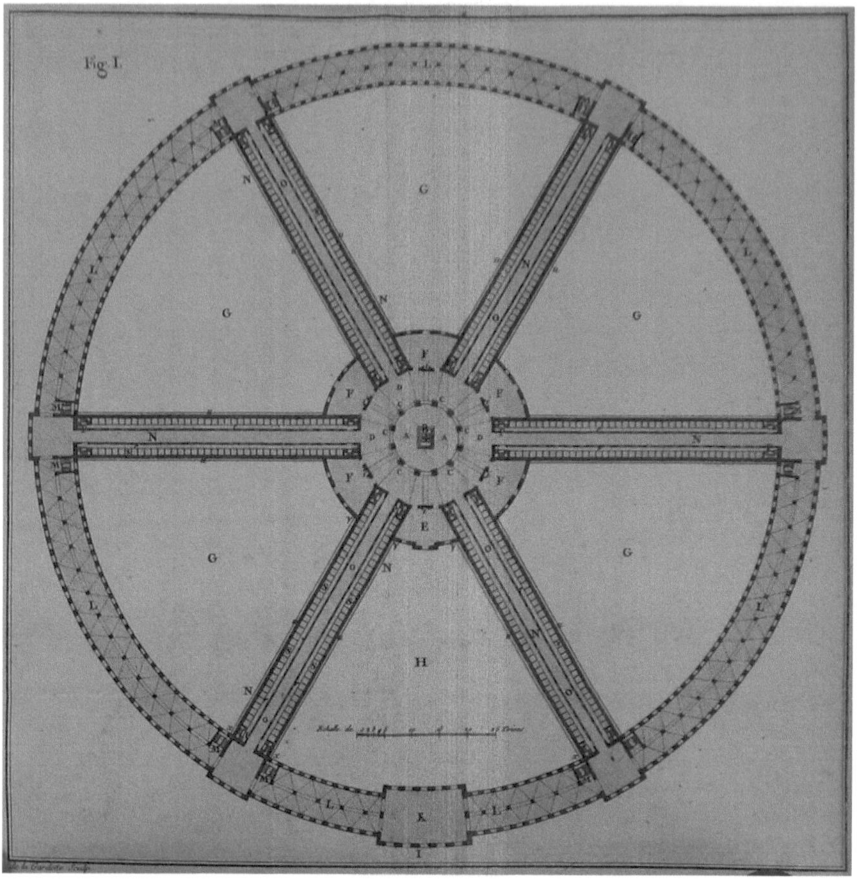

Fig. 7 Ideal hospital plan (From Petit 1774)

if given a path. He even speculated that the different orientations of the radial wings with respect to the downstream winds would give each ward its own climatic attributes, allowing them to be tailored to the environmental needs of different kinds of patients. Poyet assumed the open central court would encourage the wind flow.[43] In essence, he considered the building, seemingly so solid in the drawings, to be almost completely porous to the air.

The architectural qualities of Poyet's design—including the allusions to Roman monuments such as the Coliseum—were obvious, but its potential functionality was less clear. The opposition criticized Poyet's expectation of wind flow along the river. Critics countered that air had indeed flowed into the Hotel-Dieu's windows, despite its site on the banks of the Seine. They also saw the central court not as a funnel for vitiated air, but as a dangerous, stagnant space.[44] The revolutionary quality of the design, however, did not lie in its pneumatic performance. The revolution was that the design required pneumatic premises to be completely understood.

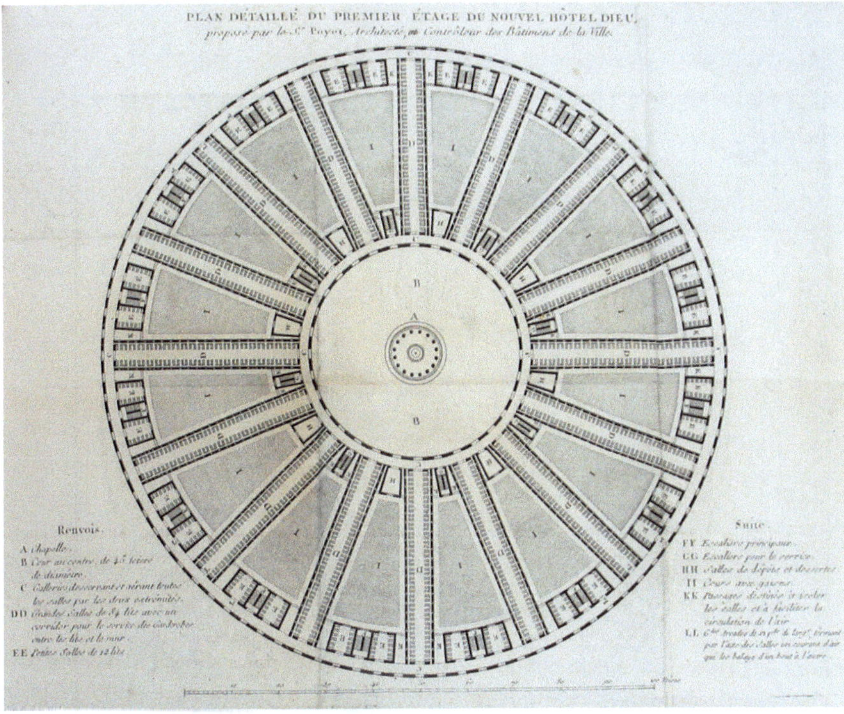

Fig. 8 Ideal hospital plan (From Poyet 1785)

Mechanical Ventilation

Architectural ventilation had a fatal flaw—all air flow was at the mercy of the winds. Mechanical ventilation, by providing its own motive force, promised complete control of both rate and pattern of air flow. Jean-Baptiste Leroy even noted that mechanical ventilation could end the argument over whether air naturally flowed upwards or downwards. Such a contrivance could make air move either direction, depending on what was needed. There were two serious drawbacks to mechanical ventilation: it was expensive, both to install and to keep running, and it was "unnatural".[45] Suspicions that mechanical ventilation somehow adversely affected the vital principle of air were powerful and lingering. Passing air through ducts was believed to remove the vital principle or to add a deleterious substance to the air. Worries about closed stoves, which were thought to expel a bad, overheated air through ducts, were particularly hard to dispel.

In 1715, John Theophilus Desaguliers, designer of an early centrifugal fan, undertook some experiments to prove that mechanically-propelled air, including air that had traveled through warmed fireplace ducts, was safe and would support life. He ran air to a bird through red-hot tubes, through cubes of heated metal, and

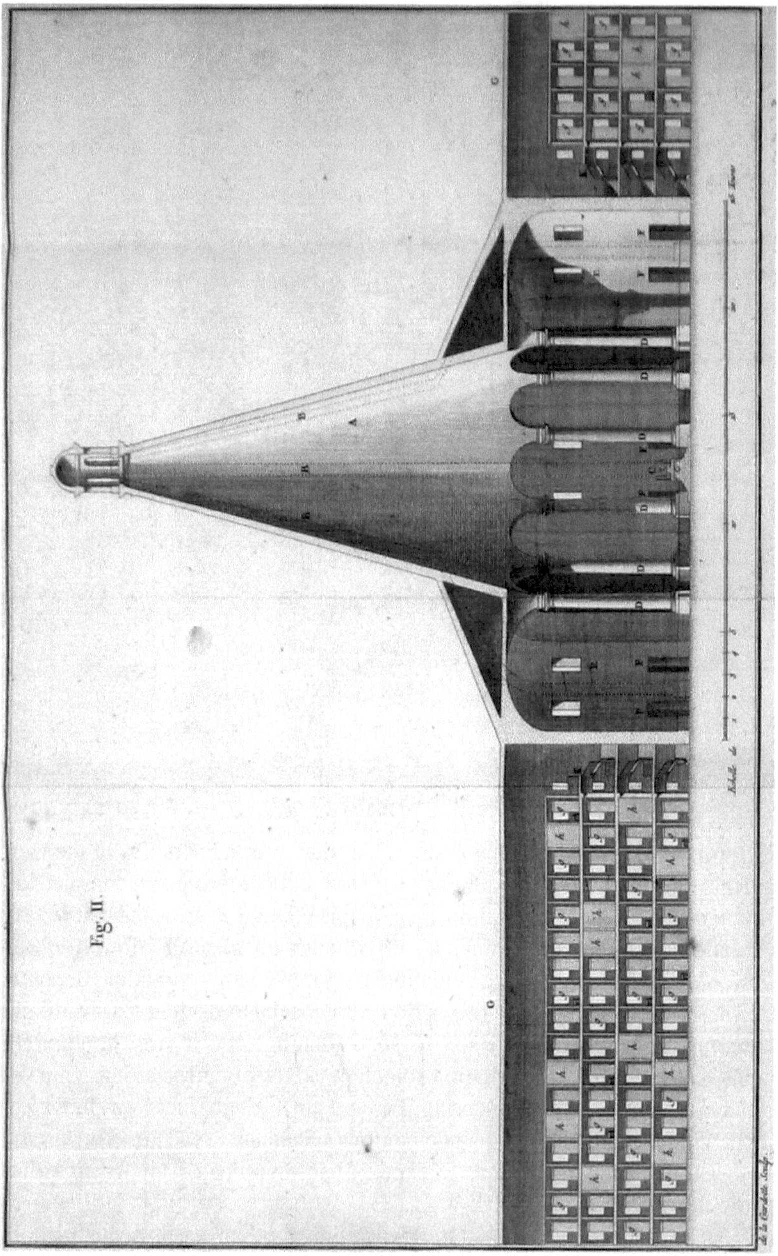

Fig. 9 Section of an ideal hospital (From Petit 1774)

Fig. 10 Elevation and section of an ideal hospital (From Poyet 1785)

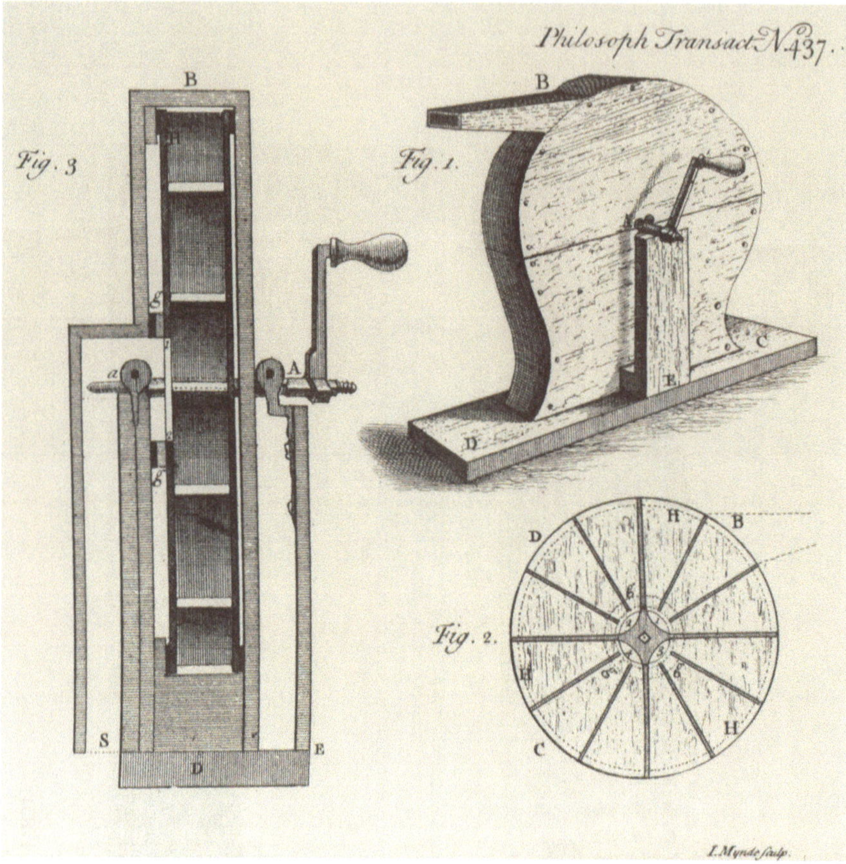

Fig. 11 Centrifugal fan (From Desaguliers 1753)

through charcoal. The bird survived many of these experiments, but not all. In one, the bird died almost instantaneously.[46] This strengthened expectations that air did indeed contain a vital principle, and that mechanical ventilation destroyed it.

The need for a power source also complicated the adoption of mechanical ventilation. The earliest ventilators relied largely on human power to pump or crank a device that moved air. Ventilators of this kind first appeared where there was desperate need and where the inhabitants were motivated (by choice or by force) to power the machines, namely in mines, ships, and prisons.

The first air movement machines were quite simple. As early as 1727, Desaguliers designed a rotary centrifugal fan to draw the foul air out of mines (Fig. 11). By 1735, he had perfected the instrument for use in sick rooms and recommended its flexible use either to draw air into a room or expel it. The year 1741 saw three more ship ventilators vie for priority: Stephen Hales' simple bellows box ventilator (Fig. 12), Martin Triewald's similar ventilator, and Samuel Sutton's system of well-placed

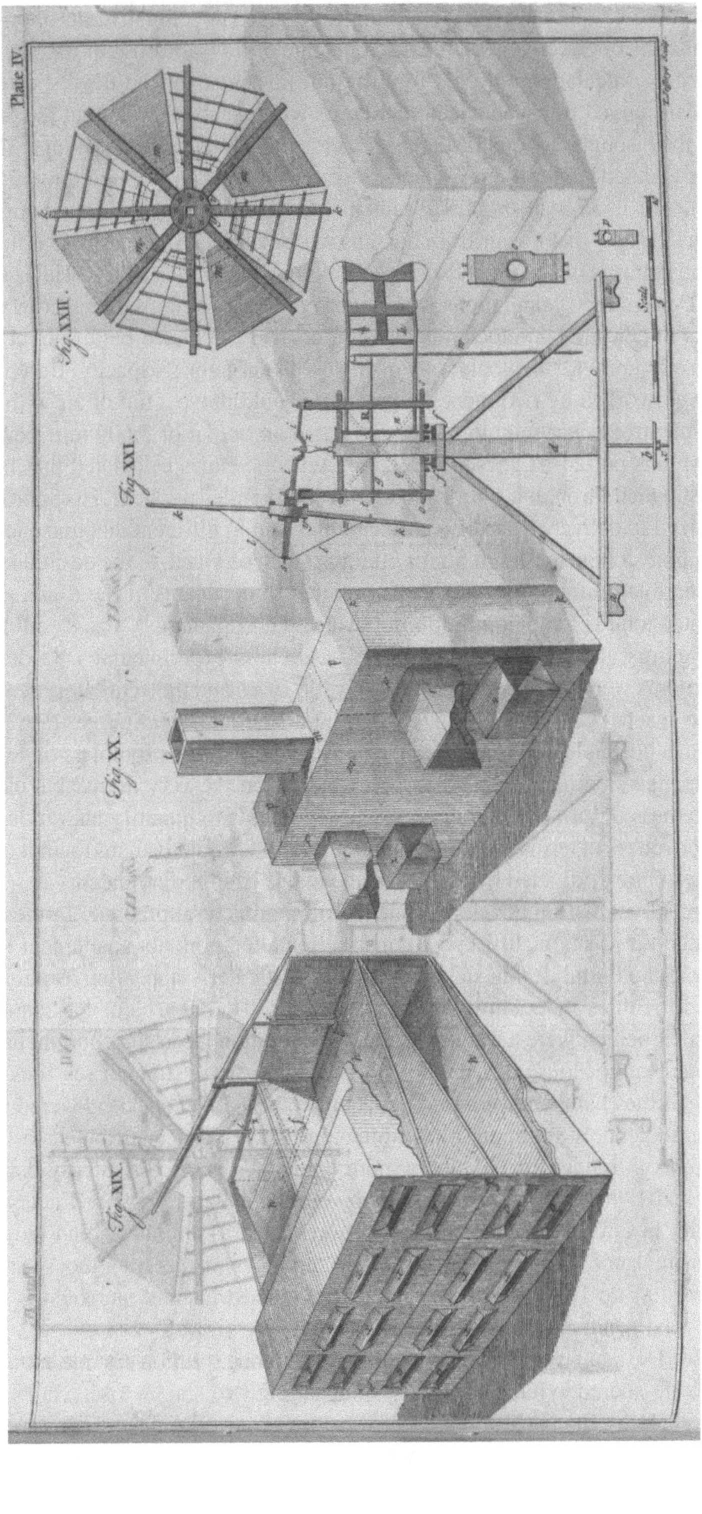

Fig. 12 Stephen Hales' Box Ventilator and his design for the Newgate Windmill (From Hales 1758)

fires and ventilation tubes.[47] The competition was judged not only in terms of priority, but also in terms of effectiveness, measured as rate of flow and ease of use.

The various calculations offered by different inventors for the flow, speed, and volume of air moved by mechanical ventilators were far from exact, but the degree to which they sought to quantify these factors, is nonetheless striking. Martin Triewald calculated his ventilator capacity on the volume of air it could move: 36,172 cubic feet of air in an hour or 21,372 tuns in 24 h. Desaguliers performed complex calculations to show that the air from his fans reached a speed of nearly 45 miles per hour (the advantages of such speeds for ventilation were left unstated).[48] Hales countered that Desaguliers' calculations were incorrect—that his fan's air speed was not the product of centrifugal velocity as Desaguliers had calculated but of direct pressure. Hales also provided calculations of his own ventilator's capacity. Based on a 12-foot lever, worked by two men, the machine would move a tun of air with each stroke. At 60 strokes per minute, Hales estimated an output of 3,600 tuns per hour and 86,400 tuns per 24 h. These were large numbers. Hales calculated that his air would be supplied through a 1-foot square duct at 25 miles per hour. His ventilators could be sized to deliver air at different velocities and in different amounts, according to the cubic volume of the room (or building) they serviced, to the desired rate of flow (perceptible or imperceptible), to the number of occupants of the room, and to the calculated total volume of air required by those occupants.[49]

The inventors also sought to measure effectiveness experimentally. In demonstrations, rooms were filled with smoke or visible gases and the ventilator then used to clear the space. Hales described an experiment where his ventilator removed the smoke from a hospital ward of 278 tun capacity in 9 min.[50] All promoters of ventilators commented—perhaps somewhat defensively—on the very noticeable change in the sweetness of air after ventilation. These attempts to quantify and to demonstrate effectiveness in terms of volume and rate of flow would be hard to understand without the context provided by the medical debates over bodies and air.

Mechanical ventilation developed outside of architectural practice. In the eighteenth century, it was typically a post-construction alteration, not an element incorporated from the initial design stages. It proved particularly appealing as a remedy for aerial difficulties in existing problem buildings. The Houses of Parliament in London, in particular, were a test case for new ventilation ideas and equipment. The hot, stuffy air in the House of Commons was notorious; the summer sessions were unbearable as the chambers were said to become ovens. Later speakers were known to lose the audience's attention, a phenomenon that Lavoisier attributed to "bad" air.[51] Attempts to render the chambers more habitable occurred regularly between the 1600s and 1800s; Christopher Wren's pyramidal exhaust ducts had not solved the problem. In effect, these truncated pyramids acted as chimneys, and they suffered from the same problem as many other chimneys of the era—they were too large. According to Desaguliers, the updraft of heated air and smoke was often defeated by a downrush of colder (heavier) air.[52]

By 1723, Desaguliers was hired to provide another solution for the House of Commons. He altered Wren's pyramids by installing two small closets, connected to the pyramids by ducts. In the closets were fire-grates. When the fires were lit before the chamber was occupied, they started an updraft that continued to exhaust

Fig. 13 Hales' Windmill on top of Newgate Prison

the air from the room, even after it was fully occupied. The expedient was successful but slow-working. The fires had to be lit well beforehand. This latter problem occurred frequently enough that Desaguliers was asked in 1735–36 to install one of his centrifugal fans into a room above the House of Commons. The fan could be used either to exhaust the air outwards or to force fresh air in. The man who turned the crank was called a ventilator. The *Gentleman's Magazine* suggested that the ventilator could draw out the longwindedness from the orators.[53]

Shortly after the Black Assizes of 1750, Newgate Prison in London installed one of the first whole-building mechanical ventilation systems. This system, designed by Hales himself, used ventilators to push air through an entire network of ducts. In 1752, the authorities installed a windmill to power the system (Fig. 13). The air flow was then regulated by sliding panels in the ducts. Hales announced the effectiveness of the ventilation in familiar terms. He claimed reduced mortality after their installation, although his results were neither substantiated nor lasting (the ventilators were soon another source of foul air, feared by the workmen who had to fix them). Nevertheless, numerous other jails also installed ventilators.[54] Each claimed success in sweeter smelling air and reduced mortality.

These ducts, blowers, and gratings were largely outside of traditional architectural expertise. They were installed, operated, and designed by external specialists, but a few architects did try their hand at incorporating these systems into their designs. John Soane was among the most adventurous, creating new heating and ventilation

Fig. 14 Interior view of John Soane's Bank of England. The blue domed structure is the closed stove which is supplied and exhausted by underfloor flues (From [Pyne and Combe] 1904)

systems for several of his buildings. In the 1790s, he designed a system of closed stoves and underfloor smoke flues in the Bank of England's stock office that was a marvel of its time (Fig. 14). Unfortunately, the system was vulnerable to worries about a new ailment, "iron cough," believed to result from breathing overheated and 'unventilated' air. The closed stoves and ducts were soon replaced with open stoves.[55]

Buildings designed with integral mechanical ventilation systems reached fruition only in the nineteenth century. With most architects still unschooled in ventilation systems and requirements, many of the earliest fully mechanically-ventilated buildings were collaborations. Like the doctor-architect partnerships of the eighteenth century, they refused to settle into peaceful patterns. The most infamous example was the collaboration between the architect Charles Barry and the ventilation engineer David Boswell Reid for the new Houses of Parliament of the 1840s. Barry initially used Reid's extensive demands for exhaust ducts as inspiration for adding a third gothic tower to his design. The collaboration, however, soon soured over conflicting claims of authority on the work site; it all ended in a venomous and career-altering lawsuit.[56] The picture would grow brighter by the late nineteenth century, but even today ventilation design remains a less-considered aspect of architectural design.

Conclusion

In the eighteenth century discoveries about air made ventilation critical to successful building design for certain building types, but left professional "ownership" of the design unclear. While "breathing room" was provided by architectural dimensions, setting the guidelines for what constituted adequate air space was the realm of science and medicine, not architecture. In other words, if architecture itself became the measure of air, it was doctors and researchers who were doing most of the measuring.

Similarly, if the promotion of air flow by the placement of openings provided an architectural means of providing air, it did so only by a transformation of traditional architectural practice. Architects drew and built the solids—columns, entablatures, walls—not the voids. If, as Jean-Baptiste Leroy would state in 1787—"interior form can only be determined by the properties of the air"—architects would have to learn to see, draw, and build the invisible architecture as well as the visible.[57] They largely did not. Nor did they become experts in mechanical ventilation. Even as mechanical systems burgeoned in new building installations, the ducts and blowers typically remained the purview of an engineer or ventilation specialist working in collaboration with the architect.

Before the pneumatic revolution of eighteenth-century science, air was clearly outside of architecture's central concerns. If new research indicated that it should be otherwise, it was an external force. Architects were not driving the research or the change, but they did not reject it. If doctors and inventors were the first to boil down the scientific findings on air into programmatic spatial terms, architects were clearly willing to engage with them, as collaborators or competitors, in the creation of novel designs. This engagement kept architects involved in the early translations of scientific findings into potentially executable (if impracticable) pneumatically-determined designs. The examples of Poyet and Soane reveal something more—a willingness to interpret and apply the measurements of science directly to the design of buildings. Throughout the various levels of architectural involvement and designs, we see the profession asserting its relevance in the face of new scientific discoveries.

Notes

1. The study of air and ventilation provides a difficult test case for architecture; it was (and still largely remains) on the fringes of traditional architectural knowledge and concerns. This fringe location, however, allows an examination of how architectural theorists and practitioners responded to external scientific discoveries that challenged rather than reinforced professional self-definitions. For an analogous examination of how experiments were incorporated into architectural knowledge and practice, see Gargiani (2003).
2. For an orientation to the material, see Bruegmann (1976); Evans (1982); Foucault (1979); Forty (1980); Greenbaum (1974, 1975); Jetter (1986); Markus (1993); Richardson (1998); Riley (1987); Rosenau (1970); Stevenson (2000); and Vidler (1987, 51–82).

3. Boyle (1660, 326–82).
4. On the fears of air, see Etlin (1977).
5. Mortality rates in this time period were calculated less frequently than today. The Hotel-Dieu, for example, did not take cumulative daily statistics. Up until the mid-eighteenth century, the institution tended to calculate their rates of occupancy, deaths, sickness, and births on the first day of every month. See Richmond (1961). Also see Wear (1992).
6. Arbuthnot (1733, 208–209).
7. See, for example, Laugier (1765, 224). The works of Benjamin Franklin, Nicolas Gauger, Count Rumford, and the Marquis de Chabannes head the list of those on improving chimney design and operation in this period.
8. Hales (1727, 134–7, 141).
9. See, for example, Lavoisier (1785, 685); Black (1777); and Priestley (1774–1777).
10. Hales (1727, 146); von Ingenhousz (1779, 35–36); and Priestley (1790, vol. 3: 268–269, 329–330).
11. See, for example, Antoine Lavoisier et al. (1780); Lavoisier (1785, 683–688); and Tredgold (1824, 69–75).
12. 2 October 1798, Latrobe (1977, vol. 2: 437–38).
13. Vitruvius (1791, see esp. 16, 181).
14. "…que les homes préfèrent toujours les choses d'éclat, et même frivoles, à celles qui n'offrent qu'un triste object d'utilité." Leroy (1787, 586). Leroy's work was read to the Academy as early as 1773 but publication was suppressed. See Greenbaum (1974). Also see Laugier (1753, 98); Duhamel du Monceau (1759, 220); Petit (1774, 1).
15. Aikin (1771, 20); Maret (1782, 50); Tenon (1996, 181).
16. Wren (1750, 213, 226). Hales (1727, 148–54); Duhamel du Monceau, (1759, 144) Howard (1777, 5, 14).
17. Huxham (1767, vol. 2: 119–120); Hales (1743, ix); Duhamel du Monceau (1759, 129–144); *Gentleman's Magazine*, 41 (1771, 43); Day (1784, 35–53); Maret (1782, 57–58).
18. Priestley (1774–1777, vol. 2, 161); von Ingenhousz (1779, xiv).
19. Arbuthnot (1733, 103); Lavoisier (1785, 572); Académie des sciences (1787).
20. Laugier (1765, 17–21); Soane (1929, 123).
21. Howard (1791, 34); Aikin (1771, 17).
22. Académie des Sciences (1786a); Tenon (1996, 179–180).
23. Laugier (1765, 16).
24. Lavoisier (1785, 685–86); Aikin (1771, 16); Lind (1777), as quoted in Lloyd and Coulter (1961, vol. 3: 219–20).
25. Priestley (1774–1777, vol. 2, 264–65).
26. Poyet (1785, 6).
27. Soane (1929, 139).
28. Chambers (1759, 69–70, 82); Soane (1929, 141); Lavoisier (1785, 685–86).
29. Lavoisier et al. (1780, 472–473). See also Duveen and Klickstein (1955).
30. Leroy (1780, 599). Also see Tenon (1996, 141, 155–56, 213–14).

31. Based on nitrous oxide's tendency to absorb oxygen, a simple test could reveal the proportion of 'good' air and 'bad' air in a given sample of air. See Priestley (1774–1777, vol. 1, 254–60). See also Schaffer (1990).
32. Soufflot to Maret in Maret (1782, 62–63). Duhamel du Monceau admired the dome for its effect on air flow, but suspected that it had been built only as decoration: Duhamel du Monceau (1759, 219).
33. Maret (1782, 30–31), and Soufflot to Maret in Maret (1782, 31).
34. Lavoisier (1785, 683–86).
35. Tenon (1996, 172–183).
36. Port (1976, 5–6).
37. Duhamel du Monceau (1759, 223).
38. These designs have been the subject of considerable study. Robert Bruegmann, Helen Rosenau, Anthony Vidler, Robin Middleton, Louis Greenbaum, Michel Foucault, and Phyllis Richmond have elucidated the historical context of these projects, although without focusing on how the aerial expectations underpinned the designs. On this aspect, see Cheminade (1993).
39. Maret (1782, 17, 38–47, 53). He described air flow as similar to the 'conical' patterns water took when flowing between the pilings of a bridge. Maret's subsequent correspondence with Soufflot and input from local engineers helped the doctor refine his design. See Lamarre (1986).
40. Leroy (1787); Académie des Sciences (1787, 8). See also Greenbaum (1974) and Vidler (1987, 51–82).
41. Petit (1774, 10–15). Maret worried that Petit's design multiplied windows but not true air exchange, and that the great funnel dome would be useless except as a generator of a large downward air current that would seep into the wards. Maret (1782, 59–60).
42. Poyet (1785, 35); Acadèmie des Sciences (1786b).
43. Poyet (1785, 1–5, 31–35).
44. Anonymous (1785, 4–5, 11); Coquéau (1786, 52).
45. Leroy (1780, 600–602). The history of mechanical ventilation is largely untold. The most comprehensive source is Donaldson and Nagengast (1994). Other works typically focus on the nineteenth century, see for example, Bruegmann (1978); Walbert (1971); Ferguson (1976); and Banham (1969).
46. Desaguliers (1763, 557–58). Desaguliers tested iron, copper, brass, charcoal, and "spirit of wine"; brass, in particular, proved mortal. This result supported the preferred use of iron in hot-air ducts. See Cohen (1900, 203).
47. Hales (1743); Sutton (1844); Triewald (1928). See also Zuckerman (1976).
48. Desaguliers (1735a).
49. See Hales (1743, 12–14, 26–30, 1758, 12–13).
50. Hales (1758, 16).
51. Lavoisier (1785, 686).
52. Desaguliers (1763, 560).
53. Desaguliers (1763, 560–561); Desaguliers (1735b, 47). On the ventilator quip, see *Gentleman's Magazine* 6 (March 1736), 132–133. The author even suggested

applying the pipe directly to the speaker's mouth, to draw the words out directly or to add words and air to the silent, rational men.

54. Hales (1758, 29–40). An initial limited human-powered system was installed in Newgate in 1750 to exhaust the bad air from the women's rooms. To inspire the laborers working the machine, Hales made the effects of their efforts visible by installing a small anemometer, a little windmill or tinkling bell to register air speed, at the duct opening. See Hales (1753) and Griffiths (1884, 1, 441). Ventilators were also installed at the Savoy Prison, Winchester, Durham, Shrewsberry, Northampton, Bristol, and Maidstone: Hales (1754, 1758, 22–26).
55. John Soane's inclusion of mechanical ventilation systems is well documented in an article by Wilmert (1993) and *Gentleman's Magazine* 57 (1787): 209. See also Bernan (1845, vol. 2, 72). This complaint was likely the result of the cast iron cockles of the stove abstracting oxygen from the air as it grew hot. This led to an odor of 'burnt air' and the complaint of iron cough, see Egerton (1968, 80).
56. Port (1976, 102–06). Also see http://www.hevac-heritage.org/victorian_engineers/reid/reid.htm.
57. Leroy (1787, 594).

Photographic Credits

Author: Figs. 1, 3, 13
Division of Rare and Manuscript Collections, Cornell University Library: Figs. 2, 4–6, 11, 12, 14
Courtesy of the National Library of Medicine, Bethesda, Maryland: Figs. 7–10

References

Académie des Sciences. 1786a. Examen d'un project de translation de Hotel-Dieu de Paris... *Histoire et mémoires de l'Académie royale des sciences*: 1–110.

Académie des Sciences. 1786b. *Rapport des commissaires chargés, par l'Académie, de l'examen du projet d'un nouvel Hôtel-Dieu*. Paris: Imprimerie Royale.

Académie des Sciences. 1787. Deuxieme rapport des commissaires chargés des projets relatifs à l'établissement des quatre hôpitaux. *Histoire et mémoires de l'Académie royale des sciences* (20 June): 1–12.

Aikin, John. 1771. *Thoughts on hospitals*. [London].

Anonymous. 1785. *Relevé des principales erreurs contenues dans la Mémoire relatif a la translation de l'Hôtel-Dieu, et examen du project du Sieur Poyet, qui est à la suite*. [Paris].

Arbuthnot, John. 1733. *An essay concerning the effects of air on human bodies*. London: J. Tonson.

Banham, Reyner. 1969. *The architecture of the well-tempered environment*. London: Architectural Press.

Bernan, Walter. 1845. *On the history of warming and ventilating rooms and buildings by open fires, hypocausts, German, Dutch, Russian, and Swedish stoves, steam, hot water, heated air, heat of*

animals, and other methods, with notices of the progress of personal and fireside comfort, and of the management of fuel. London: George Bell.
Black, Joseph. 1777. *Experiments upon magnesia alba, quick-lime, and other alcaline substances*. Edinburgh/London: William Creech/J. Murray, and Wallis, and Stonehouse.
Boyle, Robert. 1660. *New experiments physico-mechanical....* Oxford: H. Hall for Tho. Robinson.
Bruegmann, Robert. 1976. Architecture of the hospital, 1770–1870: Design and technology. Ph.D. dissertation, University of Pennsylvania.
Bruegmann, Robert. 1978. Central heating and forced ventilation: Origins and effects on architectural design. *Journal of the Society of Architectural Historians* 37(3): 143–160.
Chambers, William. 1759. *A treatise on civil architecture: In which the principles of that art are laid down*. London: for the author by J. Haberkorn.
Cheminade, Christian. 1993. Architecture et médecine à la fin du XVIIIe siècle: La ventilation des hôpitaux, de l'*Encyclopédie* au débat sur l'Hôtel-Dieu de Paris. *Recherches sur Diderot et sur l'Encyclopédie* 14: 85–109.
Cohen, I.B. 1900. *Benjamin Franklin's science*. Cambridge: Harvard University Press.
Coquéau, Claude-Philibert. 1786. *Supplément au mémoire sur la nécessité de transférer l'Hôtel-Dieu de Paris, ou Analyse du relevé des principales erreurs contenues dans cet ouvrage*. London/Paris: Desenne, Bailly, Petit.
Day, Thomas. 1784. *Some considerations on the different ways of removing confined and infectious air*. Maidstone: J. Make and sold by G. & T. Wilkie.
Desaguliers, J.T. [John Theophilus]. 1735a. A calculation of the velocity of the air moved by the new-invented centrifugal bellows of 7 feet in diameter, and 1 foot thick within, which a man can keep in motion with very little labour, at the rate of two revolutions in one second. *Philosophical Transactions* 39(437): 44–47.
Desaguliers, J.T. [John Theophilus]. 1735b. The uses of the foregoing machine, communicated in a letter to Cromwell Mortimer, M. D. R. S. Secr. Feb. 23, 1735/6, from the same. *Philosophical Transactions* 39(437): 47–49.
Desaguliers, J.T. [John Theophilus]. 1753. An Account of an Instrument or Machine for changing the Aire of the Room of sick People in a little Time, by either drawing out the foul Air, or forcing in fresh Air; or doing both successively, without opening Doors or Windows. *Philosophical Transactions* 39(437): 41–43.
Desaguliers, J.T. [John Theophilus]. 1763. *A course of experimental philosophy*. London: Printed for A. Millar, J. Rivington, R. Baldwin, L. Hawes, W. Clarke.
Donaldson, Barry, and Bernard Nagengast. 1994. *Heat and cold: Mastering the great indoors. A selective history of heating, ventilation, air-conditioning and refrigeration from the ancients to the 1930s*. Atlanta: ASHRAE.
Duveen, Denis I., and Herbert S. Klickstein. 1955. Antoine Laurent Lavoisier's contributions to medicine and public health. *Bulletin of the History of Medicine* 29: 164–179.
Egerton, M.C. 1968. William Strutt and the application of convection to the heating of buildings. *Annals of Science* 24(1): 73–87.
Etlin, Richard. 1977. L'Air dans l'urbanisme des lumieres. *Dix-Huitieme Siecle* 9: 123–134.
Evans, Robin. 1982. *The fabrication of virtue: English prison architecture, 1750–1840*. Cambridge: Cambridge University Press.
Ferguson, Eugene S. 1976. An historical sketch of central heating: 1800–1860. In *Building early America: Contributions toward the history of a great industry*, ed. Charles E. Peterson, 165–185. Radnor: Chilton Book Company.
Forty, Adrian. 1980. The modern hospital in France and England. In *Buildings and society: Essays on the development of the built environment*, ed. Anthony D. King, 61–93. Boston: Routledge & Kegan Paul.
Foucault, Michel (ed.). 1979. *Les máchines a guèrir: aux origines de l'hôpital moderne*. Brussels: Pierre Mardaga.
Gargiani, Roberto. 2003. Vers une construction parfaite: Machines et calcul de résistance de matériaux. *Matiéres* 6: 99–115.

Greenbaum, Louis S. 1974. Tempest in the academy: Jean-Baptiste Leroy, The Paris Academy of Sciences and the project of a new Hotel-Dieu. *Archives Internationales d'Histoire des Sciences* 24: 122–140.
Greenbaum, Louis S. 1975. 'Measure of Civilization': The hospital thought of Jacques Tenon on the eve of the French Revolution. *Bulletin of the History of Medicine* 49(1): 43–56.
Griffiths, Arthur. 1884. *Chronicles of Newgate*. London: Scribner and Welford.
Grimaux, Edouard. 1888. *Lavoisier: 1743–1794*. Paris: F. Alcan.
Hales, Stephen. 1727. *Vegetable staticks*. London: W. and J. Innys, T. Woodward.
Hales, Stephen. 1743. *A description of ventilators: Whereby great quantities of fresh air may with ease be conveyed into mines, gaols, hospitals, work-houses and ships, in exchange for their noxious air*. London: W. Innys, R. Manby and T. Woodward.
Hales, Stephen. 1753. An account of the good effect of ventilators, in Newgate and the Savoy Prison, by the Rev. Dr. Hales, clerk of the closet to her Royal Highness the Princess of Wales. *Gentleman's Magazine* 23: 70–71.
Hales, Stephen. 1754. A further account of the success of ventilators, &c. *Gentleman's Magazine* 24: 115–116.
Hales, Stephen. 1758. *A treatise on ventilators: Wherein an account is given of the happy effects of the several trials that have been made of them, in different ways and for different purposes*. London: Richard Manby.
Howard, John. 1777. *The state of the prisons in England and Wales, with preliminary observations, and an account of some foreign prisons*. Warrington: William Eyres.
Howard, John. 1791. *An account of the principal Lazarettos in Europe*, 2nd ed. London: J. Johnson, C. Dilly, and T. Cadell.
Huxham, John. 1767, orig. Latin ed. 1752. *Observations on the Air and Epidemic Diseases from the Beginning of the Year 1738, to the End of the Year 1748*. Trans. John Corham Huxham. London: J. Hinton and Henry Whitfield.
Jetter, Dieter. 1986. *Das europäische Hospital: Von der Spätantike bis 1800*. Cologne: DuMont.
Lamarre, Christine. 1986. Le médecin, l'architecte et le politique. *Bulletin du Centre Pierre Lèon* (no. 3–4): 43–59.
Latrobe, Benjamin Henry. 1977. *The Virginia journals of Benjamin Henry Latrobe, 1795–1798*. New Haven/London: Yale University Press for the Maryland Historical Society.
Laugier, Marc-Antoine. 1753, repr. 1977. *An Essay on Architecture*. Trans. Wolfgang and Anni Herrmann. Los Angeles: Hennessey & Ingalls.
Laugier, Marc-Antoine. 1765. *Observations sur l'Architecture*. Paris: Desaint.
Lavoisier, Antoine. 1785. Alterations qu'eprouve l'air respire. In *Oeuvres*. Paris: Ministre de l'Instruction Publique.
Lavoisier, Antoine, et al. 1780. Rapport sur les prisons. *Histoire et Mémoires de l'Académie royale des sciences*: 409–424.
Leroy, Jean-Baptiste. 1780. Mémoire sur quelques moyens simples de renouveler l'air des endroits dans lesquels il ne circule pas, ou dans lesquels il ne circule qu très-difficilement; & sur les applications qu'on peut en faire. *Histoire de l'Académie royale des sciences*: 598–603.
Leroy, Jean-Baptiste. 1787. Précis d'un ouvrage sur les hôpitaux. *Memoires de l'Acadèmie des Sciences*: 585–600.
Lind, James. 1777. *Holograph Pharmacopoeia of Haslar Hospital: Regulations respecting the physical patients*. London: Pharmaceutical Society of Great Britain.
Lloyd, Christopher, and Jack L.S. Coulter. 1961. *Medicine and the Navy: 1200–1900*. Edinburgh/London: E. & S. Livingstone.
Maret, M. 1782. Mémoire sur la construction d'un hôpital, dans lequel on détermine quel est le meilleur moyen à employer pour entretenir dans les infirmeries un air pur & salubre. *Nouveaux mémoires de l'Académie de Dijon pour la partie des sciences et arts*: 25–67.
Markus, Thomas A. 1993. *Buildings & power: Freedom and control in the origin of modern building types*. London/New York: Routledge.
Monceau, [Henri] Duhamel du. 1759. *Moyens de conserver la santé aux équipages des vaisseaux: avec la manière de purifier l'air des salles des hôpitaux; et une courte description de l'Hôpital Saint Louis, à Paris*. Paris: H. L. Guerin & L. F. Delatour.

Petit, Antoine. 1774. *Mémoire sur la meilleure manière de construire un hôpital de malades*. Paris: Louis Cellot.

Port, M.H. (ed.). 1976. *The Houses of Parliament*. New Haven/London: Yale University Press.

Poyet, Bernard. 1785. *Mémoire sur la nécessité de transférer et reconstruire l'Hôtel-Dieu de Paris suivi d'un projet de translation de cet hôpital.*

Priestley, Joseph. 1774–1777. *Experiments and observations on different kinds of air*. London: Printed for J. Johnson.

Priestley, Joseph. 1790. *Experiments and observations on different kinds of air, and other branches of natural philosophy, connected with the subject*, 2nd ed. Birmingham: Thomas Pearson.

[Pyne, W.H.], and [Combe W.]. 1904. *The microcosm of London: Or London in miniature*. London: Methuen.

Richardson, Harriet (ed.). 1998. *English hospitals, 1660–1948*. Swindon: Royal Commission on the Historical Monuments of England.

Richmond, Phyllis Allen. 1961. The Hotel-Dieu of Paris on the eve of the revolution. *Journal of the History of Medicine and Allied Sciences* 16: 335–353.

Riley, James C. 1987. *The eighteenth-century campaign to avoid disease*. New York: St. Martin's Press.

Rosenau, Helen. 1970. *Social purpose in architecture: Paris and London compared, 1760–1800*. London: Studio Vista.

Schaffer, Simon. 1990. Measuring virtue: Eudiometry, enlightenment, and pneumatic medicine. In *The medical enlightenment of the eighteenth century*, ed. Andrew Cunningham and Roger French, 280–318. Cambridge/New York: Cambridge University Press.

Soane, John. 1929. *Lectures on architecture: As delivered to the students of the Royal Academy from 1809 to 1836 in two courses of six lectures each*. London: Sir John Soane's Museum.

Stevenson, Christine. 2000. *Medicine and magnificence: British hospital and asylum architecture, 1660–1815*. New Haven/London: Yale University Press.

Sutton, Samuel. 1844. *An historical account of a new method for extracting the foul air out of ships*. London: J. Brindley.

Tenon, Jacques. 1996. *Memoirs on Paris Hospitals*. Trans. of 1788 ed. Canton: Science History Publications.

Tredgold, Thomas. 1824. *Principles of warming and ventilating public buildings, dwelling-houses, manufactories, hospitals, hot houses, conservatories, etc*. London: J. Taylor.

Triewald, Marten. 1928, 1st ed. 1734. *Short Description of the Atmospheric Engine Published at Stockholm*. Trans. Are Waetland. London: Newcomen Society.

Vidler, Anthony. 1987. *The writing of the walls*. Princeton: Princeton Architectural Press.

Vitruvius. 1791. *The architecture of M. Vitruvius Pollio: Translated from the original Latin, by W. Newton, architect*. London: James Newton.

von Ingenhousz, John. 1779. *Experiments upon vegetables: Discovering their great power of purifying the common air in the sun-shine, and of injuring it in the shade and at night*. London: Elmsly and H. Payne.

Walbert, Benjamin L. 1971. The infancy of central heating in the United States: 1803 to 1845. *Association for Preservation Technology* 3(4): 76–87.

Wear, Andrew. 1992. Making sense of health and the environment in early modern England. In *Medicine in society: Historical essays*, ed. Andrew Wear, 119–148. Cambridge/New York: Cambridge University Press.

Wilmert, Todd. 1993. Heating methods and their impact on Soane's work: Lincoln's Inn Fields and Dulwich Picture Gallery. *Journal of the Society of Architectural Historians* 52(1): 26–58.

Wren Jr., Christopher. 1750. *Parentalia*. London: T. Osborn and R. Dodsley.

Zuckerman, Arnold. 1976. Scurvy and the ventilation of ships in the Royal Navy: Samuel Sutton's contribution. *Eighteenth-Century Studies* 10(2): 222–234.

James "Athenian" Stuart and the Geometry of Setting Out

David Yeomans, Jason M. Kelly, and Frank Salmon

A characteristic feature of the neoclassical attitude to Greco-Roman architecture that ran from the middle of the eighteenth century to the middle of the nineteenth has long been held to be the minute surveys of ancient buildings that were undertaken and published during that period. Ultimately inspired by Antoine Desgodetz's *Les édifices antiques de Rome* (1682), measured surveys of antique buildings across the Mediterranean world became a staple part of architectural and antiquarian study from the 1750s, especially in relation to the growing interest in Greek architecture. The British were especially assiduous in framing these surveying activities as part of a discourse about "truth" (as Robert Wood put it in 1753) and "accuracy", a term used by James Stuart in the preface to the first volume of *The Antiquities of Athens* in 1762.[1] Later eighteenth- and early nineteenth-century architects appear to have accepted that these surveys did indeed represent ever-more refined attempts to establish absolute sets of dimensions for the great monuments of Athens and Greece. In his eleventh lecture to Royal Academy students, for example, first delivered in 1815, John Soane spoke unquestioningly of "the accurate and laborious

D. Yeomans (✉)
Independent Scholar, 24, Winchester Close, Banbury 0X16 4FP, UK
e-mail: mail@davidyeomans.co.uk

J.M. Kelly
Department of History, Indiana University, 15202 Dan Patch Ct., Carmel, IN 46032, USA
e-mail: jaskelly@iupui.edu

F. Salmon
Department of History of Art, University of Cambridge, Cambridge, UK
e-mail: fes11@cam.ac.uk

A. Gerbino (ed.), *Geometrical Objects: Architecture and the Mathematical Sciences 1400-1800*, Archimedes 38, DOI 10.1007/978-3-319-05998-3_12,

representations of Stuart and [Nicholas] Revett, who measured those proud remains of ancient glory to the thousandth part of an inch."[2] This tradition reached its culmination in the extremely fine measurements made in Athens in the 1840s by Francis Cranmer Penrose, who described the purpose of his 1851 *Investigation of the Principles of Athenian Architecture* as having been "to fill up what had been left imperfect by Stuart and Revett."[3]

Since that time, architectural historians have tended to accept at face value the claims of protagonists from Stuart to Penrose that their surveys tended towards an ultimate goal of absolute accuracy of measurement, locating them within a framework of Enlightenment scientific research that contrasts with broader and perhaps more Romantic contexts of coeval history and anthropology. However, in 1956, Jacob Landy, writing in the journal *Archaeology,* pointed out that while the measurements in *The Antiquities of Athens* were indeed "more accurate than those of any previous publication," Stuart and Revett were also "not concerned with as complete a presentation of the facts as they pretended," being "burdened with literary, mythological and historical allusions."[4] While we should indeed now see *The Antiquities of Athens* as a work conditioned by the cultural circumstances of the later eighteenth century, there is another dimension to the question of accuracy that has not yet been studied and that this chapter explores. Recent research on the case of Stuart's survey of the Tower of the Winds in Athens has shown, thanks to the fortuitous survival of a proof plate showing the plan of the roof, that at least one of the dimensions appearing on the version finally published was not measured at all but calculated trigonometrically.[5] It is possible that such an intervention amounts to no more than a pragmatic way of providing data that Stuart realized he had not collected in Athens when he was working on the surveys in London many years later, but the same plate also suggests that Stuart was systematically double checking all measurements in order to ensure geometrical coherence.

We cannot be certain, then, that the figures offered in *The Antiquities of Athens* by Stuart—and perhaps by others involved in the same type of pursuit—represent measurements physically made. The process Stuart followed in preparing this plate of the Tower of the Winds was not a simple one of transcribing dimensions taken in the field onto the image but one that involved some computation, and this realization now presents the historian of neoclassical architecture with significant problems. We know that architects in ancient Greece and Rome would have used geometrical rules to design their buildings, and we are now learning that men like Stuart, by measuring those early buildings that survived, sought to understand those rules, both to ensure the consistency of their reconstructions and to use them in their own designs.[6] However, the process of surveying an existing structure is by no means commensurate with that of setting it out in the first place, since some dimensions are effectively concealed by the fabric of the building itself. Further still, methods appropriate for drawing on the smooth surface of a drawing board may be quite different from those appropriate for the staking out of the plan in the field or the marking of stone by the mason. This situation raises a number of related conundra: How did Stuart take measurements in the field? How did they get translated to published form? What assumptions did he make about Greek setting out, and how did these assumptions color his measurements and his reconstructions?

This paper probes these questions by close consideration of cases taken from Stuart's surviving papers. Of the large number of his field notebooks in circulation in the early nineteenth century, only one integral version is now known: the "Edinburgh Notebook" in the Laing Collection of the University of Edinburgh.[7] This source provides useful data in the form of Stuart's field notes and musings on two large classical buildings: the Temple of Rome and Augustus at Pola in Istria (early first century) and the Theater "of Bacchus" (actually of Herodes Atticus, c.162) in Athens. We have also been able to examine problems of geometry at the smaller scale of a building component, the Ionic capital, thanks to the survival of a number of loose sheets on which Stuart analyzed the problem of the setting out of the volutes.

Whereas the surviving sheets on which Stuart studied the Ionic capital are loose and difficult to date, it seems reasonable to assume that the Edinburgh Notebook is typical of the way he kept his records whilst in the field and resident in Athens. It was essentially a commonplace book, used to record a variety of notes on different subjects. There are descriptions of different parts of Athens and other places that he visited, a few drawings, and many calculations (Fig. 1). The notes usually begin on the right-hand page and only sometimes continue on the left. This doubtless reflects Stuart's method of working in the field, holding the book in the left hand or resting it on the right knee, and using the right hand side of an opening.[8] It is not always clear what dimensions were measured nor how accurately. In his description of some buildings, he left gaps for the dimensions that he presumably intended to measure at a later date, or which he was content to delegate to Revett. We know something of their working practice, as Stuart mentions measurements with a chain (presumably with a trained assistant) that Revett would later check with a rod, which he specified, "will be more accurate."[9] The measurements Stuart recorded in his notebooks are by no means straightforward, as Joseph Woods found when editing the fourth volume of the *Antiquities* in 1816: "The following list of heights of these buildings are given by Stuart; they are not always consistent with the figures on the sketches, nor do I always understand the exact application."[10] Whereas Woods thought "it would be best to give them just as I found them," Stuart, as we have seen at least in the case of the Tower of the Winds, went to considerable effort to make them consistent when preparing his work for publication.

The Temple of Rome and Augustus at Pola

The Edinburgh Notebook contains memoranda on the temple at Pola, the Roman city in Istria to which Stuart and Revett made an excursion from Venice in 1750 while waiting to travel to Athens (which they did in 1751). Their work at Pola thus stands as something of a rehearsal for the methods they would deploy when in Greece. There is a small plan showing the general scheme of the temple and then two larger sketches (fol. 14), all of which can be combined for convenience into Fig. 2. The arithmetic accompanying Stuart's sketches is simply the addition of the various measurements that were made to obtain overall lengths and widths.

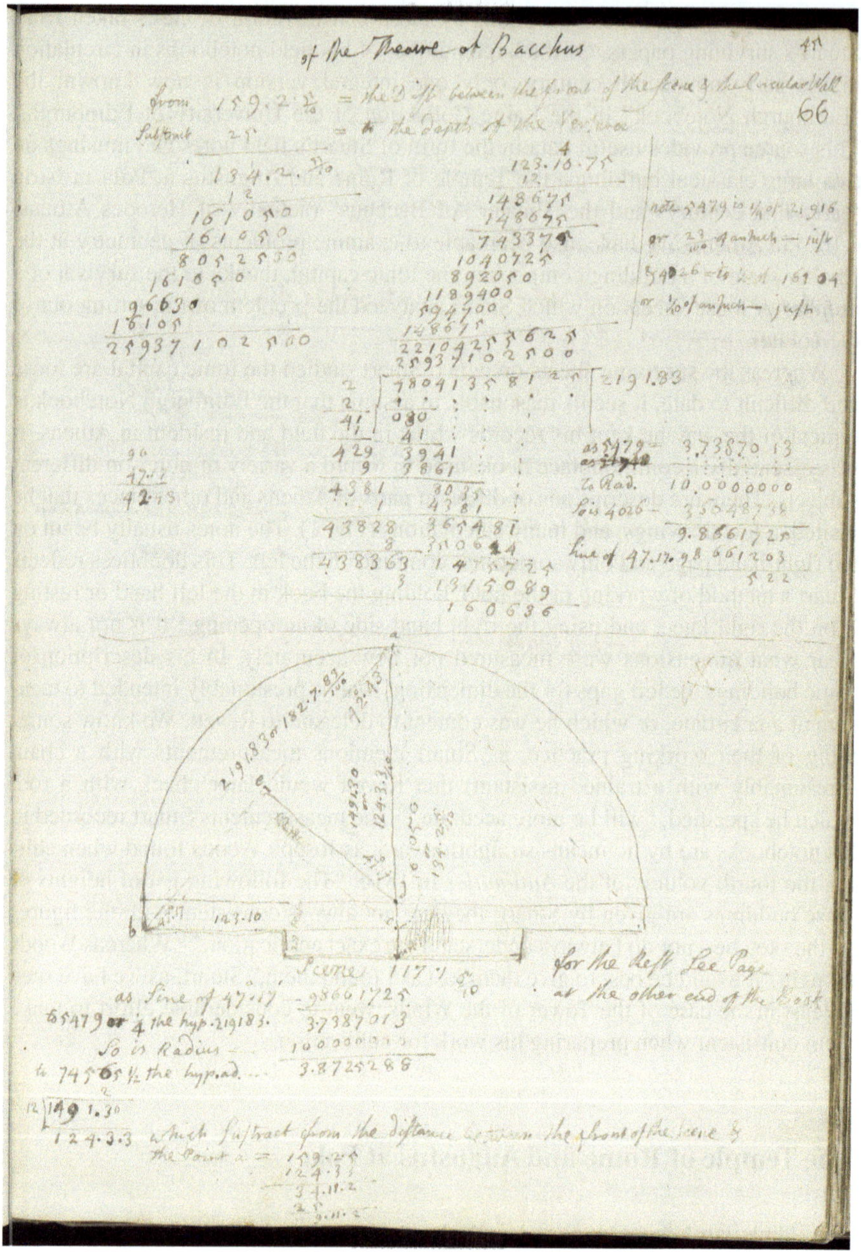

Fig. 1 James Stuart's "Edinburgh Notebook," fol. 66v, Laing Collection, University of Edinburgh

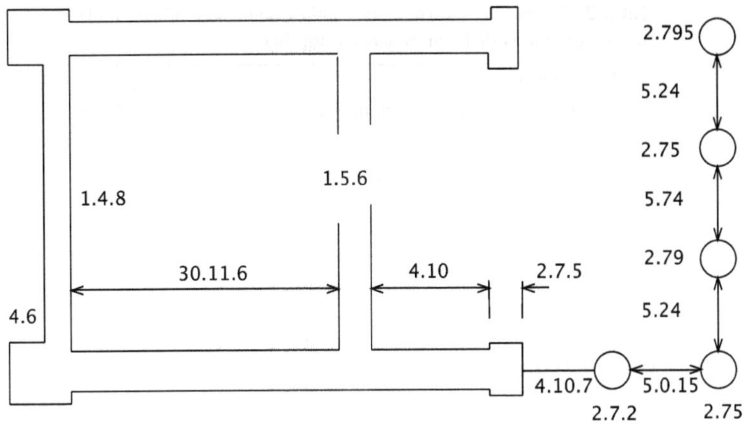

Fig. 2 Plan of the Temple of Rome and Augustus at Pola with inscribed measurements, based on sketches and field notes in the Edinburgh notebook

Table 1 James Stuart's arithmetic for the dimensions of the temple at Pola, from the Edinburgh Notebook (fol. 14)

		4.	6	
	1.	4.	8	
	30.	11.	6	
	1.	5.	6	
	4.	10.		
	2.	7.	50	
	4.	10.	70	++22 7. 05
	2.	7.	20	
	5.	0.	15	
	2.	7.	50	/100
Entire length of the temple	56.	9.	65	

Reproduced here are the measurements for the length of the temple, in which Stuart has made a division between the cella and the pronaos (Table 1).

The figures are in English feet, inches, and decimals of an inch up to the hundredth, the standard Stuart had adopted when surveying the Obelisk of Psammetichus II in Rome, probably in 1748, for his first major archaeological publication *De Obelisco Caesaris Augusti* of 1750.[11] For the Athens expedition, the authors were able to achieve this level of accuracy with the aid of a brass yard rule engraved by the foremost mathematical instrument maker of mid-eighteenth century London, John Bird. Other than a chain and a compass, this was the only piece of surveying equipment we know them to have had.

The arithmetic for the parts of the Pola temple is correct, but changes were introduced between the Notebook and the published version that appears in Chapter II, plate III of the fourth volume of the *Antiquities*. This was published after Stuart's

Table 2 James Stuart's arithmetic for the dimen-sions of the temple at Pola, from the Edinburgh Notebook (fol. 14)

To the length of the cell	30	11.	6
Add the thickness of the wall at the front	1.	5.	6
	32.	4.	2
Subtract the width of the cell including the walls	26.	1.	85
	06.	2.	35
			4
	24	9.	4

death, and Woods's introduction to the volume indicates that he was working from Stuart's drawings. The value 1.4.8 for the thickness of the end wall of the cella in the Notebook was reduced by just over an inch to 1.3.6, while the overall length of the building was increased by just under an inch to 56 ft, 10.6 in. There is nothing to indicate why these changes were made. The most likely explanation is that Woods was working from more than one version of the plan and needed to make a choice between them. Whatever the reason, this maneuver certainly made the correct arithmetic of the Notebook record incorrect on the published plate. Woods almost certainly noticed the discrepancy, but he evidently considered it more important—as he stated—to maintain editorial neutrality than to try to reconcile the numbers.

A second area of Stuart's concerns raised by his Notebook observations involves the issue of proportional relationships. Although Stuart was to make a clear statement in the introduction to the first volume of the *Antiquities* that he had avoided any system of design based on modules (and to imply that previous surveys had made errors because of the preconceptions that modules brought with them), that did not mean he never considered the possibility of proportional relationships. On the same page as the calculations above appears a set of other figures, where Stuart seems to be checking possible relationships that might have determined the actual dimensions. The following example, taken from folio 14, shows Stuart engaging with Vitruvius on this question. Disregarding the Roman author's modular starting point (using the diameter of the column as the generator of the design), he concentrated instead on the relationship of length to width of the cella (Table 2).

Stuart was checking that the Pola temple was in line with Vitruvius's prescription (Book 4, Chapter 4) that "the cella itself will be longer by one fourth than its width, including the wall in which the doors are to be located."[12] The dimensions evidently did not match Vitruvius's recommendations, at least in terms of a common module. A presumed module of 6 ft 2.35 in. would produce wall dimensions about 5 % shorter than those built. A second set of calculations shows Stuart trying to relate the length of the cella to that of the whole, also presumably to compare the result with Vitruvius's recommendation. Here, too, Stuart would have noticed a significant deviation from the text. As built, the cella is closer to 4 parts of 9 than the 5 parts in 8 that Vitruvius allows.[13] The notes and calculations made at Pola thus stand as evidence of the approach Stuart intended to adopt at Athens, which was, as he put it in the 1751 "Proposals" for the *Antiquities,* to analyze buildings "by pointing out the relation they may have to the Doctrine of Vitruvius."[14]

The Theater of Bacchus

The notes for the Theater of Bacchus are divided between the front and the back of the Edinburgh Notebook, probably reflecting work done at different dates during Stuart and Revett's stay in Athens, which lasted from 1751 to 1754. The plate with the plan of the theater was eventually published in volume two of the *Antiquities,* which appeared in 1789–90 under the editorship of William Newton (Stuart having died in 1788).

In Athens, Stuart and Revett were confronted with a ruin. The remains of the *frons scaenae* and the flanking *versurae*—the tower-like buildings that provided access to the scene from the corners of the hemicycle—were all that remained of the structure. A grass-covered hollow lay where the seating had been cut into the hillside, but none of the tiers remained intact, almost certainly because the site had been quarried for building stone in the centuries since the classical era. The ensemble can be seen in Stuart's perspective view, which also features Revett in the right foreground at work drawing the masonry of the *frons scaenae* (Fig. 3). As an added difficulty, the surveyors were unable to excavate inside the ruin because of its proximity to an Ottoman garrison, though some digging behind the *versura* wall was permitted.[15] The Notebook records Stuart's aim in surveying the site. He hoped to elucidate Vitruvius's description of the setting out of Greek and Roman theaters, improving on the accounts of earlier editors, in particular Claude Perrault and Daniele Barbaro:

> The Theater of Bacchus is so ruined that only the front of the scene, the versurae & the exterior circuit appear above ground naked of ornament and the upper parts entirely ruined. The pulpitura above [lies more than] 16 feet below the present surface of the ground. Yet what remains may serve to explain Vitruvius better than all his commentators[.] Lett [sic] us see his words & comparing the designs of Barbaro[,] Perault & theatre of Bacchus observe which agrees best with his description and documents[.][16]

Of the Vitruvius commentators mentioned here, Stuart does not seem to have made further use of Barbaro. The published text makes reference to the Italian translation of Marchese Berardo Galiani, which superseded Barbaro when it appeared in 1758. The Perrault translation, however, was much more important. It was on this French source that Stuart depended for his understanding of Vitruvius. The Notebook contains a transcription and translation of this section of the text (Book 5, Chapter 8) for quick reference. As Stuart reports, Vitruvius differentiates between the layout of Greek and Roman theaters. In his published comments, Stuart appeared to recognize that the date (actually second-century AD) and identity of the builders was open to question. Nevertheless, for the purposes of the reconstruction, he proceeded on the assumption that what Vitruvius "had said concerning the Greek Theaters [was] applicable to this building."[17]

Stuart's first task in describing the theater was to obtain true dimensions from the surfaces that he could directly measure. A schematic cross-section in the form of a sketched triangle shows two of these: the descending slope of the seating and its height at the back (Fig. 4). Neither of these measurements was easily obtained.

Fig. 3 Nicholas Revett drawing the *frons scaenae* of the Theater "of Bacchus" (Herodes Atticus) (From Stuart and Revett 1762–1830)

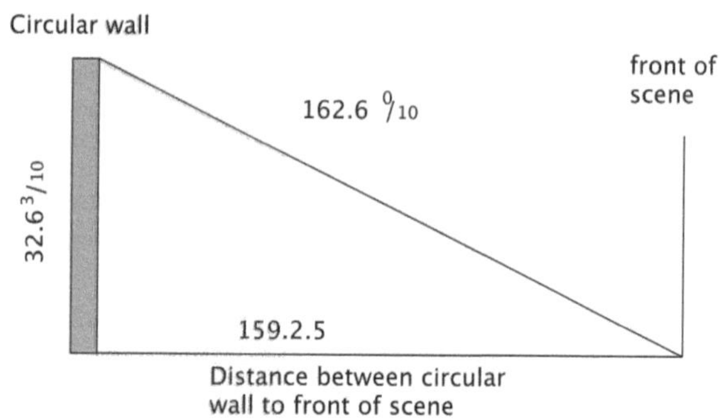

Fig. 4 Schematic cross-section of the Theater of Bacchus, based on Stuart's fieldnotes in the Edinburgh notebook

The slope was hardly constant, as is evident from the published view. Likewise, with the theater built into the hillside, the height of the seating could only be measured from the side. This was the area they had to excavate, which Stuart referred to in his notes as the "corridor going from the wings of the Theater to the Orchestra or gate of the Versura."[18] With these two dimensions, Stuart was able to find the horizontal depth of the theater in plan from the *frons scaenae* to the back wall. For this, he simply used Pythagoras. The Notebook contains a clear calculation for that, where the squares of the two distances 162 ft 6 in. (the measurement along the

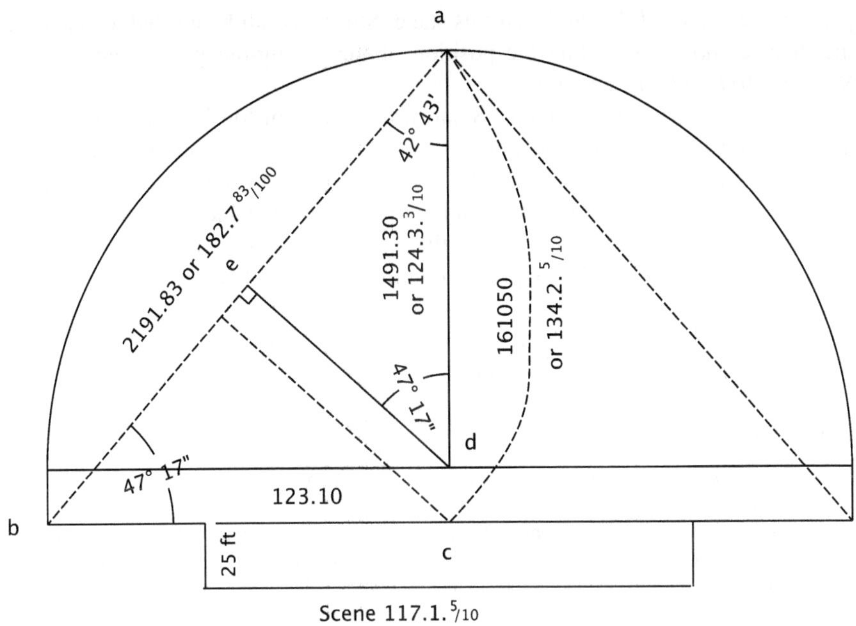

Fig. 5 Schematic plan of the Theater of Bacchus, based on fol. 66v of the Edinburgh notebook (Fig. 1)

slope) and 32 ft 6 in. (the height of the seating at the back wall) are reckoned and the second subtracted from the first. The square root of the result is the horizontal distance to the wall. Stuart extracted this root by hand, a calculation that fills a considerable portion of the page. This is the first of several instances where it is evident that Stuart used derived dimensions rather than those directly measured.

According to Vitruvius, the design of theaters was based on a semi-circle touching the far edges of the *versurae*, but with a center some distance before the *frons scaenae*. If Stuart was to demonstrate the use of this principle in the design of the Theater of Bacchus, it would be necessary to find the centre of that circle. Figure 5 simplifies the plan that Stuart sketched in his notebook (Fig. 1, fol. 66v) to find the point *d* and the accompanying dimensions. His first step was to measure the width of the theater from the far edges of the *versurae* and halve it to give *cb* as 123 ft 10 in. He also measured the depth of the *versurae* from the *frons scaenae*: 25 ft. These were the only directly measured values that Stuart used for this reconstruction. The rest were derived from trigonometric operations on the plan. The length of the perpendicular center line *ac* was established by subtracting the depth of the *versurae* from the horizontal value determined in the previous calculation (Fig. 4: 159 ft 2.5 in.). Applying Pythagoras, Stuart could then calculate the length of the hypotenuse of the triangle *abc* and from this, the sine and the opening of the angle *bac*. Because the hypotenuse was a chord of the large semi-circle formed by the back wall of the seating, a line bisecting that chord (*ed*) would pass through the required point at the center of the circle. At this stage, Stuart had all he needed to calculate the distance *ad*, thus finding the position of the setting-out point suggested by Vitruvius ($ad = bd = ae/\sin ade$).

In carrying out his calculations, Stuart used a table of logsines, which allowed him to avoid tedious multiplication and to perform the calculations with simple addition. As it would have been simpler to do the calculations involved with a table of tangents, we must assume that he did not have one available. The other point to note here is that the table of logarithms that he was using provided seven figures in the mantissa, which would have given him more accuracy than he needed.

Armed with this information, we can check Stuart's measurements against those generated by Vitruvius's method. The most readily verifiable dimensions were the size of the orchestra circle, the width of the scene, and the depth of the *versurae*. Stuart explained Vitruvius's construction in the text accompanying the published plan of the theater (Fig. 6):

> On this we must observe, that the exterior wall is the portion of a circle, the centre of which being found, it will follow from the precepts of Vitruvius, if we suppose what he has said concerning the Greek Theater applicable to this building, that the extent of the Procenium, with the situation and dimension of the Orchestra, may be determined. For the distance *a.b.* from the centre *a.* of the exterior circle, to the front of the Scene *C.B.D.* will be the radius of a lesser concentric circle, in which three squares being inscribed, after the manner he has directed, the side of the square *e.f.* [sic *g.f.*] nearest to the Scene and parallel to it, will then mark the limit of the Procenium, and the remaining part of the circle, if we do not mistake Vitruvius, will form the space assigned by him to the orchestra; within which space, I am persuaded, the Pulpitum or Logeum projected at least as far as to the centre *a.* for I cannot imagine, that the actors were confined to the narrow space assigned by this scheme to the Procenium.[19]

Vitruvius's instructions, interpreted here largely via Perrault, call for the radius of the orchestra circle to determine the other elements. The first step, as shown in simplified form on the right side of Fig. 7, is to inscribe a square into the orchestra circle. As Stuart relates, extending the side parallel and closest to the *frons scaenae* provides the forward edges of the proscenium, the raised platform for performers. The width of the scene is obtained by setting the radius of the orchestra circle at the outer edge a' and rotating it until it meets the line of the proscenium at h. The depth, finally, of the *versurae* is determined by rotating the same radius around the upper corner of the inscribed square e' until it meets the orchestra circle at c'. It should be clear that the side of the square $a'h = ab/\sqrt{2}$ and the total width of the *frons sceanae* $= ab(2+\sqrt{2})$. Likewise, the depth of the *versurae* (H in Fig. 7) will be $ab - ab.\sin 15°$ or $0.74ab$.

Recognizing these relationships, we can check to see whether Stuart would have been able to confirm Vitruvius's method by a simple calculation. Using the published figures, which differ slightly from those in the Notebook, Stuart and Revett record the distance *ab* as 35 ft 0.3 in. On that basis, the width of the scene should be about 119 ft 6 in. In fact, Stuart and Revett have it as only 117 ft 1.5. The depth of the *versurae* is marked as 25 ft 4.5 in. Calculated from *ab*, however, it would be 7 in. longer. In both cases, the discrepancy is about 2 %. Although the published plan suggests a close correlation between the built remains and Vitruvius's setting-out method, Stuart's measurements do not bear it out.

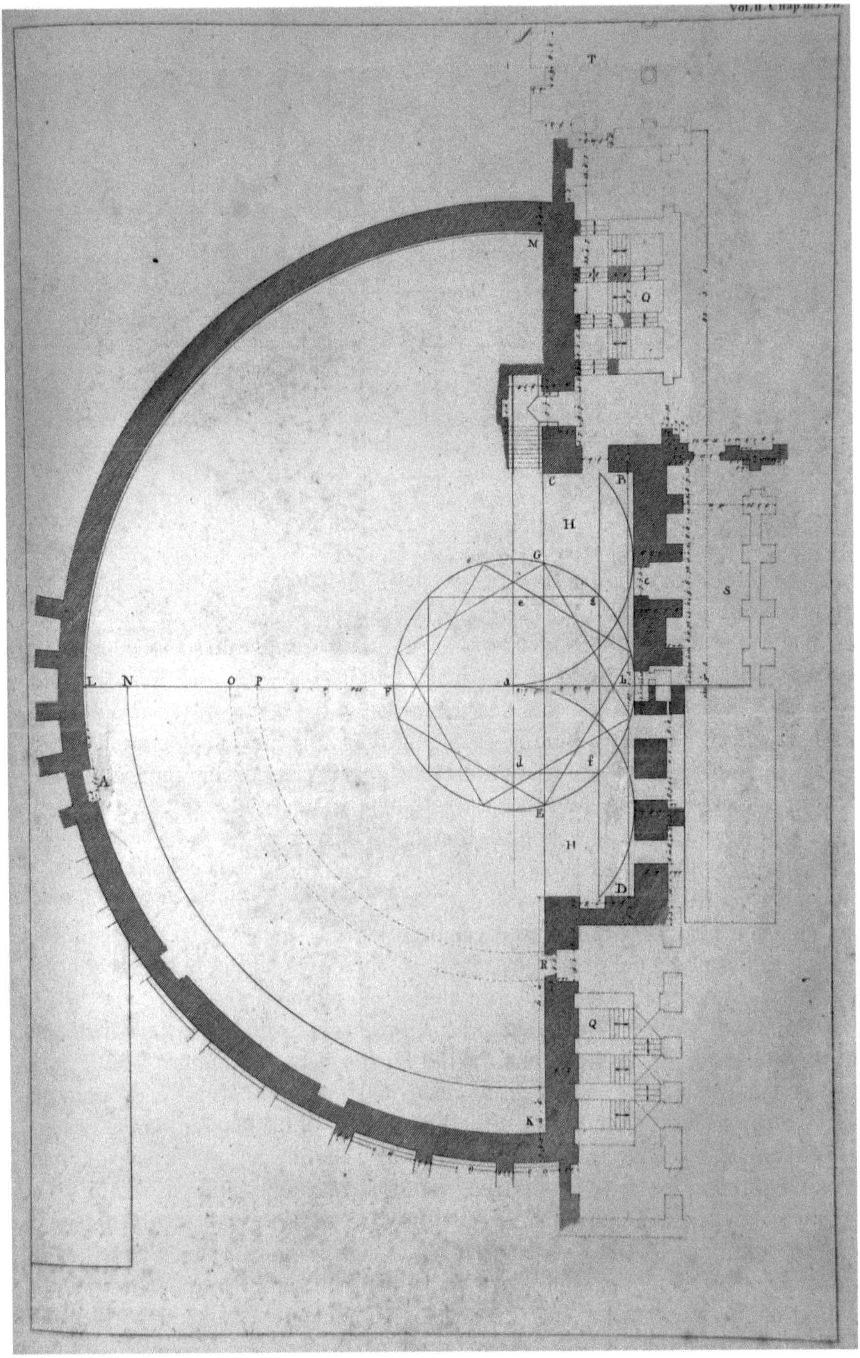

Fig. 6 Plan of the Theater of Bacchus, with reconstruction of Vitruvius's setting out method (From Stuart and Revett 1762–1830)

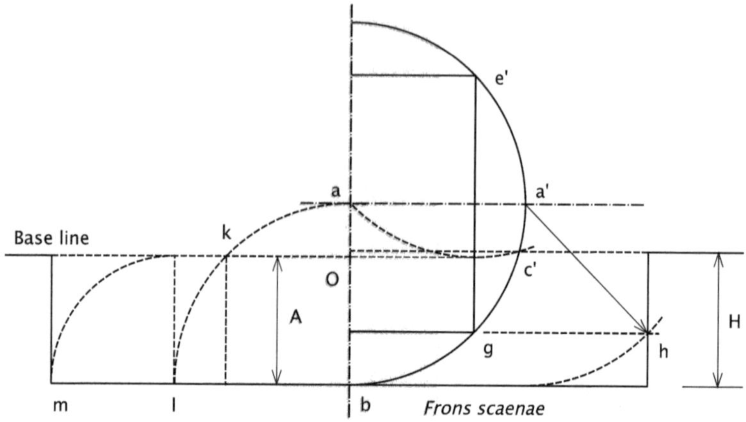

Fig. 7 Setting out the Greek theater. The right side shows a simplified version of Vitruvius's setting out method, which involves inscribing a square in the orchestra circle, and the left a simpler and more feasible method of achieving the same results

Had he carried them out, these calculations might have given Stuart some reason to doubt Vitruvius's authority as a source, at least for the Theater of Bacchus. In fact, there are other reasons to discount Vitruvius's suggested procedure. His approach, with whirling squares and many extraneous lines and found points, appears little more than a drawing board exercise. What is missing from Vitruvius (and hence from his commentators, including Stuart) is any consideration of the practical operations necessary to transform the geometry of design into setting out marks on site using stakes and string lines. In the first place, it would be extremely difficult to set out the theater using the center of the circle as a starting point, and it is difficult to imagine that builders would have begun there. Not only is *a* itself a found point, it also requires several intermediate steps to inscribe the square before determining the location of the proscenium, the *frons scaenae*, and the *versurae*. Rather than unfolding the plan from a point, a builder would surely begin with the longest lines in order to ensure that the building fit on the site.

Could Stuart have recognized the practical problems with Vitruvius's method? Perhaps not: aside from the Roman author's still-powerful authority both as a designer and a source for ancient design methods, Stuart was at this early point in his career still an inexperienced architect. The left side of the diagram in Fig. 7 suggests a simpler alternative, with the more believable starting point *O*. It assumes that the depth of the *versurae* (*A*) is derived, not from a complicated construction, but from the radius of the orchestra circle. *Ok* equals *Ob* and *ab* equals *kb*, the diagonal of the square drawn on *Ob*. The length *bl* is the same diagonal rotated towards the *frons scaenae*. The point *m* is obtained by adding the length *Ob* to *l*. This method produces almost the same layout that Vitruvius recommends. The width of the *frons scaenae*, $ab(2+\sqrt{2})$, is the same and the depth of the *versurae* very nearly so, but this operation is much simpler and also more workable at full scale.

The Ionic Volute

The volute of the Ionic capital presented Stuart and Revett with a series of problems that, for modern scholars, illuminate the complexities of Enlightenment measurement. The problems stemmed from the fact that their aims as both architects and archaeologists were often at cross-purposes. The conflicting epistemologies that spurred their research provoked questions that, while distinct to the modern researcher, were complementary or even integrated in the minds of Enlightenment researchers. The first question was methodological: how should one measure the Ionic volute? If the purpose of their research was simply the empirical description of classical architecture, then their approach would be quite different from that of Desgodetz, who sought to uncover the hidden language of classical architectural proportion. The second question was practical, with application both to archaeology and architecture. How could Stuart and Revett work back from their on-site measurements and observations to deduce the working methods of ancient architects and masons? Their interest here was not purely academic. The reintroduction of the Greek Ionic order formed part of the "improvement" Stuart hoped to see in later eighteenth-century architecture and to which he himself was to make great contributions in his own designs of the 1770s and 1780s. In this sense, his studies of Greek Ionic capitals had greater relevance in the context of neoclassical design than did the public building types we have been able to study here.

On the subject of the Ionic order, Stuart and Revett found their predecessors' works incomplete. While Vitruvius discussed the Doric and Ionic styles at some length, the moderns had virtually ignored them in favour of the Corinthian. The reason for this was the centrality of Rome to the mental landscape and travel itineraries of early modern scholars, artists, and antiquarians. Palladio's *I Quattro Libri*, for example, the central text of British architectural classicism, was replete with examples of the Corinthian, but included only one description of an Ionic temple: the Temple of Manly Fortune, now known as the Temple of Portunus (Fig. 8). Even Desgodetz, to whom Stuart and Revett looked with nearly uncritical admiration, failed them here. Stuart dismissed all three of the examples that Desgodetz examined: the Temple of Manly Fortune, the Theater of Marcellus, and the Colosseum (Amphitheater of Vespasian). To Stuart, following Fréart, the Temple of Manly Fortune was "ill wrought" and "covered with Stucco." Although it was the best surviving example of the Ionic order in Rome, its features were "not only incorrect, but they are likewise so decayed, that the original form and projection of these Mouldings cannot now be ascertained."[20] In contrast, Greek examples of the Ionic, like the temple on the Illissos, were "simple", "elegant", "well executed", and "among those Works of Antiquity which best deserve our Attention."[21]

Such statements provide important insight into the influence of Vitruvian rationalism that dominated many early modern architectural debates. Stuart and Revett worked from the assumption, then prevalent, that ancient Greek architects abided by the strict laws later transcribed by Vitruvius in *De Architectura*. The Greco-Roman architects who followed the Greeks borrowed their proportional principles, corrupting them into the variations that could be seen in Rome, such as the "incorrect" Ionic

Fig. 8 The Temple of "Fortuna Virilis" (Portunus), Rome (From Ware 1738)

capitals of the Temple of Manly Fortune. That said, Stuart and Revett were not entirely satisfied with rationalist dogma. True to the empiricism that then dominated British intellectual circles, both men recognized the need to survey the remaining examples of Ionic architecture in Athens. It is here that their assumptions and desires worked at cross-purposes. On the one hand, the architects-in-training sought an abstract system of beauty, preferably one based on Vitruvius's system of modular proportion. Furthermore, their role as architects encouraged them to inquire into the practicalities of ancient architectural practice. There was, on the other hand, the physical evidence of the buildings themselves. These measurements, as it turned out, did not allow easy rationalization, much less conversion into convenient rules-of-thumb.

Measuring the Volute

The setting out of the volute was a problem that had engaged architects long before Stuart and Revett. A close approximation to an Archimedean volute—which uncoils at a constant width—can be obtained by unwinding a cord from a cylinder, but what

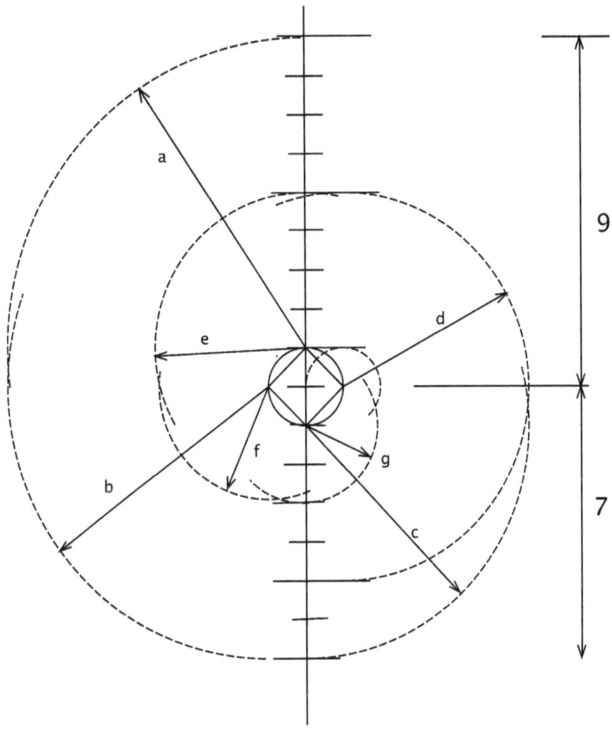

Fig. 9 Vitruvius's method for setting out the spiral of the Ionic Volute (From *De Architectura*, Book 3, Chapter 5.5–6)

was more usually wanted was a logarithmic volute, which widens as it uncoils. Moreover, the volute had to be drawn inwards within a block of stone of known size. Once the centre of the volute and the size of the oculus have been found, the method usually involves drawing a series of arcs of diminishing radius, each subtending 90°. Vitruvius describes a method of this sort (Book 3, Chapter 5), with the point of the compasses stepped round a square inscribed inside the oculus. However, the figure that originally accompanied his account was lost, so subsequent architects had to imagine several unstated elements of his method. For example, if one were to draw ever-shorter radii from a fixed centre for each arc, the diminution of the scroll would not be smooth and each new arc would create a noticeable break. Moreover, a literal translation of Vitruvius's text resulted in an Archimedean volute of only two revolutions (Fig. 9).[22]

Renaissance architects were well aware of these deficiencies, and using the Roman remains as guides, they created improved and often sophisticated systems for laying out volutes. The most important refinement to Vitruvius's method involved manipulating the centers of the arcs to give a smooth transition from one to the next. Sebastiano Serlio provided a simple solution in 1537. His adaptation stayed very close to Vitruvius's text, with semicircular arcs plotted from points set within the oculus and along the vertical axis of the volute.[23] Further refinement came from a

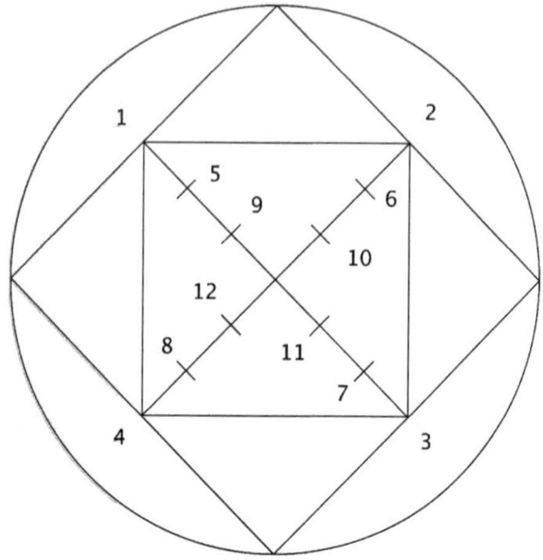

Fig. 10 Method for setting-out the compass points in the oculus of the Ionic volute (From Salviati 1552)

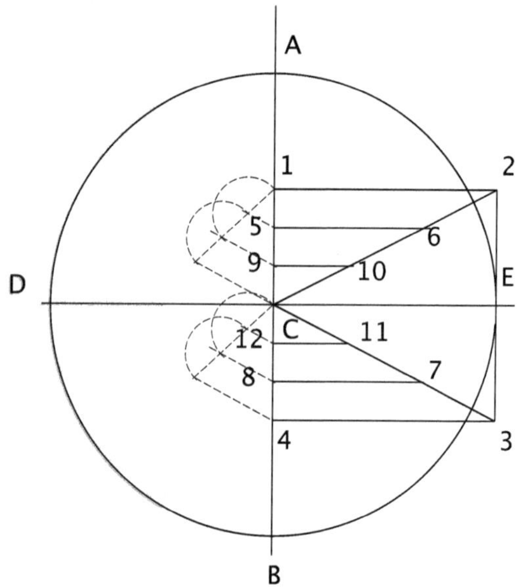

Fig. 11 Method for setting-out oculus centers (From Goldmann 1649)

text by Giuseppi Salviati in 1552. His technique—simplified and popularized by Philibert de l'Orme, Giacomo Barozzi da Vignola, and Andrea Palladio—placed the centers of the arcs along diagonals at 45° to the vertical axis (Fig. 10). The points could be found geometrically, inscribing squares within the oculus and dividing the diagonals into thirds.[24] Their approach remained popular into the eighteenth century, having the sanction of both Fréart and Perrault.[25] However, many practicing architects, especially in Britain, found Nicolaus Goldmann's seventeenth-century solution preferable, as it minimized the breaks between the spiraling arcs (Fig. 11).[26]

According to William Chambers, Goldmann's technique was the best, because the arcs "have their radii… in the same straight line; so that they meet, without forming an angle: whereas in that of De l'Orme, the radii never coincide; and consequently no two of the curves can join, without forming an angle."[27] For Stuart, these researches must have created an almost unrealizable expectation of what the Ionic volute was or should be. If the Greeks did possess the secrets of ancient architecture, they too must have understood the volute not merely as a decorative form, but as a highly elaborated and coherent geometrical construction.

The proper setting out the Ionic volute concerned architects both before and after Vitruvius. The two problems faced by Stuart and Revett were more specific. They had to determine how to measure particular examples *in situ* and how to draw them for the plates in their book. Fortunately, we know something of their methods in confronting both of these tasks, thanks to the survival of several of Stuart's preparatory sketches, notes, and drawings in the British Library and in the Drawings Collection of the Royal Institute of British Architects.[28] The former group of papers includes field notes of the measurements for the volutes they found in Athens, along with some accompanying calculations. The latter archive contains a number of different recipes for finding a series of centers from which curves could be drawn through their measured points.

The most practical method of measurement would entail climbing a ladder and measuring with calipers, while calling out the results to an assistant on the ground. This procedure is perhaps easier described than executed, for one, because it requires the two collaborators to agree beforehand on a set series of points to measure. The field drawings suggest the use of just such a formula. What Stuart and Revett measured were the distances from the center at which the spiral intersected with the vertical and horizontal axes, as well as with diagonals struck at 45° intervals. In this respect, the field notes provided the basis for the plates of the published work, which appear in much the same form. The method may have been inspired by the setting-out technique of Guillaume Philandrier (1544), whose construction called for pre-set lengths measured out along diagonals in the same manner (Fig. 12).[29]

Four of the field drawings can be associated with a single temple, that of Minerva Polias, corresponding to the western part of the Erechtheion and specifically its north-facing portico. The numbers recorded on the drawings match or nearly match the published plate (Vol. II, Ch. II, plate IX), with each of the sketches reporting slightly different data (Table 3). The published figures give the radii of the volute at successive points measured from the center, and two of the drawings (fols. 63 and 65) almost exactly match these figures. However, a third drawing (fol. 64) accords with only some of the published values. It appears that the others are running dimensions taken from the outside toward the center; these are not shown on the published plate but can be deduced from it. The change suggests that Stuart and Revett began their surveying with this drawing, but altered their method of measurement mid-course, shifting to the center of the volute only at the 135° mark. The figures on fol. 68 also appear to be running dimensions from the outside, but with less general agreement with the published figures. These discrepancies may reflect slightly different measurements of the same capital or—perhaps less likely given the minute

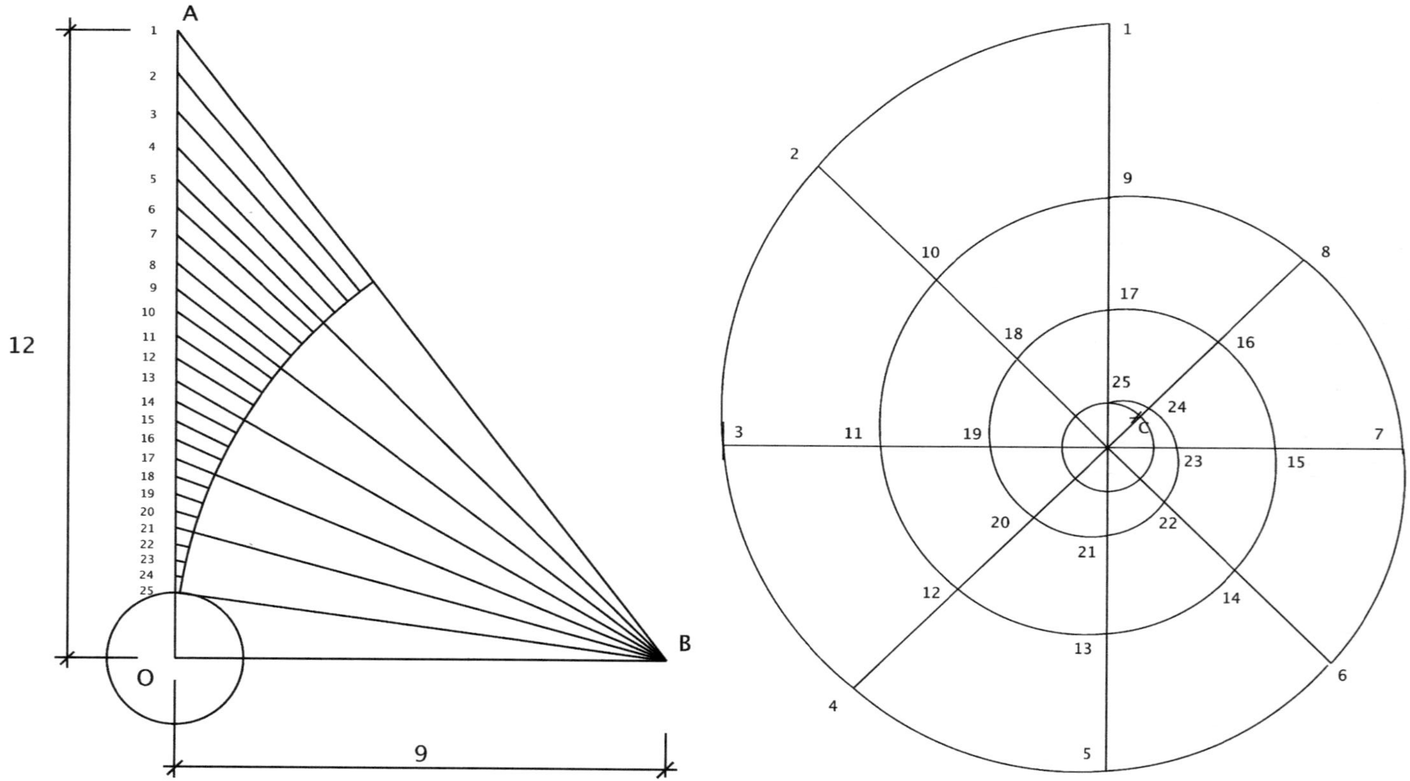

Fig. 12 Method for setting-out the Ionic volute, according to Guillaume Philandrier (1544). A preparatory drawing produces line segments that are then transferred to the oculus at 45° intervals

Table 3 Comparison of the measurements for the Ionic Volute of the temple of Minerva Polias, (Vol. II, Ch. II, Plate IX) with Stuart's field notes in the British Library: Additional Manuscripts 22153, fols. 63–65, and 68

	Plate IX	Toward center	f.63	f.65	f.64	f.68
0°	11.8	5.734	11.8	11.8		6.034
	6.066	8.024	6.06	6.066		9
	3.776	10.375	3.1	3.1		10.501
	1.425	11.8	no data?	no data?		no data?
45°	11.2	**5.45**	11.2	11.2	**5.4**	5.4
	5.75	**8.35**	5.374	5.374	**8.3**	8.3
	2.85	**11.2**	2.834	2.834	**11.15**	11.15
	–					
90°	9.725	**4.615**	9.725	9.725	**4.15/4.612**	4.615
	5.11	**7.484**	5.11	5.11	**7.234**	7.234
	2.241	**8.341**	2.4/2.491	2.404	**8.3**	8.3
	1.384	9.725			9.725	9.725
135°	–		no data	no data		8.42
	8.475	4.041	no data	no data	8.475	4.434
	4.434	6.341	no data	no data	4.434	2.314
	2.134	8.475	no data	no data	2.154	no data?
180°	7.35	3.7	7.35	7.35		3.7
	3.65	5.925	3.65	3.65		5.8
	1.425	7.35				1.35
225°	6.766	3.106	6.766	6.766		3.4
	3.66	6.766	3.366	3.366		6.766
	–					
270°	6.4	3.134	6.4	6.4		3.134
	3.266	5.016	3.266	3.266		4.925
	1.384	6.4	no data?	no data?		6.4
315°	6.25		6.25	6.25		3.1/3.225
	3.15		3.15	3.15		6.25

The figures given on the plate (column 1) are running dimensions *from* the center. Column 2 gives the running dimensions calculated *towards* the centre. The figures in *bold* very nearly match those from fol. 64, which suggests that Stuart and Revett began the survey with this drawing, before altering their method of measurement for the others

differences—another capital from the same set of columns. It is also worth pointing out the curious prevalence of dimensions with three decimal places, normally an impossible level of accuracy for direct measurement. Such figures could have been generated arithmetically, but the drawings do not suggest the use of any intermediate constructions or calculations. One possible explanation, in the absence of any other, is that the surveyors used calipers and a diagonal scale to obtain such fine readings.

Stuart was interested here, as with the Pola Temple and the Theater of Bacchus, in reconciling his own measurements with the prescriptions of the *De Architectura*. Although he published no attempt along these lines, at least one of Stuart's surviving manuscripts shows him actively searching for correspondences. Fol. 61 of the British

Table 4 James Stuart's arithmetic for columns from the Minerva Polias temple (the Erechtheion), from the Edinburgh Notebook, fol. 61

18. 5.05	Height of the columns of Min. Polias
..........12	
9) 221050	(24.56+2.0.56 diameter at bottom of columns
8) 2456	
3.07	
· 2. 3.60	
9) 27.63	Length of abacus
3.07	=1/9
6.14	=2/9 which subtracted from the number below it
27.63	
21.49	
Or feet 1.9.49	Distance by calculation from centre to centre of the eye of the volute.

Library papers records the following calculations, which pertain to Vitruvius's recipe for setting out the Ionic capital (Table 4).

Here Stuart first calculated the diameter of the base, making it—per Vitruvius—1/9 the height (Book 4, Chapter 1.8). He then calculated the length of the abacus, the uppermost slab of the capital. Although this particular column is shorter than 19 ft, he used the author's prescription for a column taller than 25 ft, for which the abacus was to be "as wide as the bottom of the column with one eighth added on (Book 3, Chapter 5.7)."[30] He then checked the result (2 ft, 3.6 in.), dividing it by nine to produce a repeat of the Figure 3.07. The length of the abacus determines the distance between the centers of the two volutes, the latter shorter than the former by 2/9.

It is difficult to know what to make of these calculations. In the first place, the figures do not correspond to the columns of the north porch of the Minerva Polias temple. They seem, rather, to match—but only partially—the dimensions of the engaged columns on its enclosed western flank (Vol. II, Ch. II, plates XI and XII). The lower column diameter is given there as 2 ft, 0.55 in. and the abacus length as 2 ft, 3.6 in., that is, within .01 and .03 in. of the values in the calculations. The column height, however, is far greater than that recorded on the published plate, which registers at a mere 17 ft, 7.5 in., including the capital. Given these circumstances, it is possible that the calculation represents an attempt to determine a potential column height and its corresponding abacus by working backwards from some of its other dimensions. That might also explain Stuart's departures from Vitruvius, not only for the length of the abacus, but also for the distance between the volute centers. The ratio for the latter of 7/9 the abacus length is not given by Vitruvius, but it does result in a value very close to the capital's actual measurements (1 ft, 9.49 in. versus 1 ft, 9.68 in., a difference of about .2 in.).[31] Stuart seems to have been applying a rough Vitruvian logic to see which dimensions were related and which not.

Reconstructing Greek Volutes

Measuring existing capitals was one problem. Reproducing those capitals in visual form was quite another. In essence, Stuart had to reconstruct the method that the original craftsman had used to set out each volute, finding in the process the correct centers for all the arcs and ensuring that they fit the measured points. His aim here was not merely antiquarian. If Stuart had any hope of accurately reproducing the volutes for publication, so that the engraved curves actually corresponded to the recorded measurements, he had to find the original setting out procedure. For this task, however, he faced a significant obstacle: he was unaware of the techniques with which Greek architects worked and had, if anything, an overdeveloped view of their geometrical complexity. A brief review of what we know today about these techniques reveals two factors that may have aided Stuart in his own reconstructions. First, there were multiple ways that the ancients constructed the volute; some were very elaborate, but others were much simpler. Second, efficiency in the building process was often just as or more important than geometrical coherence.

Classical architects were on site to provide direction to the craftsmen. In all cases, these architects would have negotiated budgets and general designs (*syngraphai*) with their patrons—often with the city-state itself. The lead architect determined the non-essential elements of design on site. He could determine the decorative details, such as the volute, only after constructing the base of the temple, most importantly the stylobate and intercolumniations. These dimensions determined the proportions of the upper elements. Once the ground plan of the temple was established, the architect was responsible for overseeing specific elements of the design, providing *paradeigmata*, or templates, for his craftsmen to copy. In the case of the Ionic capital, a craftsman under the supervision of the architect would have created a wood, clay, or possibly stone model of the volute. This prototype served as the basis from which all the capitals would then have been carved. Making it was a simple task of transferring its outline, via calipers, to new stone blocks.[32] The *paradeigmati* would have had much the same function as workshop drawings do today. For a repetitive element such as a volute, they had the advantage that the complex process of setting it out only had to be undertaken once.

Modern archaeologists have discovered several examples of *paradeigmati*. The Temple of Apollo at Didyma, for example, houses an extensive set of full-scale "blueprints," as Lothar Haselberger has described them.[33] He identified thin inscriptions on the walls and floors as construction drawings for elements of the temple's architecture. Other massive *paradeigmati* have been found in Pergamon, Priene, Baalbeck, and Rome, among other locations.[34] Likewise, prototype *paradeigmati* for small details have also been uncovered, including examples of Ionic capitals. One exemplary specimen, now at the University Museum, Berne, comes from an unknown location in Greece and still includes the inscribed vertical axis, vertical and horizontal tangents, as well as 11 compass points within the oculus. Thomas

Loertscher has analyzed this example in detail, and his conclusions point to a setting-out system not described by early modern architectural writers.[35]

In addition to *paradeigmati*, surviving elements of ancient architecture also point to the working methods of ancient architects. Unfinished capitals from Priene and Didyma suggest at least two of the ways that ancient architects designed their volutes. The first method was used at the Temple of Apollo at Didyma (ca. 330 BCE), designed by Paeonius of Miletus and Daphnis of Ephesus. Several of these examples include the craftsmen's inscribed guidelines, including a series of intersecting lines to divide the oculus into eighths.[36] These capitals were likely the *paradeigmati* for the other capitals at Didyma to which craftsmen could have turned as a reference. While a detailed investigation of the Didymaean volute is wanting, it is probable that the centers rest along the diagonals, as there would be little need to inscribe them otherwise.

Another surviving fragment is from the Temple of Athena Polias at Priene (ca. 350–330 BCE), studied by Gorham P. Stevens in 1931.[37] Although the volutes on this capital are finished and the method of laying out erased, Stevens was able to reconstruct 63 points along the spiral. Applying Euclid's theorem that only one arc—with, of course, only one centre—can pass through any three points, he discovered the system used for creating the volute, which was set out from 16 centers along the diagonals to the vertical axis. Similar to Vignola's method, the Priene capital was laid out with a square inscribed within the oculus, the corners bisected by 45° radii. Each diagonal was divided into 16 units, with the centre of each arc steadily rotated towards the centre, in fact, partly following an Archimedean spiral.

Stuart's task was the same as that later taken up by Stevens, namely to find the centers from which an existing volute has been set out and to determine from these the sequence in which the arcs were drawn. For the first of these tasks, he probably used the same method that Stevens did, by bisecting two or more chords in each arc and extending perpendiculars from them. This elementary procedure, known since Euclid, was also an implicit part of Guillaume Philandrier's method of volute construction. Once the centers were located, however, a recipe for connecting them still had to be found. This problem was not at all straightforward, and Stuart met it with only mixed success. The first volute that he and Revett published was that of the unnamed Ionic temple on the Illissos river (destroyed in 1778), included in the first volume of the *Antiquities* (Ch. II, plate VII). Their measurements must have been reasonably accurate, for they served as the basis of a convincing reconstruction by their friend and colleague Stephen Riou. Riou had travelled with the two men to Athens and, on his return, worked out a method for laying out the temple's volute, transforming the published dimensions into a system of modular parts (Fig. 13). He published the reconstruction in 1768.[38]

Riou's achievement may have given Stuart an unjustified confidence. As he began preparing his notes for the second volume—which contained the volutes of the several temples in the Erechtheion—Stuart struggled to make sense of the measurements they had taken. We are able to follow his attempts thanks to the survival of a number of his papers in the RIBA Drawings Collection. These diagrams are accompanied by written notes, consisting of numbered sequences of steps for laying

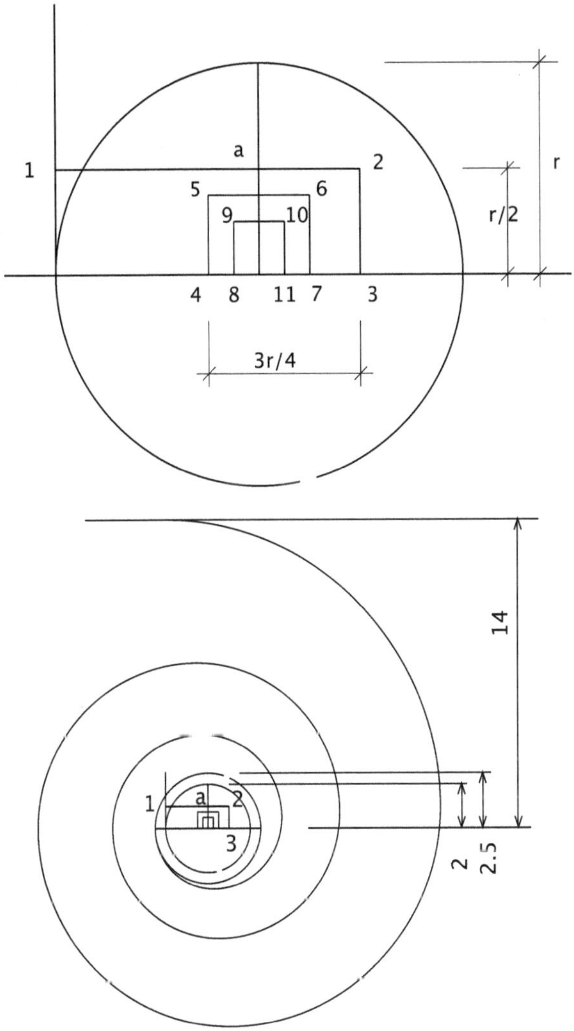

Fig. 13 Reconstruction of the volutes from the unnamed Ionic temple on the Illissos river (now destroyed) (From Riou 1768)

out the construction in the eye of the volute. The centers of the arcs are numbered, and corresponding numbers are to be found on the volutes themselves, indicating the limits of each arc. The latter are also found on some of the published plates, marked with a small asterisk. Their presence suggests that the engraver followed one or more of Stuart's recipes to draw the volutes. However, it is not clear why he might have retained numbers on the arcs without the construction diagrams in the eye of the volute, for the former are meaningless without the latter.

What stands out about Stuart's reconstruction attempts is how cumbersome, impractical, and hard to follow they are. The unidentified construction reproduced in Fig. 14, for example, requires 17 different steps with a compass and ruler, and

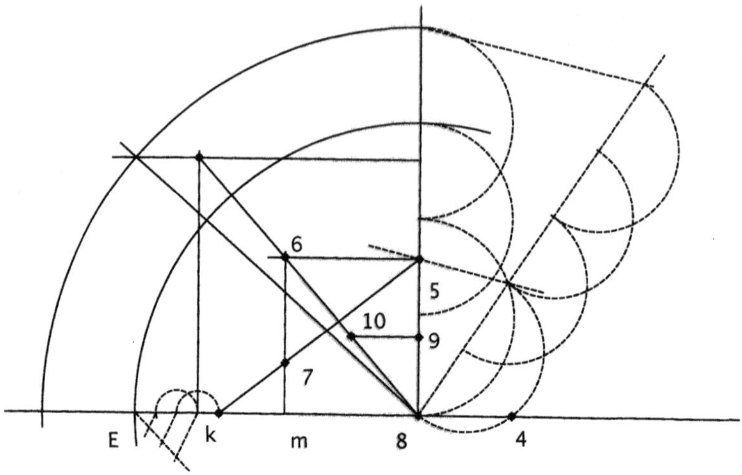

Fig. 14 Stuart's attempt to reconstruct the setting out of the volute of an unidentified Ionic capital, from his papers in the RIBA Drawings Collection, SD 93/4/3

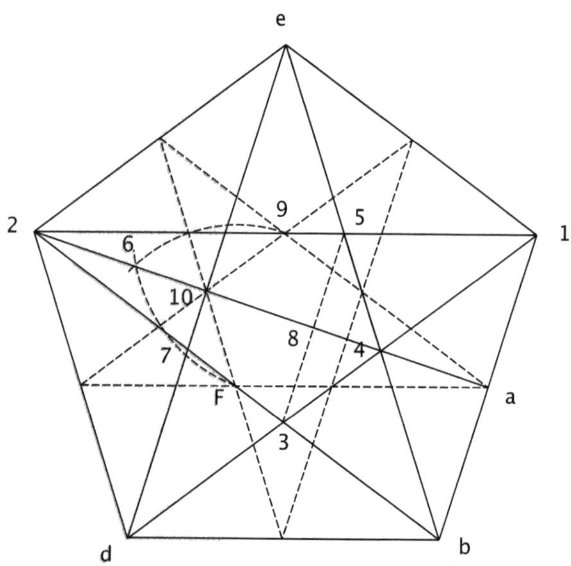

Fig. 15 Stuart's attempt to reconstruct the volute of the engaged columns of the "Temple of Minerva Polias" (the western flank of the Erectheion), from his papers in the RIBA Drawings Collection, SD 93/4/3

even one with a protractor. The fine dotted lines are used to make the numerous divisions called for in the recipe. Even more curious is that the prescribed compass centers follow no obvious pattern for their sequence or placement in the eye. A second diagram for the volutes on the engaged columns of the western flank of the Erechtheion is only marginally more sensible. The construction takes the recognizable—if complicated—form of pentagons nestled in five-pointed stars (Fig. 15).

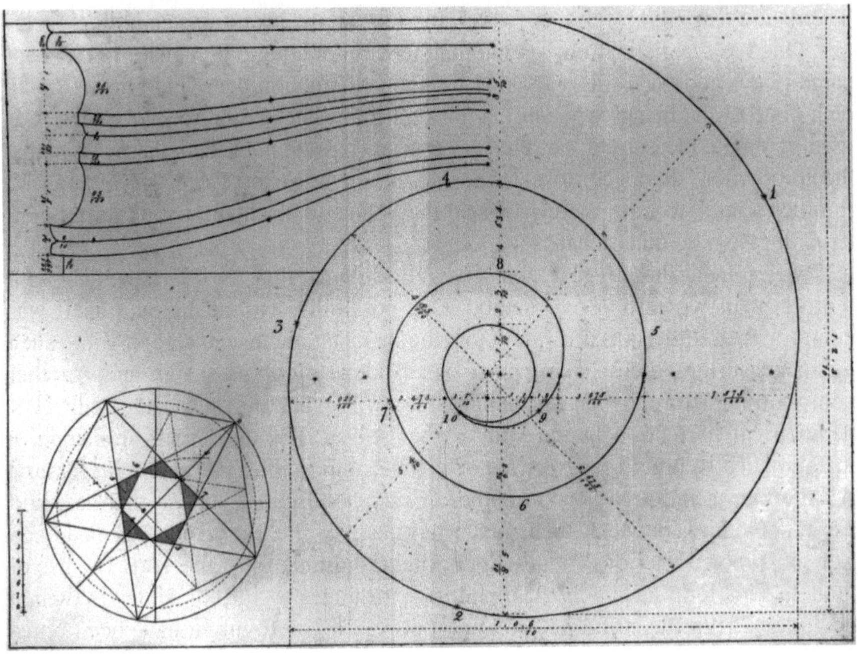

Fig. 16 Pentagonal diagram (*lower left*) based on Stuart's attempted reconstruction in Fig. 15, as published in Stuart and Revett (1762–1830)

In principle, such shapes could help to form arcs of 72° (360/5), diminishing from one pentagon to the next at a regular rate. Yet the construction does not follow this process. Instead, the compass centers leap randomly about the whole construction, from the outer points of the larger star and back and forth between the larger and smaller inscribed pentagons. In general, the prescribed radii diminish as they progress, which is what one would be looking for. The distance 2–3, for example, is shorter than the distance 1–2 and 3–4 is smaller than 2–3, but Stuart does not stick to this pattern. Moving the compass point from 4 to 5 produces no reduction in the radius of the curve. Moreover, the curious step that leads to point 6 appears to have no logic whatsoever, for it results in an *increased* radius. William Newton, the editor of this second volume of the *Antiquities*, published the construction (Fig. 16), but it is unlikely that the engraver actually used it to produce the accompanying volute. A third method, for the capital of the "Temple of Erectheus", or the eastern portico of the Erechtheion, is even more complex.[39] In the first place, it requires the construction of a heptagon, which is not strictly possible with a ruler and compass alone. Even apart from this difficulty, the recipe is so convoluted that it resists attempts to follow it, even with the aid of the corresponding sketch.

How can we explain Stuart's thinking here? None of these procedures is credible as the method by which the original volutes were set out, and it is difficult to understand why Stuart felt it necessary to look for and propose such unlikely recipes. One has the distinct impression that he was not able to fit a curve to the measured points and was simply looking for any construction that would do so. In the end, none of the manuscript constructions for the volutes in the second volume was used. A simple measure of the spirals as engraved on the published plates shows that their proportions do not conform to the measurements given. The engraver appears, in other words, to have merely labeled the volutes with Stuart's values after constructing them by other means.[40]

Two explanations suggest themselves for this inability to square the data with a more plausible setting out method. One possibility is that the data itself was compromised. Indeed, it is difficult to imagine inaccuracies *not* creeping into such a fine measuring process taking place on a ladder 18 ft above the ground. Another contributing factor may have been that Stuart was expecting an unrealistic level of geometrical perfection in the volutes themselves. For the Greek craftsman, a mathematically precise curve was, of course, impossible, but also unnecessary. All the mason required was a number of points sufficiently close that he could carve something between them that would satisfy the eye. Both Stevens's and Lörtsch's reconstructions suggest approximate spirals of just this sort. The use of models and templates—*paradeigmata*—introduces a further stage in which deviations from the strict mathematical form of a volute might have been introduced: in the copying process. Even if Stuart's measurements were accurate, in other words, they may have included too many variations—introduced between the original setting-out drawing and the process of carving—to allow them to be fitted to a regular geometrical construct. Indeed, Stuart and Revett's own figures suggest that they were not measuring perfect spirals. Four volutes are illustrated in the *Antiquities*, and to check their "accuracy", we can plot the radii of the volutes at successive points on a graph (Fig. 17). A geometrically precise volute would produce a smooth curve, but this does not appear. The graph for each volute dips and rises unevenly.

It is worth noting that Stuart could have easily avoided these difficulties, both for himself and his engraver. Philandrier's method showed how to draw a volute by first determining the radials from the center toward a series of points set out at 45° intervals along the spiral. The system was well-known and had recently been republished by Abraham Swan in a popular practical handbook.[41] Working backward from his own measurements, Stuart could have used Philandrier's method to draw a volute composed of successive 45° arcs. This approach would have no doubt entailed significant drawbacks. The arcs would not lie on a continuous curve, but would have been subtly "broken" from one to the other. More importantly, once the centers of the arcs were found, the method gave no recipe for connecting them. Philandrier's method, in other words, would have been adequate for the engraver, but it gave no way of replicating the volutes at different scales for use in a practical design context. This seems to have been enough to dissuade Stuart.

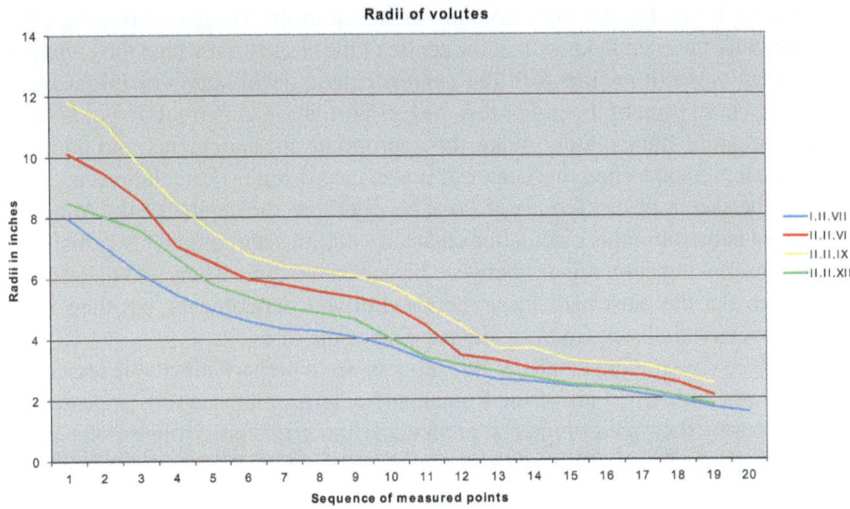

Fig. 17 Graph of the radii of volutes, from Stuart and Revett (1762–1830)

Conclusion

The original survey drawings that survive represent only a small proportion of the measurements that Stuart and Revett made during their sojourn in Greece, but they are enough to give us a good idea of their working method. Whereas the published dimensions are presented as polished, unproblematic, and absolute, we can now see that the reality was somewhat messier. The field notes often show not only minor differences between the recorded measurements and those on the published plates, but also small variations between multiple repeated measurements. These variations suggest that a process of selection and correction took place both in the field and while preparing the published work. It is now also clear that some calculation was involved in obtaining these figures. Indeed, measurements within one thousandth of an inch were usually possible in no other way. The "accuracy" that so impressed the architects of the nineteenth century in effect disguised a process marked by trial-and-error, figures derived from others, and subjective judgment.

Stuart made it clear in his introduction to the first volume of the *Antiquities* that his purpose was simply to record the dimensions of the monuments as accurately as possible and not to impose upon his measurements any preconceived theory of design. He took care to proceed by "purposely forbearing to mention Modules, as they necessarily imply a System." On several occasions, however, he could not avoid making assumptions about the forms of the buildings and the way in which they were set out for construction. Indeed, the large number of Stuart's notebook calculations rather suggests that he made strenuous attempts to check for such proportions. When faced with a structure as ruined as the Theater of Bacchus, for example, Stuart needed to know that the centre of the circle from which the *pulpitum* was set out was not in line with the *versurae*, as a naïve surveyor might have supposed, but in front of them. For this, he had Vitruvius and Perrault to guide him, so that he could immediately make the appropriate measurements and adopt a geometrical method to find this center. It is also clear from his calculations accompanying the sketch of the Temple at Pola, his check on the capital of the Minerva Polias, and numerous other calculations that have not yet been completely explained, that he was making comparisons between his measured dimensions and theories of proportion. For the most part, these remained private experiments, but they were nonetheless essential steps in his understanding of the ruins.

One of the overarching results of this process was a perhaps inevitable lessening of Vitruvius's authority. Few of the Roman author's recommendations appeared to be borne out by their measurements, particularly those for Ionic volutes. Nor were modern methods from Serlio onwards satisfactory either. Worse, they appear to have misled Stuart toward an unrealistic expectation of the kinds of constructions that the Greeks actually used. Did he not see how improbable his own attempts were to find a method that fitted the measured points? Despite the occasional failure, Stuart and Revett's work entailed important insights. Their intimate contact with and their scrupulous measuring of the architecture itself must have convinced them that not only could no modular system be strictly applied, but that architectural practice in ancient Greece was more varied than they and their contemporaries often assumed. This realization is best represented not by the dimensions in their final, published form, but rather by the manuscript field notes, which record how they measured and made sense of them.

Notes

Jason M. Kelly contributed to the section of this paper that deals with the Ionic volute. Frank Salmon wrote the introduction and edited the whole. David Yeomans wrote the remainder of the paper and produced most of the illustrations. We would like to thank Prof. Charles Goldie for his comments on the geometry of the Ionic volute and Anthony Gerbino for his advice and patience.

1. See Wood (1753, (a)*r*): the "principal merit of works of this kind is truth" and James and Revett (1762–1830, vol. 1, vii): "we determined to avoid Haste, and System, those most dangerous enemies to accuracy and fidelity."
2. See Watkin (1996, 641).
3. Penrose (1851, ix).
4. Landy (1956, 255, 258, and 259). For a more recent account of Stuart and Revett that locates the debate about "accuracy" in cultural terms see Kaufman (1989, 74).
5. See Salmon (2006, 107–17).
6. Stuart's Temple of Winds at Mount Stewart, Co. Down, of 1782–85, for example, follows the principal dimensions of the Athenian original very closely.

7. Laing Mss., La.III.581, Edinburgh University Library. There is also a sketchbook in the collection of the Royal Institute of British Architects (SKB/336/2). Joseph Woods, who edited the fourth volume of *The Antiquities of Athens* (1816, ii), recorded that he had been handed 54 of Stuart's manuscript notebooks.
8. Much as we see in the Stuart's drawing of himself studying the Erechtheion in Stuart and Revett, *The Antiquities of Athens,* 2: plate II.
9. Edinburgh notebook, fol. 73v. For further detail on their methods and equipment see Salmon (2006, 131–32).
10. Stuart and Revett (1762–1830, vol. 4 (1816), 29).
11. Stuart and Revett (1762–1830, vol. 1, vii). For a full account, see Salmon (2006).
12. Vitruvius (1999), 58.
13. He began by dividing the total length of the building by 50. Taking the result of that calculation (13.6), he then divided the length of the cella by a figure close (but not exactly equal) to it (13.8), obtaining 27 as the result. Even aside from Stuart's intentional fudging, the arithmetic here is incorrect. The answer should have been 28.1, but neither result would have agreed with Vitruvius's recommendation of 5/8.
14. See Salmon (2006, 124).
15. This is explained in Stuart's published account, Stuart and Revett (1762–1830, vol. 2, 23). The break in the site notes possibly reflects a period of failed negotiations with the Turkish authorities.
16. Edinburgh Notebook, fols.165v and 167r. The editions to which he refers are: Vitruvius (1556) and Vitruvius (1673, 2nd ed. 1684). These pages of the notebook also reproduce the French text of Perrault's Vitruvius, which Stuart translated to obtain the section quoted here.
17. Stuart and Revett (1762–1830, vol. 2, 24).
18. Edinburgh Notebook, fol. 65v. Stuart's dimensions were also recorded in a cross-section of the theater: Additional MSS 21153, fol. 72, British Library.
19. Stuart and Revett (1762–1830, vol. 2, 24).
20. Stuart and Revett (1762–1830, vol. 1, ii, fn. A). Also see Fréart de Chambray (2005, orig. ed. 1650, 91–2) for a similar judgment.
21. Stuart and Revett (1762–1830, vol. 1, ii and 7).
22. Carpenter (1926, 253).
23. Serlio (1537). On this subject, see Losito (1993) and Andrey and Galli (2004, 33–36).
24. See Salviati (1552); Vignola (1572, pl. 20, 1999).
25. Fréart de Chambray (2005, 110) and Perrault (1722, 69–74, pl. IV).
26. First published in Goldmann (1649).
27. Chambers (1791, 53).
28. British Library, Additional Manuscripts 22153, fols. 61–68 and RIBA Drawings Collection, SD 93/4, fols. 1–7.
29. See Andrey and Galli (2004, 37–38).
30. A sheet in the RIBA Drawings Collection, SD 93/4 reproduces this calculation in the form of a small sketch.

31. The latter figure does not appear on the published plates but can be deduced from them by subtracting from the whole capital width (36.834 in.) the distances from the centers to the outer edges of the volutes (7.575 on each side).
32. See Coulton (1977, 53–58); Petronōtēs (1972); and Coulton (1976).
33. See the following articles by Lothar Haselberger (1980, 1983, 1985, 1991).
34. See Haselberger (1994); Kalayan (1971); Koenigs (1983); and Schwandner (1990). For a more extensive bibliography, see Wilson Jones (2000, 249).
35. Loertscher (1989). Also see Haselberger (1997, 89–92).
36. Haselberger (1985).
37. Stevens (1931).
38. Riou (1768, 34–5, pl. 9).
39. RIBA Drawings, SD 93/4/3
40. The carefully drawn volute in Vol. III for the Ionic colonnade near the Lantern of Demosthenes (Ch. XI, Pt. 1) provides a similar case. The diagram, produced by Willey Reveley from Stuart's surviving notes, depicts a logarithmic volute with the centers of the arcs based on diminishing squares. The dimensions given for the volute, however, correspond only fitfully to the illustration, and several values are simply missing. We must conclude that Reveley was presented with poorly recorded field notes that were ultimately impossible to interpret. He appeared to be aware of the discrepancies but decided to let them stand, on the justification that "Mr Stuart has left no memorandum on the subject of these disagreements." Stuart and Revett (1762–1830, vol. 3 [ed. Willey Reveley], vii).
41. Swan ([1745], pl. VIII).

Photographic Credits

Edinburgh University Library: Fig. 1
Author: Figs. 2, 4, 5, 7, 9–15
Faculty of Architecture and History of Art, University of Cambridge: Figs. 3, 6, 16
Cambridge University Library: Fig. 8

References

Andrey, Denise, and Mirko Galli. 2004. Geometric methods of the 1500s for laying out the ionic volute. *Nexus Network Journal* 6(2): 31–48.

Carpenter, Rhys. 1926. Vitruvius and the Ionic order. *American Journal of Archaeology* 30(3): 259–269.

Chambers, William. 1791. *A treatise on the decorative part of civil architecture*, 3rd ed. London: J. Smeeton.

Chambray, Roland Fréart de. 2005. *Parallèle de l'architecture antique avec la moderne suivi de Idée de la perfection de la peinture*. Paris: École nationale supérieur des beaux-arts.

Coulton, J.J. 1976. The meaning of Anagrafeus. *American Journal of Archaeology* 80(3): 302–304.

Coulton, J.J. 1977. *Greek architects at work: Problems of structure and design*. London: Elek.
Goldmann, Nicolaus. 1649. Vitruvii voluta ionica hactenua amissa: Restituta a Nicolao Goldmanno. In *M. Vitruvii Pollionis De architectura libri decem*, ed. Johannes de Laet, 265–272. Amsterdam: Ludovicum Elzevirium.
Haselberger, Lothar. 1980. Werkzeichnungen am Jüngeren Didymeion–Vorbericht. *Istanbuler Mitteilungen* 30: 191–215.
Haselberger, Lothar. 1983. Bericht über die Arbeit am Jüngeren Apollontempel von Didyma–Zwischen-bericht. *Istanbuler Mitteilungen* 33: 90–123.
Haselberger, Lothar. 1985. The construction plans for the Temple of Apollo at Didyma. *Scientific American* 253(6): 126–132.
Haselberger, Lothar. 1991. Aspekte der Bauzeichnungen von Didyma. *Revue Archéologique* 1: 99–113.
Haselberger, Lothar. 1994. Ein Giebelriss der Vorhalle des Pantheon: Die Werkrisse vor dem Augustmausoleum. *Mitteilungen des Deutschen Archäologischen Instituts, Römische Abteilung* 101: 279–308.
Haselberger, Lothar. 1997. Architectural likenesses: Models and plans of architecture in classical antiquity. *Journal of Roman Archaeology* 10: 77–94.
Kalayan, Haroutune. 1971. Notes on assembly marks, drawings, and models concerning Roman period monuments in Lebanon. *Annales archéologiques arabes syriennes* 21: 269–273.
Kaufman, Edward. 1989. Architecture and travel in the age of British eclecticism. In *Architecture and its image: Four centuries of architectural representation*, ed. Eve Blau and Edward Kaufman, 59–85. Montreal: Centre canadien d'architecture [Canadian Centre for Architecture].
Koenigs, Wolf. 1983. Die Athenatempel von Priene. *Istanbuler Mitteilungen* 33: 134–176.
Landy, Jacob. 1956. Stuart and Revett: Pioneer archaeologists. *Archaeology* 9: 252–259.
Loertscher, Thomas. 1989. Voluta constructa: Zu einem kaiserzeitlichen Volutenkonstruktions-modell aus Nordafrika. *Antike Kunst* 32: 82–103.
Losito, Maria. 1993. La ricostruzione della voluta ionica vitruviana nei trattati del Rinascimento. *Mélanges de l'École Française de Rome: Italie e Méditerranée* 105(1): 133–175.
Penrose, Francis Cranmer. 1851. *An investigation of the principles of Athenian architecture*. London: Longman & Co., J. Murray.
Perrault, Claude. 1722. *A treatise of the five orders of architecture*. London: John James.
Petronōtēs, Argyrios. 1972. *Zum Problem der Bauzeichnungen bei den Griechen*. Athens: Dodona Verlag.
Philandrier, Guillaume. 1544. *In decem libros M. Vitruvii Pollionis de architectura annotationes....* Rome: G. A. Dossena.
Riou, Stephen. 1768. *The Grecian orders of architecture*. London: J. Dixwell for the Author.
Salmon, Frank. 2006. Stuart as antiquary and archaeologist in Italy and Greece. In *James "Athenian" Stuart 1713–1788: The rediscovery of antiquity*, ed. Susan Soros, 107–117. New Haven/London: Yale University Press.
Salviati, Giuseppi. 1552. *Regola di far perfettamente col compasso la voluta jonica et del capitello ionico et d'ogni altra sorte*. Venice: F. Marcolini.
Schwandner, Ernst-Ludwig. 1990. Beobachtungen zu hellenistischer tempelarchitektur von Pergamon. In *Hermogenes und die hochhellenistische Architektur*, ed. Wolfram Hoepfner, 93–102. Mainz: Zabern.
Serlio, Sebastiano. 1537. *Regole generali di architettura sopra le cinque maniere degli edifici*. Venice: Francesco Marcolini.
Stevens, Gorham P. 1931. The volute of the capital of the Temple of Athena at Priene. *Memoirs of the American Academy in Rome* 9: 135–144.
Stuart, James, and Nicholas Revett. 1762–1830. *The antiquities of Athens: Measured and delineated*. London.
Swan, Abraham. 1745. *The British architect*. London: R. Sayer.
Vignola, Giacomo Barozzi da. 1572. *Regole delli cinque ordini*, 2nd ed. Rome.
Vignola, Giacomo Barozzi da. 1999. *Canon of the Five Orders of Architecture*. Trans. Branko Mitrovic. New York: Acanthus Press.
Vitruvius. 1556. *I dieci libri dell'Architettura*, ed. Daniele Barbaro. Venice: F. Marcolini.

Vitruvius. 1673. *Les dix livres d'architecture de Vitruve*, ed. Claude Perrault. Paris: J. B. Coignard.

Vitruvius. 1999. *Ten books on architecture*, ed. Ingrid D. Rowland and Thomas Noble Howe. Cambridge: Cambridge University Press.

Ware, Isaac. 1738. *The four books of architecture by Andrea Palladio*. London: for R. Ware.

Watkin, David. 1996. *Sir John Soane: Enlightenment thought and the Royal Academy lectures*. Cambridge: Cambridge University Press.

Wilson Jones, Mark. 2000. *Principles of Roman architecture*. New Haven/London: Yale University Press.

Wood, Robert. 1753. *The ruins of Palmyra*. London.

Index

A. Gerbino (ed.), *Geometrical Objects: Architecture and the Mathematical Sciences 1400-1800*, Archimedes 38, DOI 10.1007/978-3-319-05998-3,

MIX
Papier aus verantwortungsvollen Quellen
Paper from responsible sources
FSC® C105338

If you have any concerns about our products,
you can contact us on
ProductSafety@springernature.com

In case Publisher is established outside the EU,
the EU authorized representative is:
Springer Nature Customer Service Center GmbH
Europaplatz 3, 69115 Heidelberg, Germany

Printed by Libri Plureos GmbH
in Hamburg, Germany